WITHDRAWN

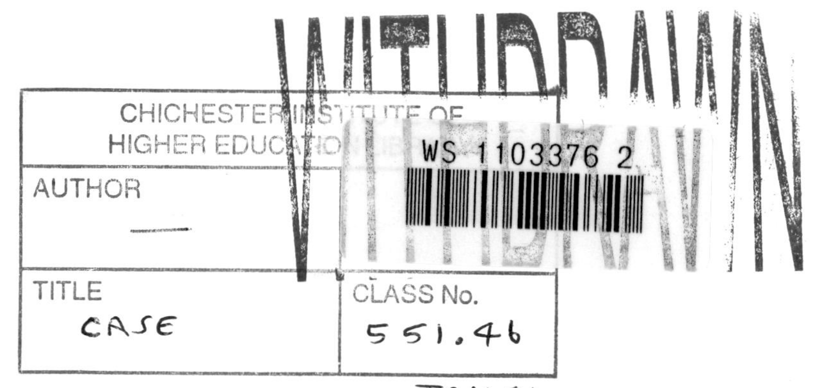
WITHDRAWN
CHICHESTER INSTITUTE OF
HIGHER EDUCATION
WS 1103376 2
AUTHOR
TITLE
CASE
CLASS No.
551.46

THE OCEANOGRAPHY COURSE TEAM

Authors
Joan Brown
Angela Colling
Dave Park
John Phillips
Dave Rothery
John Wright

Editor
Gerry Bearman

Design and Illustration
Sue Dobson
Ray Munns
Ros Porter
Jane Sheppard

This Volume forms part of an Open University course. For general availability of all the Volumes in the Oceanography Series, please contact your regular supplier, or in case of difficulty the appropriate Pergamon office.

Further information on Open University courses may be obtained from: The Admissions Office, The Open University, P.O. Box 48, Walton Hall, Milton Keynes MK7 6AA.

Cover illustration: Satellite photograph showing distribution of phytoplankton pigments in the North Atlantic off the US coast in the region of the Gulf Stream and the Labrador Current. (*NASA and O. Brown and R. Evans, University of Miami.*)

CASE STUDIES IN OCEANOGRAPHY AND MARINE AFFAIRS

PREPARED BY AN OPEN UNIVERSITY COURSE TEAM

PERGAMON PRESS
OXFORD · NEW YORK · SEOUL · TOKYO

in association with

THE OPEN UNIVERSITY
WALTON HALL, MILTON KEYNES MK7 6AA, ENGLAND

U.K.	Pergamon Press plc, Headington Hill Hall, Oxford OX3 0BW, England
U.S.A.	Pergamon Press, Inc., Maxwell House, Fairview Park, Elmsford, New York 10523, U.S.A.
PEOPLE'S REPUBLIC OF CHINA	Maxwell Pergamon China, Beijing Exhibition Centre, Xizhimenwai Dajie, Beijing 100044, People's Republic of China
KOREA	Pergamon Press Korea, KPO Box 315, Seoul 110-603, Korea
JAPAN	Pergamon Press, 5th Floor, Matsuoka Central Building, 1-7-1 Nishishinjuku, Shinjuku-ku, Tokyo 160, Japan

---

First edition 1991

Library of Congress Cataloging in Publication Data

Case studies in oceanography and marine affairs/prepared by an Open University Course Team.—1st ed.
Includes bibliographical references and index.
1. Oceanography. 2. Marine resources. 3. Maritime law. I. Open University. Oceanography Course Team.
GC11.2.C37 1991
551.46—dc20   91-16914

British Library Cataloguing in Publication Data

Case studies in oceanography and marine studies.
(Open University Oceanography)
I. Open University II. Series
551.46

ISBN 008 036376 8 (Hardcover)
ISBN 008 036375 X (Flexicover)

Jointly published by the Open University, Walton Hall, Milton Keynes, MK7 6AA and Pergamon Press plc, Headington Hill Hall, Oxford OX3 0BW.

Designed by the Graphic Design Group of The Open University.

Printed in Great Britain by BPCC Wheatons Ltd, Exeter

CONTENTS

CHAPTER 4 THE ARCTIC OCEAN: ICE, OIL AND SOVEREIGNTY

CHAPTER 5 THE GALÁPAGOS: ISLANDS OF VARIETY AND CHANGE

APPENDIX 1 TABLES I-III

APPENDIX 2 SUMMARY OF THE UN CONVENTION ON THE LAW OF THE SEA

APPENDIX 3 THE COD WAR IS NOT ONLY ABOUT FISH

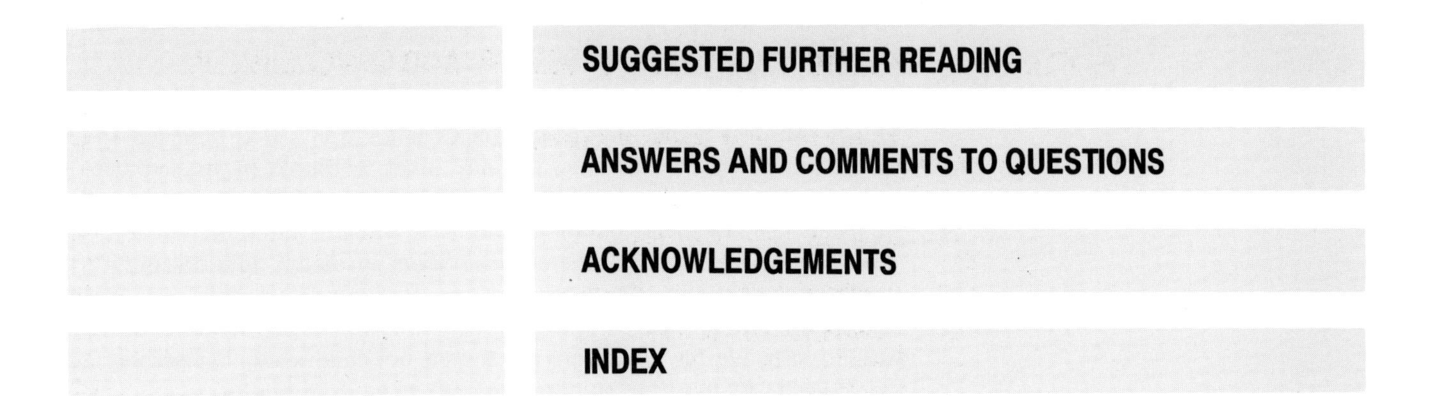

ABOUT THIS VOLUME

This is one of a Series of Volumes on Oceanography. It is designed so that it can be read on its own, like any other textbook, or studied as part of S330 *Oceanography*, a third level course for Open University students. The science of oceanography as a whole is multidisciplinary. This Volume differs from others in the Series in that it does not draw specifically upon any one or other of the 'traditional' scientific disciplines. The first part outlines the international legal framework within which marine affairs are (or should be) conducted. The second part consists of two case studies illustrating the legal and/or multidisciplinary scientific aspects of selected oceanographic environments.

Chapter 1 summarizes the many ways in which the seas and oceans constitute a global resource that must be properly managed if conflicts of interest and use are to be reconciled, and if lasting damage to marine ecosystems is to be avoided. Chapter 2 describes the long and tangled history of attempts to establish a legal system (a Law of the Sea), whereby marine affairs may be managed for the benefit of all. It takes us up to the late 1970s. Chapter 3 outlines the evolution of the present international legal regime, through the 1980s and into the early 1990s.

Chapter 4 is concerned with the Arctic Ocean, much of which is permanently covered with ice. The geographical setting of the Arctic Ocean, and its importance to the Inuit, the indigenous people, mean that perceptions of 'who owns what' have developed in an unusual way. Until relatively recently, there seemed to be no pressing need for the legal situation regarding the waters and sea-bed of the Arctic to be clarified; however, as its sub-sea resources have become increasingly valuable and techniques of exploration and exploitation have improved, all this has changed. Taking hydrocarbons as our main example, we look at the practical difficulties of extraction and transportation and at the associated legal issues, particularly regarding rights of passage. Finally, we consider how global warming could transform the navigability of Arctic waters and hence perhaps also their legal status.

The Galápagos Islands, discussed in Chapter 5, are famous for their peculiar endemic flora and fauna, both terrestrial and marine. These are largely the result of the islands' oceanographic setting, far from land and in the seasonally changing current system of the eastern equatorial Pacific. The Galápagos Islands also experience dramatic interannual changes: their equatorial position means that they are hit hard by the climatic perturbation known as El Niño. We look briefly at how the marine flora and fauna of the islands were affected by the particularly severe El Niño of 1982–83, and see how tectonically uplifted massive corals provide an insight into past climatic conditions. Since 1986–87, the marine life of the Galápagos has been to some extent protected through being in a marine nature reserve. We look at how this reserve came to be set up, and its role as a marine *resources* reserve rather than simply as a conservation area.

You will find questions designed to help you to develop arguments and/or to test your own understanding as you read, with answers provided at the back of this Volume. Important technical terms are printed in **bold** type where they are first introduced and defined. Note, however, that several of these terms are defined in previous Volumes in the Series and may not necessarily be redefined here.

While reading this Volume, and Chapters 1 to 3 in particular, you should bear the following points in mind:

1 In international negotiations, treaties, agreements, etc., relating to the Law of the Sea, the unit of lateral distance employed is the nautical mile and the unit employed for depth is usually the metre. This hybrid system has been adopted in this Volume because to do otherwise would (a) create problems in quoting from the original documents and (b) destroy the conventional whole-number system of maritime limits (e.g. the 12 (nautical)-mile limit would become the 22.224-km limit, or, at best, the 22.2-km limit). The term 'nautical mile' is generally abbreviated to 'mile'; in other words, 'mile' should always be understood to refer to the nautical mile and not to the shorter statute mile used on land. Where the abbreviation 'm' is used, it refers to metres and *not* to miles. The (nautical) mile used here is the International Nautical Mile, the primary *numerical* definition of which is 1.852 km. The UK nautical mile (*not* used here) is slightly longer, being defined numerically as 6 080 feet, which converts to 1.853 km. Note also that 1° latitude = 60 International Nautical Miles.

2 Unless otherwise stated in a specific context, the information in Chapters 2 and 3 on Law of the Sea refers to the legal position during peacetime. In wartime, the rules of international law, however widely agreed and accepted beforehand, are liable to be ignored.

CHAPTER 1 MARINE RESOURCES AND ACTIVITIES

'Running through the web of English history one perceives the connecting thread of maritime interest and occupation interwoven with the national life, and at all times affecting the national policy.... The sea must be 'kept.' That has been the maxim and watchword of national policy throughout the ages....'

T.W. Fulton (1911) *The Sovereignty of the Sea*, Blackwood & Sons.

This introductory Chapter seeks to provide a general survey of those properties of the sea that make it a resource—something of material value to humanity. The term 'resource' is applied in the broadest sense, to include any use to which the sea is put—activities such as navigation, waste disposal and research as well as the more conventional resources like fisheries and sea-bed minerals.

The management of marine resources and activities has become increasingly complex as new technologies have developed and new nations have appeared—especially in the decades following World War II. Protection of the marine environment is now a matter of global concern.

1.1 THE SEA ITSELF

The largest and most 'intangible' resource is the sea itself. What we may call marine *space* and its influence upon cultural and national identities and attitudes have provided much material for political geographers. Is it better to be an island, a coastal state or a land-locked state? The question can be endlessly debated, especially when you consider the extremes: the concentration of land-locked states in Africa with no direct access to marine resources on the one hand; and on the other, the island states of the Pacific, which have large marine resource potentials but inadequate economic bases to develop them independently.

In the space of a few decades, the relationship between the developed countries and the less developed countries has undergone a transformation. For one thing there are now many more of the latter, so that in international bodies such as the United Nations they can easily outvote the developed countries. Secondly, the growing desire of many less developed countries to industrialize has focused attention on the importance of both physical and biological resources and hence strengthened the determination of such nations to bargain over the resources available.

A nation's shoreline length, and hence the area of coastal sea over which it can legitimately claim some jurisdiction (the territorial sea, defined in Chapter 2), bears no relation to the area of the country, its population or its needs, however defined. Moreover, two countries with equal-length shorelines could well, and often do, find themselves with quite different resource potentials. A strong feeling has developed among the less developed nations that an attempt should be made to iron out some of these inequalities by ensuring that the poorer countries receive some financial benefit from the resources of the deep oceans. Moreover, the less developed countries now have more power to back up this view, partly because of their voting strength at the United Nations, partly because some of them have resources of their own that are required by the developed nations, and partly because of a more sympathetic attitude generally towards the reduction of economic differentials between states.

In short, marine 'space' by itself is of little consequence to any nation unless it can be put to some use, even if only as a protective barrier. And the sea has many more uses than that.

1.1.1 TRANSPORT AND COMMUNICATIONS

Sea-going ships probably evolved from river craft of the earliest civilizations (Figure 1.1). Probably the earliest known economy based on seaborne trade was that of Minoan Crete, which became established between about 1700 and 1400 BC. Mycenaeans, Phoenicians, Greeks and

Figure 1.1 Excavated and re-assembled model of a funeral ship of the Pharaoh Cheops, buried alongside the pyramid at Giza, *c.* 2600 BC. It is a ceremonial version of river craft commonly in use at that time.

Romans subsequently increased the scale of marine trading (and warfare) throughout the Mediterranean and beyond. Similar activities were in progress during approximately the same period in the western Pacific. Maritime trade has played a large part in establishing the political geography of the world that we know today. In fact, the boundaries of many of the world's greatest empires were established in lands discovered by mariners in great feats of endurance and navigation.

Technological advance has not diminished the importance of seaborne trade—rather the reverse, if anything, especially for bulk commodities such as petroleum, coal, iron ore, grain, and fruit, to name but a few. Between 1945 and the 1980s, total world-wide seaborne tonnage increased from around 500 million tonnes to ten times that amount. Air travel has led to a drastic reduction in maritime passenger traffic since 1945, but air transport remains uncompetitive for bulk cargo.

Military use

Warships have been a part of the oceanic scene for thousands of years, but maritime warfare entered a new phase with the development of the submarine, the first effective use of which was in the American Civil War. An even more significant development took place in 1955 with the introduction of the nuclear-powered submarine, able to stay submerged for long periods. Such a submarine can rest on the sea-bed for many months without refuelling or surfacing.

Mishaps to such submarines in deep waters is a cause for concern, because speedy recovery is difficult, if not impossible. The loss of a Soviet nuclear submarine in the Barents Sea in 1989 brought the known number of nuclear power plants lost on the sea-bed to ten. There is danger of radioactive contamination of the marine environment, though this is more localized than that caused by the testing of nuclear weapons during the 1950s and 1960s. More recently, the oceans provided a useful area of open space over which to test ballistic missiles.

Submarine cables

Submarine electric telegraphy was first discussed as a theoretical possibility in 1795, although its practical implementation had to await the discovery of a material (gutta-percha) suitable for insulating conducting cable against chemical attack and the penetration of water. Experiments were carried out with cables laid under the Rhine in 1847, under Kiel Harbour in 1848 and under part of the English Channel in 1849. The first commercial cable was laid on the bed of the Channel between Dover and Calais in 1850, but it had a useful life of only a few hours because of breakage. A second cross-Channel cable was more successful and was opened for public use in 1851. The first Atlantic cable was laid between Ireland and Newfoundland in 1857–58 but only lasted for $2\frac{1}{2}$ months and 732 messages. The second, laid in 1866 by Brunel's *Great Eastern* (Figure 1.2), was more successful. Since then, all the world's seas and oceans have been crossed by telegraph, and later telephone cables. Rubbish and sunken vessels apart, cables were the first permanent artefacts on the sea-bed and thus represented the first major new use of the oceans for thousands of years.

Their importance for communications has declined with the development of satellite technologies, but submarine cables will continue to have their

Figure 1.2 The *Great Eastern*, Isambard Kingdom Brunel's 'great iron ship', from which the Atlantic telegraph cable was laid in 1866.

uses, both for the transmission of electrical power (e.g. between France and the United Kingdom) and for communication (a fibre optic telephone link between the United Kingdom and North America was laid in the early 1990s).

1.2 LIVING RESOURCES

Fishing is probably humanity's oldest marine activity, and is still one of the most important. We use many types of marine plants and animals, mainly for food but also for other materials such as oils and pharmaceuticals. The kinds of animals caught for food in any one region will depend on their availability and the demand for them, which used to be governed primarily by such factors as regional taste and the extent of other food resources, but is now commonly influenced by the size of the export market. Along shores and in some estuaries and coastal waters, certain molluscs (e.g. scallops, clams, cockles, mussels and oysters) and crustaceans (e.g. lobsters, crabs, prawns and shrimps) provide limited but lucrative catches. Most of these live in the intertidal zone or on the sea-bed, and may be gathered by hand or caught in baited traps. Some molluscs (e.g. mussels and oysters) grow attached to solid surfaces, and can therefore be grown on long lines or moveable frames and easily collected (Figure 1.3(a),(b)). Also, fish farming of high value species has increased greatly since the 1970s (Figure 1.3(c)).

(a)

(b)

(c)

Figure 1.3 Aquaculture (mariculture) in Loch Etive, Scotland. (a) and (b) mussel farming using a long-line system; (c) fish enclosures for salmon farming.

Fish are most abundant and accessible in continental shelf seas and shallow coastal waters (Figure 1.4). Those that spend most of their time near the bottom are termed **demersal fish** and may be round (e.g. cod, haddock) or flat (e.g. sole, halibut). Those that spend at least some time near the surface (e.g. anchovy, herring, mackerel, tuna, sardines) are called **pelagic fish**.

1.2.1 FISHING METHODS

Fish are caught in many ways. Otter or beam trawls are commonly used for cod and plaice, which tend to escape upwards when disturbed, and are trapped because the upper edge of the mouth of the net is trawled in advance of the lower part. Purse-seine nets are often used to take those pelagic fish that form shoals, such as herring (Figure 1.5). Also used for this purpose are drift nets, which just hang from floats and may be up

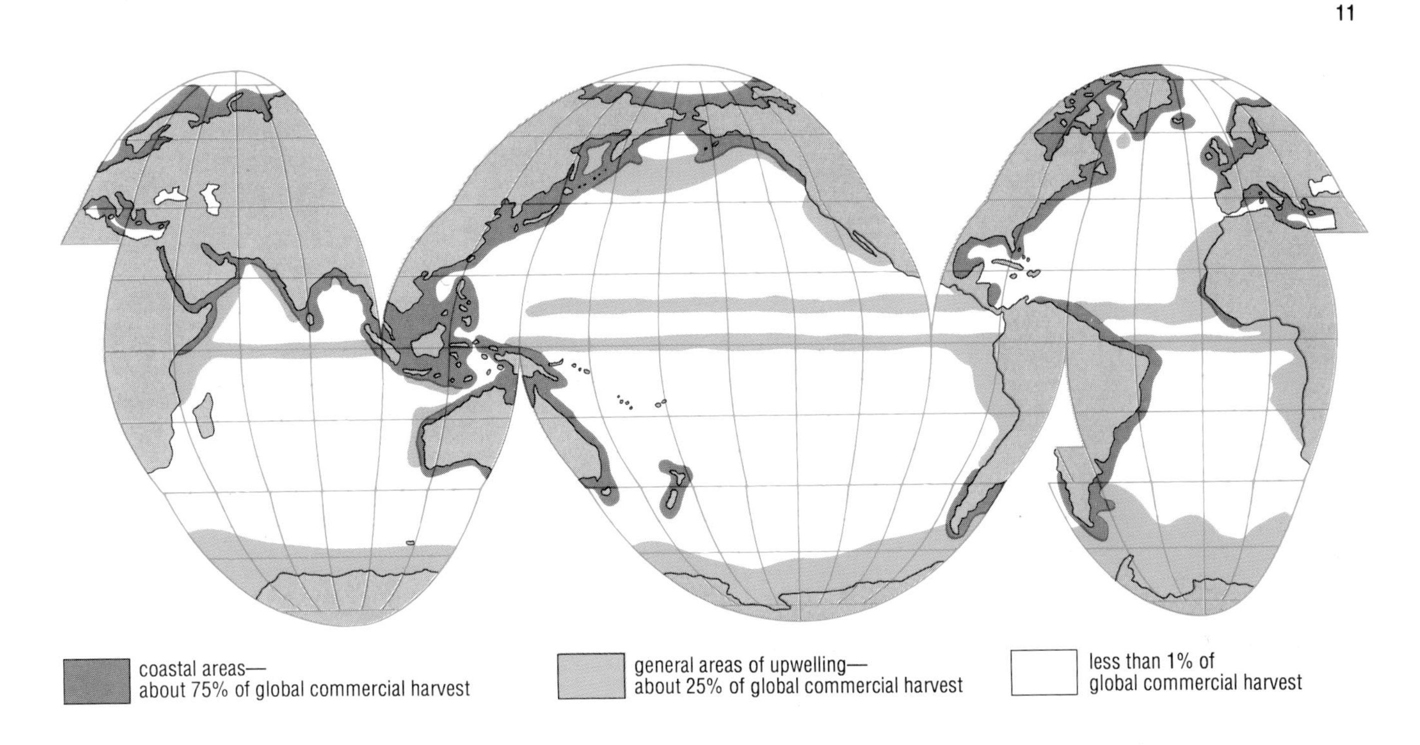

Figure 1.4 The distribution of the world's fisheries, in particular coastal areas and those characterized by areas of **upwelling**.

to 50 km long. This method began to attract much adverse publicity and international protest during the 1980s, because of the shift from visible nets made of hemp, cotton or flax, which eventually rot and sink if not recovered, to invisible plastic filament nets, which do not rot and can float indefinitely if not recovered. Such nets have become an international environmental problem because of their widespread and often careless application, not only in coastal waters but also in the deep sea, and the problem extends from Europe to the Pacific. According to

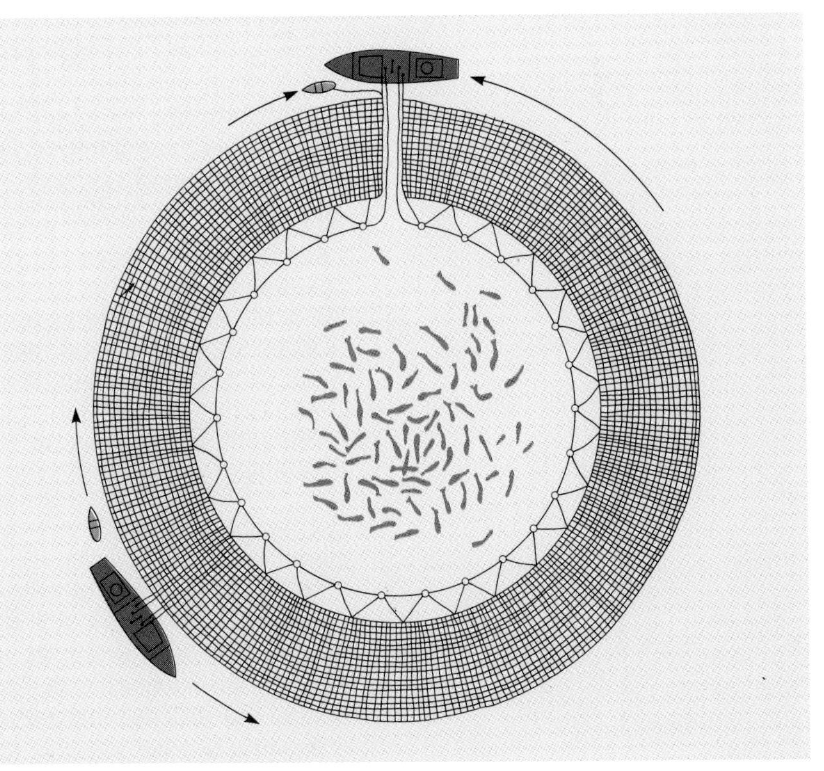

Figure 1.5 The purse-seine net. Moving in a circular path, the boat unwinds a long net until the circle is closed. The rim of the net is kept on the surface by floats while the bottom sinks. Fish inside the net are trapped when a 'purse string' draws the bottom of the net closed.

some authorities, these nets have killed off more seabirds and many more sea mammals than all oil spills put together. In the Pacific, for example, Japanese, Taiwanese and Korean fishing vessels alone have routinely deployed tens of thousands of kilometres of plastic filament nets, entangling and killing tens of thousands of dolphins, sharks, turtles, seabirds and other animals, and threatening some species with extinction. In 1989, the New Zealand Prime Minister called the nets a 'wall of death' in a speech to the UN General Assembly; the Japanese Fisheries Association opposed any attempt to prohibit the use of drift nets until 'scientific evidence' showed the practice to be harmful, but then bowed to international pressure and agreed to stop using drift nets, at least temporarily.

Mackerel are commonly caught using lines of baited hooks, up to 24 km long, a technique also used for cod where the bottom is too rough for trawling, and for deep-water pelagic fish such as tuna (also caught with drift nets nowadays). New deep-water fisheries have been developed for stocks of krill (a small crustacean, the staple food of baleen whales in the Southern Ocean) and cephalopods, which live at intermediate depths (e.g. squid round the Falkland Islands).

1.2.2 OVERFISHING

Until about a hundred years ago the oceans had always seemed to represent an infinite source of food, but technological advances in the 19th century led to devastating onslaughts on some fishing grounds. During the 1880s, for example, steam trawlers were introduced into the North Sea. Not only were they relatively independent of the weather, but they also carried more efficient fishing gear than the old sailing smacks. By the turn of the century, at least 600 such ships operated from British harbours, and other European countries possessed similar fleets. Expansions of fishing fleets also occurred in North America and Japan. Catches began to decline. For example, during the last decade of the 19th century the **catch per unit effort** (the average catch per vessel per fishing period) in the North Sea fell by a third, and catches began to fail completely in some other fishing grounds. But as catches declined prices rose, making it economically—if not environmentally—viable to go on fishing.

Not only has our ability to catch fish developed over the past hundred years, but the total demand for fish has also increased, for direct consumption as well as for fish meal and fertilizer. For example, from the 1950s to the 1980s, the total global catch of marine fish, including shellfish but excluding whales, rose from about 20 million tonnes a year to about 70 million tonnes, and then levelled out at 70–80 million tonnes.

Partial solutions to the problem of overfishing lie in fish farming (aquaculture or mariculture of both fish and shellfish, *cf.* Figure 1.3), and in exploiting non-traditional species, such as those living at mid-depths (*c.*1 000 m) over the continental slope; but this may require consumer re-education in some countries ('this fish may *look* 'orrible, but it's very tasty') if it is to be economically successful. However, catches of traditional species must still be limited, and this causes political problems. The uneven distribution of fish in the seas and oceans, the fact that some countries are economically more dependent on fish than others, and scientific disagreements on the precise definition of overfishing are just a few of the potential sources of conflict.

In order to resolve such conflicts and to minimize overfishing, international agreements to limit catches will be necessary, and these agreements should be based on scientific data relating to the numbers, rate of reproduction and ecology of the fish species concerned. Solutions may, however, depend more on national attitudes than on basic science.

1.2.3 THE PROBLEMS OF REACHING AGREEMENT

Obviously, the more fishing activity there is and the greater the number of nations involved, the harder it will be to achieve universal agreements, not least because the newly independent and often underdeveloped countries entering the field will not necessarily share the political, social and economic assumptions of the traditional maritime nations. For example, a newly independent ex-colonial nation with rich offshore fishing grounds and acutely aware of its own relative poverty is unlikely to take kindly to the continued exploitation of its resources by a rich highly developed power, however much the latter may regard those fishing grounds as its traditional preserve. Nor, as it begins to develop its fishing industry, will the new nation welcome suggestions from traditional maritime nations that the time has come to regulate the use of fishing grounds, even if regulation can be justified according to scientific criteria.

The increased scale and variety of fishing techniques has arisen in part from the uneven distributions of fish in both space and time—a natural circumstance to which different nations have responded in different ways. Distant-water fishing fleets and factory ships are obvious examples of such responses, as is the ability of the Japanese to roam the world for tuna. Peru, on the other hand, was for a time the world's foremost fishing nation in terms of fish landed and fishmeal produced, by exploiting the large anchovy resources just off its coast. The technology used was relatively modest but nevertheless enabled the Peruvians to raise their annual anchovy catch from practically nothing in 1950 to more than 12 million tonnes in 1970. Unfortunately, the greater the diversity of fishing activity the more difficult it is likely to be to frame common rules. Moreover, the ability of certain advanced nations to exploit a wider variety of fish off the shores of a greater number of other nations is an obvious spur to those other nations to protect their economies. All this may contribute to overfishing as well as raising the temperature of international relations.

Persistent overfishing will lead to more or less permanent diminution of the fish stock and perhaps ultimately to extinction. Unfortunately, however, theories about overfishing and the associated concept of **sustainable yield** (the rate at which fish may be caught without permanently depleting the stock—i.e. without overfishing) are ambiguous; there is usually disagreement on what they mean in numerical terms for any given species. Although the need for resource management and conservation is now widely accepted, there are often differences of opinion about how to optimize such activities, as subsequent Chapters will show.

Different countries have different degrees of dependence upon fish and thus react quite differently when natural and/or political events conspire to change the patterns of fish exploitation. One way of assessing relative dependences upon fish would be simply to rank nations in terms of the annual tonnage of fish caught. However, this would not represent a good measure of dependence, for it takes no account of relative populations, levels of investment, standards of productivity, savings on foreign exchange, the sometimes seasonal nature of employment in the fishing industry, and so on. Iceland, for example, has traditionally caught less fish in total than Britain; but practically its whole economy is based on fish, whereas if all fish were to disappear tomorrow Britain would be severely inconvenienced but hardly ruined.

The subject of whaling is frequently in the news and it is a long-running saga. In the days of sail, whalers hunted only the five species of 'right' whale which could be overhauled by rowing boats and floated after death (Figure 1.6). Hunted with savage intensity, these whales were probably saved from extermination only by a sharp decline in the whale-oil market late in the 19th century. The introduction of faster

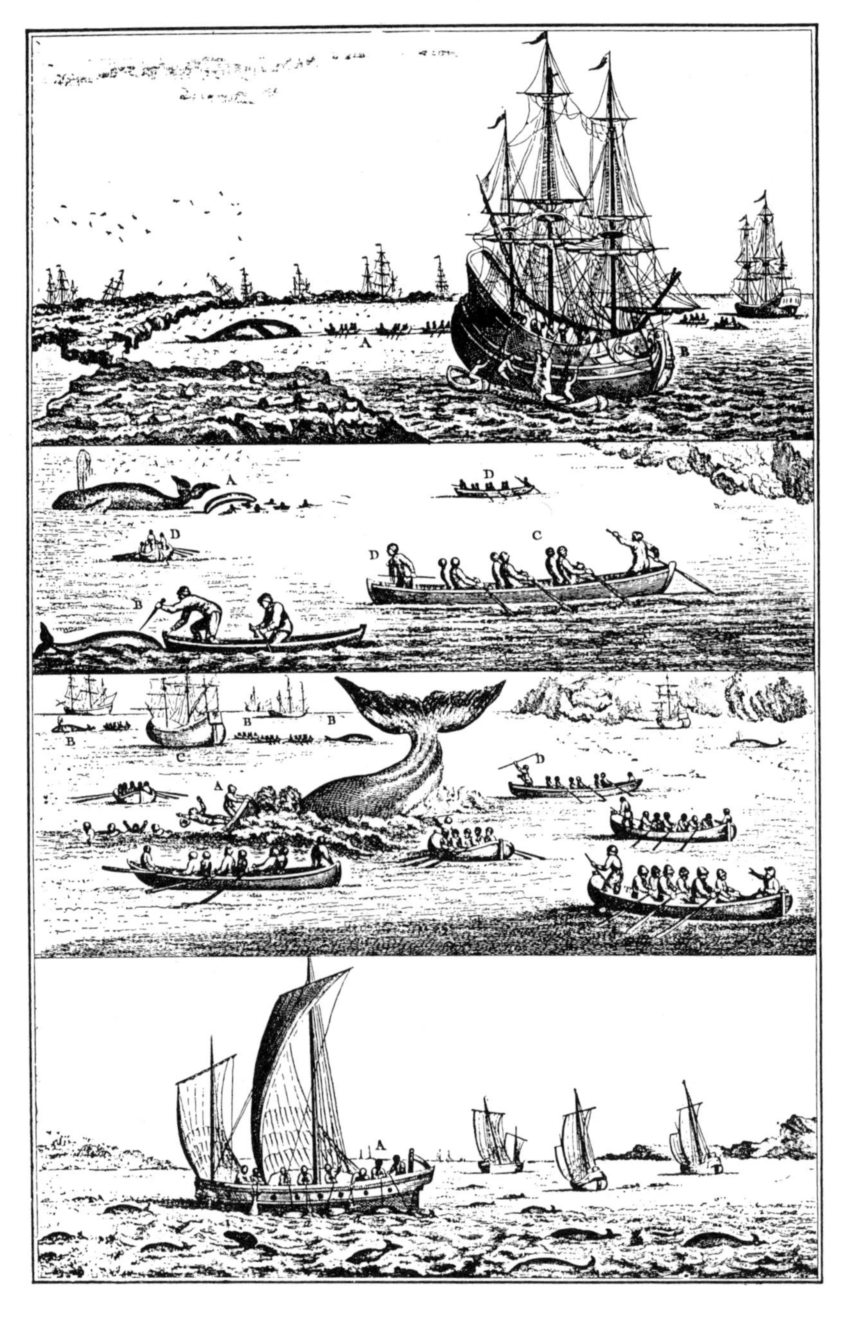

Figure 1.6 17th-century Dutch whalers operating off Spitsbergen.

ships and more effective techniques, such as the explosive-headed harpoon and the use of compressed air to make dead whales float, added another seven species of great whale to the number that were hunted. The International Whaling Commission (IWC) was set up in 1946 (entering into force in 1949), in an attempt to preserve whale stocks by setting annual catch quotas for each species. Despite the continuing decline of stocks, annual IWC meetings and UN conferences and conventions, and repeated attempts to impose a total moratorium on whaling in the succeeding decades, some nations are still killing significant numbers of whales, often using scientific research as the excuse.

1.2.4 POLLUTION

Until recently, it was legitimate in practice if not in theory to regard the oceans as an infinite sink for the absorption of anthropogenic wastes, where everything from old cars to radioactive wastes could be discarded and forgotten. But this is no longer so. The vast increase in the quantity and variety of wastes entering the oceans has become apparent at the most mundane level with the accumulation of plastics, not only along coasts but even out in mid-ocean—and plastics have been in general use only since World War II. Near Dover on the south coast of England, for example, there are accumulations of polystyrene 'sands', the disintegration products of coffee cups and other disposable utensils discarded from cross-Channel ferries and other ships.

Oceanic pollutants include toxic chemicals carried down to the sea by rivers; industrial gases and particles and radioactive fall-out from the atmosphere; wastes of all kinds (toxic, organic, radioactive) dumped directly into the sea (Figure 1.7 (a)) or indirectly as fall-out from incinerator ships (expected to be phased out by the mid-1990s); oil deposits ejected from tankers and oil wells both accidentally and

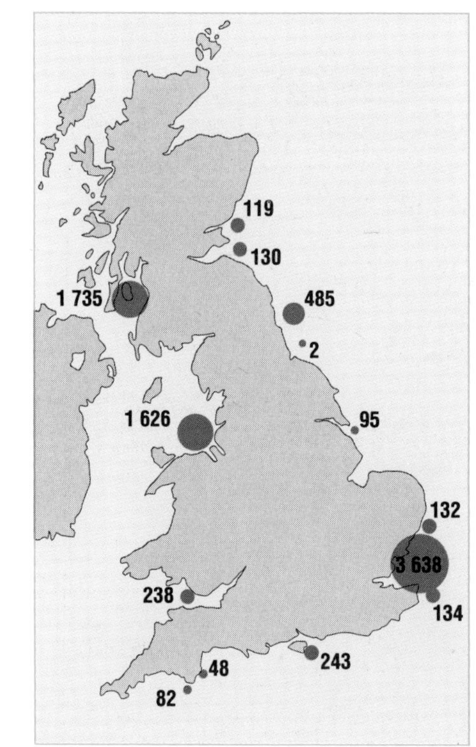

(a)

(b)

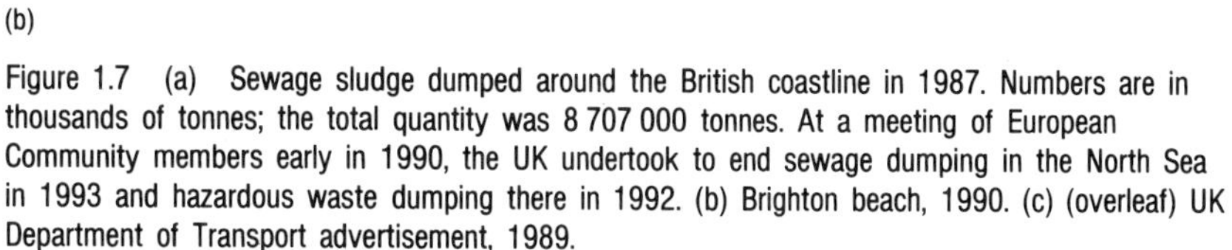

Figure 1.7 (a) Sewage sludge dumped around the British coastline in 1987. Numbers are in thousands of tonnes; the total quantity was 8 707 000 tonnes. At a meeting of European Community members early in 1990, the UK undertook to end sewage dumping in the North Sea in 1993 and hazardous waste dumping there in 1992. (b) Brighton beach, 1990. (c) (overleaf) UK Department of Transport advertisement, 1989.

OVER THE SIDE IS OVER!

The disposal of any plastic into the sea is prohibited.

Other forms of garbage may be discharged only under strictly controlled conditions.

Dunnage, lining and packing material which float can only be disposed of at sea more than 25 miles from land.

Food wastes and all other garbage (including paper products, rags, glass, metal, bottles and crockery) cannot be discharged within 12 miles of land unless it has first been passed through a grinder. Even so, the minimum distance from land at which discharge is permitted is set at 3 miles.

Reception facilities for garbage from ships are available at ports and harbours.

Disposal of garbage is governed by the provisions of The Merchant Shipping (Prevention of Pollution by Garbage) Regulations 1988.

The Merchant Shipping (Reception Facilities for Garbage) Regulations 1988.

In force from 31 December 1988.

OVER THE SIDE IS OVER!

Now you can't do this.

Litter spoils beaches, for everyone.

Plastic waste maims and kills.

Chemicals and their containers are dangerous.

Litter that injures, especially children.

Discarded plastic is a danger to ships.

Seal photo by Charles W Fowler Propellor photo by Ambassador Marine Ltd deliberately; and sewage entering through pipelines. You can no doubt think of other examples. Local catastrophes (e.g. oil spills and the loss of leaky drums containing radioactive or toxic chemicals) that have occurred in the past (Figure 1.7(b)) are unlikely to decrease in frequency or magnitude; though efforts to reduce the quantities of waste entering the sea are increasing in some countries (Figure 1.7(c)).

The control of oceanic pollution requires knowledge about:

1 The pathways by which pollutants reach the oceans.

2 The rates of input and dispersal of pollutants in the oceans.

3 The forms in which pollutants occur and how they behave in oceanic systems.

Estimates of the flux of natural and artificial materials into the natural environment enable us to place pollution in perspective.

QUESTION 1.1 Many human activities merely increase the amounts or concentrations of natural substances in the environment. So why should we look upon such activities as causing pollution? Consider (a) inert material such as dredged sediments from harbour clearance operations and (b) organic wastes of domestic and agricultural origin.

Heavy metals such as cadmium, chromium, copper, lead and zinc all occur naturally in seawater, albeit in very low concentrations. Natural processes can cause concentrations of many of these metals to reach levels in marine organisms that are many tens of thousands of times greater than in the ambient seawater. Heavy metals are discharged into the sea from industrial complexes along many of the world's coastlines, significantly increasing the ambient concentrations to levels that may be toxic to life.

Table 1.1 The sources and modes of transport to the oceans of major marine pollutants. A, most released directly into the oceans; B, most released within the boundaries of sovereign territories; C, all released within the boundaries of sovereign territories.

Pollutant	A	B	C	Chief mode of transport to oceans
Mercury		✓		atmosphere
Lead			✓	rivers, atmosphere
Pesticides			✓	atmosphere
Polychlorinated biphenyls (PCBs)			✓	atmosphere
Perchlorethylene			✓	atmosphere
Petroleum	✓	✓		atmosphere
Radioactive elements		✓		atmosphere, rivers
Heat		✓		atmosphere, rivers

Note: Lead, pesticides and PCBs are becoming less common as marine pollutants through stricter legislation controlling their use, introduced in the 1980s. PCBs and other halogenated hydrocarbons are insoluble in water but soluble in fats and oils, including the fatty tissue of animals and birds, where they can accumulate with adverse effects on life and reproductive cycles.

Table 1.1 shows some of the chief marine pollutants, their sources and the pathways by which they reach the oceans. The almost complete lack of entries in column A makes the message clear: most marine pollutants originate on the continents and reach the oceans via the atmosphere and, to a lesser extent, rivers. E.W. Seabrook Hull and Alfred W. Koers, when at the Woodrow Wilson International Center for

Scholars (United States), put it this way:

> '... marine pollution cannot be separated and isolated from global pollution generally. Indeed it is an avoidance of reality, which will assure failure of any such effort, to turn one's back on the land, to ignore the atmosphere and then to try to develop effective means for the international control of pollution of the marine environment alone. Ocean pollution cannot be controlled unless the release of the materials that pollute the oceans can be controlled. These are generated mainly by activities entirely within the boundaries of states and they are transported primarily by the atmosphere, secondarily by rivers and only tertiarily by specific acts of man. Thus, ocean pollution control requires the control of human polluting activities everywhere. It is as simple as that.'

The detailed impact of marine pollution will depend on:

1 The reactions that pollutants take part in.

2 The extent to which pollutants accumulate in different parts of the oceanic environment.

Some anthropogenic pollutants have no natural counterparts (e.g. DDT, PCBs—polychlorinated biphenyls). Many others differ from their natural counterparts only in their chemical form (or **speciation**), but may then *behave* quite differently because of their greater availability to marine organisms. For example, until the 1960s, only inorganic forms of mercury were thought to be involved in marine processes; but in 1963 an outbreak of crippling and fatal poisoning in the Minamata Bay area of Japan was traced to locally caught fish accumulating organic mercury compounds from industrial effluents. Organic mercury compounds were in fact subsequently found also to occur naturally—but natural concentrations are not high enough to cause toxic effects.

QUESTION 1.2 Mercury is a particularly controversial element, for it is volatile, highly toxic and easily absorbed by marine organisms, as the Minamata Bay disaster showed.

Here are some figures:

Annual world industrial production	$\sim 9 \times 10^3$ t yr^{-1}
Estimated natural river-borne flux (in solution)	$\sim 4 \times 10^3$ t yr^{-1}
Estimated loss from the Earth's crust direct to atmosphere (by sublimation/evaporation)	$\sim 2 \times 10^4$ to 2×10^5 t yr^{-1}

(a) Do you think that the mercury content of the world's oceans is controlled by natural or artificial processes?

(b) What is probably the chief pathway for mercury to the oceans?

The coastal zone is particularly vulnerable to human disturbance. It can be affected by environmental changes inland as well as in the sea. Simply damming a river, for example, can greatly reduce the flow of freshwater, nutrients and sediments, with profound effects on both sediment movement and ecosystems in the coastal zone. The building of the Aswan Dam on the Nile completely destroyed the important delta fishery in the eastern Mediterranean. Coastal ecosystems are rich and varied and their diversity can be progressively eroded by urbanization, industrial development, and their effluents. The world's wetlands, which are transitional coastal regions comprising various combinations of tidal flats, salt-marshes, mangrove swamps, islands and inlets, are especially threatened by industrialization and reclamation. They are important wildlife habitats and frequently provide irreplaceable nursery grounds for economically important marine species.

Pollution knows no boundaries and the atmosphere is an international pathway (Table 1.1). Thus, significant amounts of polychlorinated biphenyls (PCBs) have been found in bottom sediments as far apart as the Baltic, the Mediterranean and the Sargasso Sea. Ash particles in Sargasso Sea sediments probably also come via atmospheric fall-out from coal-fired power stations in North America. Fall-out from the Chernobyl nuclear reactor explosion and fire in 1986 was found in the waters and sediments of the Mediterranean within a matter of weeks.

Deep-sea sludge dumping

Figure 1.8 represents a fairly typical situation: most dumping at sea occurs in relatively shallow water not far offshore. Some dumping also occurs at shallow depths in open waters. Either way, there are food-chain pathways for both toxins and pathogens that can lead rapidly back to humans, as most fish are caught in continental shelf waters.

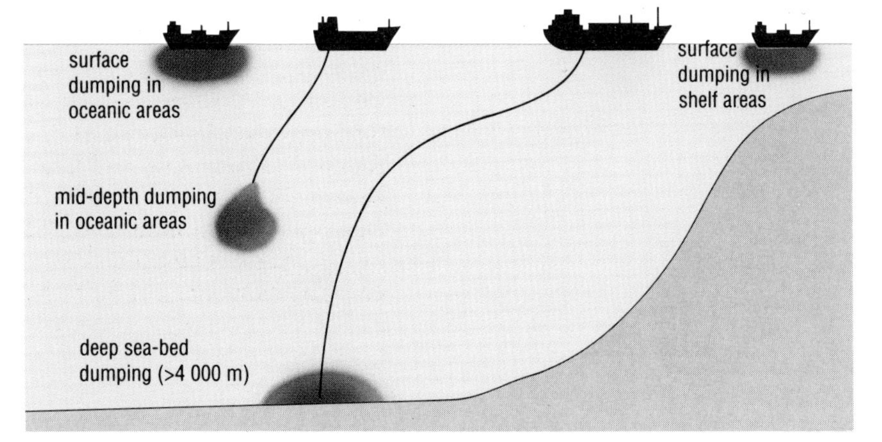

Figure 1.8 Sludge dumped at different locations and depths in the ocean (not to scale).

Disposal in the water column at mid-depths in the open oceans (Figure 1.8) is less likely to damage planktonic production in sunlit surface waters (the **photic zone**), but there may still be pathways back to species caught for human consumption, because some animals migrate through considerable vertical distances. In addition, there is growing interest in harvesting deeper water species (Section 1.2.2). Dispersal through the water column by currents, internal waves and turbulence is also likely—as in the case of shallow-water dumping in open water.

The deep ocean is not the quiet, still environment that it was once thought to be. There are currents, eddies and even unpredictable **abyssal storms** down there. Nonetheless, the turnover time for deep ocean water is hundreds of years, so dispersal of material dumped on the deep sea-bed (Figure 1.8) is unlikely to reach species that affect humans to any significant extent—especially as the relatively high particulate content of sludge will ensure that it spreads out as a sort of abyssal turbidity flow. It will therefore be largely confined to the *benthic boundary layer* (the bottom hundred metres or so of the water column), abyssal storms notwithstanding. Though it should be noted that this somewhat anthropocentric conclusion takes no account of potential damage to the species that *do* live on the deep sea-bed. Any such operations must be monitored, to check, for example, that the **biochemical oxygen demand** of organic matter in the sludge does not reduce dissolved oxygen concentrations to levels that can have widespread adverse biological and geochemical impacts.

1.3 PHYSICAL RESOURCES

Seawater has been a physical resource since prehistoric times. The extraction of salt by solar evaporation must have been among the first chemical processes carried out by early humans. The distillation of freshwater from the sea has also been practised for many centuries: while besieging Alexandria, Julius Caesar's army was supplied with water from a distillation apparatus. Freshwater, salt and other chemicals are still extracted from seawater on a commercial scale today (Table 1.2).

Table 1.2 The production of chemicals from seawater.

Commodity	Approx. total annual world production (10^6 tonnes)	Approx. annual production from seawater (10^6 tonnes)	Approx. percentage of total
Water (H_2O)	350	270*	75
Salt (NaCl)	140	40	30
Magnesium compounds (MgO)	10	0.5	5
Magnesium (Mg)	0.2	0.1	50
Bromine (Br)	0.1	0.03	30

*Water produced by desalination, mainly from seawater, but also from saline lakes.

QUESTION 1.3 Examine Table 1.2.

(a) Which product is obtained in greatest quantity from seawater, and in what regions would you expect this to be important?

(b) For which industrial chemical does extraction from seawater provide the largest proportion of the total production?

Virtually every coastal country has at some time produced sea salt commercially and many still do, either by industrial processes or by solar evaporation. Most of them are labelled on Figure 1.9, which also shows places where the other products listed in Table 1.2 are obtained. In pre-industrial times, salt was produced almost entirely for human consumption, but it is now used mainly to manufacture chemicals. In the USA, for example, the chemical industry takes two-thirds of the salt produced, and about half of that goes to make chlorine and sodium hydroxide. Magnesium hydroxide precipitated from seawater is used to produce magnesium and its compounds. Bromine is released from seawater as a gas and then condensed to liquid.

However, desalination can have important ecological side-effects. The removal of freshwater from seawater leaves highly concentrated brines to be disposed of. They are much denser than seawater, so if released from pipes on the sea-bed, for example, they simply flow along the bottom, mixing only slowly with the surrounding normal seawater. This could easily destroy bottom fauna and flora, and organisms in the water above could also suffer, either directly or through the destruction of their food

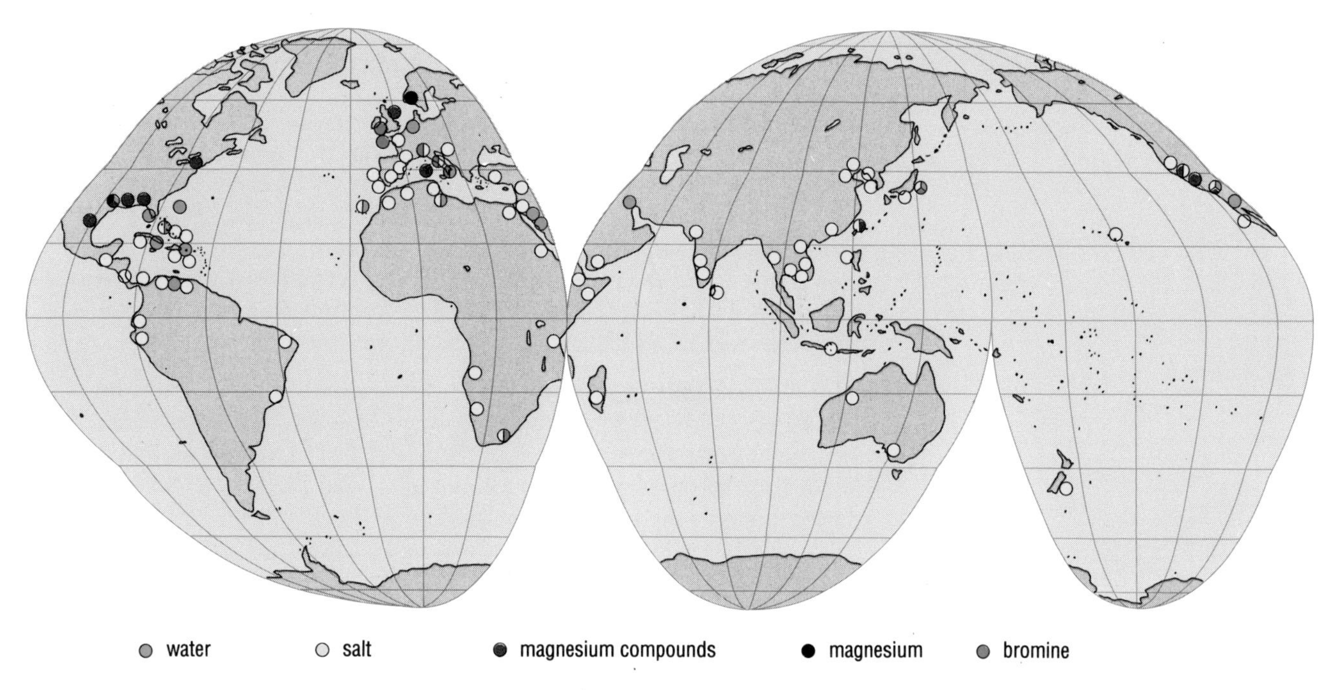

Figure 1.9 Centres for producing chemicals from seawater.

supplies. Waste brines must therefore be dispersed as widely as possible, and selecting the most effective method requires detailed knowledge of water movements near the processing plant.

Could anything else be done with the waste brines?

Combined desalination and salt-extraction plants might be an economic proposition. Of course, the brines will contain greatly increased concentrations of other industrially important elements and it may be economic to extract some of them—certainly more economic than processing 'raw' seawater.

Most of the naturally occurring elements have been detected in seawater, the great majority at extremely low concentrations (in the order of parts per *billion* (10^{-9})), which makes extraction direct from seawater an uneconomic proposition. Measured concentrations of many trace elements in seawater have decreased progressively over the years, as marine chemists have improved their analytical techniques and eliminated contamination. For example, until about the 1930s, the accepted average concentration of gold in seawater was 65 milligrammes per litre ($65\,\mathrm{mg\,l^{-1}}$). By the 1950s, this had gone down to $4 \times 10^{-6}\,\mathrm{mg\,l^{-1}}$ (four-thousandths of a microgramme per litre, a reduction of some seven orders of magnitude); and this value was accepted until measurements during the 1980s indicated an average as low as $2.5 \times 10^{-8}\,\mathrm{mg\,l^{-1}}$, suggesting that there are barely 40 000 tonnes of gold in the whole of the oceans.

The oceans yield many other kinds of physical resources, especially from the continental shelves. New methods are also being developed to recover minerals from the deep oceans, but the extraction of minerals in international waters presents the usual complex legal and political problems of ownership, as we shall see in Chapter 3. Interest in harnessing the energy of waves, tides and currents has been sharpened since the 1970s by the realization that excess carbon dioxide produced by burning fossil fuels and by deforestation may be causing climatic changes which could destabilize global ecological processes.

1.3.1 CONTINENTAL SHELVES

The continental shelves are geologically part of the adjacent continents, so their physical resources are not necessarily bounded by the shoreline. There is great variety in the form and structure of continental margins, depending upon their geological and plate-tectonic history. Of special interest are thick sequences of sediment forming many continental shelves, some of which may contain oil, gas and other minerals.

Not all sediments of continental shelves are marine: some have been deposited by rivers or represent drowned beaches or coastal plains, for coastlines are ephemeral features. Sea-level has varied by as much as 200 metres in the past couple of million years, and there have been greater fluctuations in the more distant geological past.

Surface resources

These occur in many places (Figure 1.10) and comprise various sedimentary deposits resulting either from physical sorting processes or from chemical precipitation. Sands and gravels are the most abundant surface deposits on continental shelves, being composed of common rock-forming minerals; in many places they consist principally of quartz. They provide reserves of **aggregate** for concrete, as planning controls increasingly restrict supplies of land-based deposits. The sand and gravel resources of the US continental shelf alone are estimated to be in the order of $1\,400 \times 10^9$ tonnes, 800×10^9 tonnes of them round the Alaskan coast, though only about 5 million tonnes are extracted by the US each year. Britain takes some 12–15 million tonnes annually, chiefly from the North Sea, while Japan extracts about 85 million tonnes of sands and gravels from its continental shelf each year.

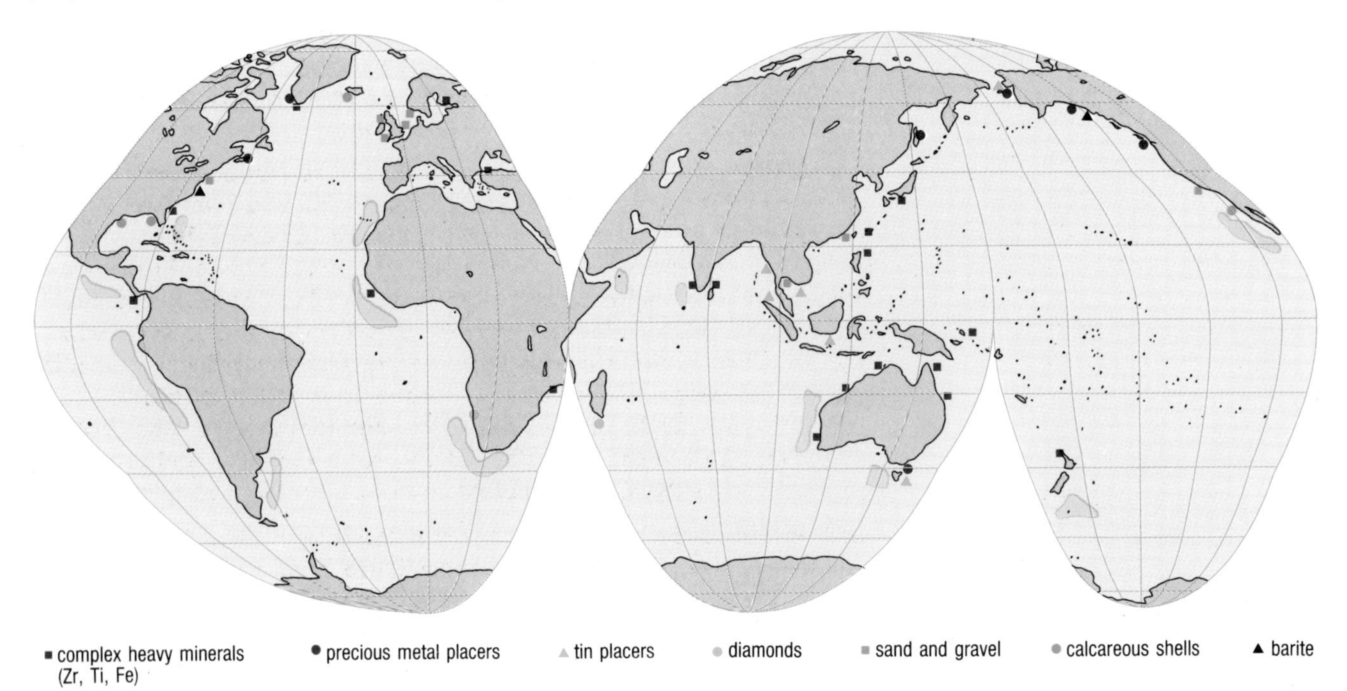

Figure 1.10 The approximate locations of identified aggregates, placer deposits and phosphorites.

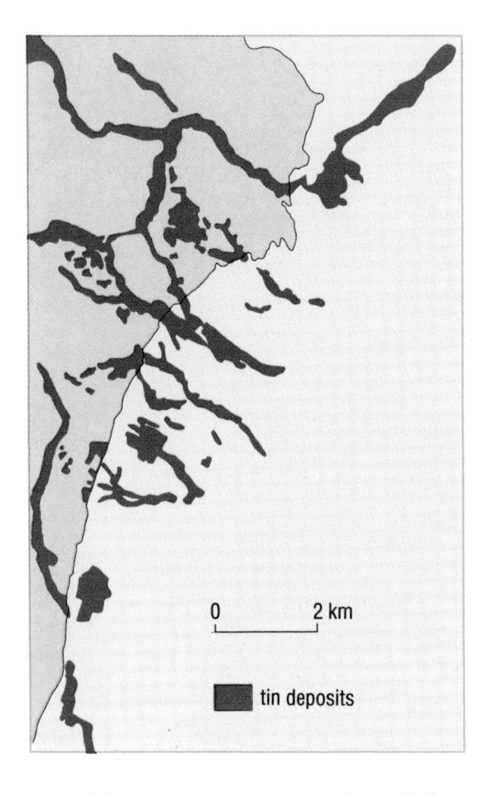

Figure 1.11 Placer deposits of cassiterite (SnO_2) formed in streams draining Singkep Island in Indonesia, at a time of low sea-level during the Pleistocene. When the sea-level rose, they were partly submerged, but locating them was fairly simple because they are extensions of old stream channels. The deposits are now worked by offshore dredging.

In so-called **placer deposits**, various combinations of complex heavy minerals, tin ore (cassiterite), precious metals and diamonds (Figure 1.10) have been concentrated into workable accumulations by currents and waves that have winnowed out most of the less dense mineral particles. Some placers are submerged fluvial and shoreline deposits formed at times of lower sea-level (Figure 1.11).

From Figure 1.11, can you suggest a simple method of delineating potential areas of offshore placer deposits?

As the submarine deposits may be continuations of placer concentrations in rivers that were partly drowned by the rising sea-level, the occurrence of coastal placer deposits may well be a pointer to offshore concentrations.

Calcareous shell deposits (Figure 1.10) can provide important sources of calcium carbonate (limestone), for example on volcanic islands lacking land deposits. In lower latitudes, where the water is warm enough for reef formation, limestone from coral can also be an important resource. Another type of chemically precipitated mineral deposit is **phosphorite** (Figure 1.10) in accumulations consisting of crusts, nodules and pellets of calcium phosphate, initially formed by primary biochemical processes and then subjected to physical and chemical modification on the sea-floor. Exploitation of phosphate deposits of continental shelves is not yet economic because large reserves of higher grade exist on land. They do represent a large potential resource, however, estimated at 8×10^{10} tonnes—enough for 100 years at present rates of consumption. Chemically precipitated concentrations of *barite* ($BaSO_4$) are also locally abundant on continental shelves (Figure 1.10), but are not yet worked on any scale.

Sub-surface resources

Many mineral deposits beneath continental shelves are also extensions of underground deposits on land. Examples include the coalfields of north-eastern Britain, where workings extend beneath the North Sea, and the tin lodes of south-west England, which were worked below the coastal waters off Cornwall in the 19th century.

Even petroleum fields, representing the most important sub-surface resources of the continental shelves, frequently have their landward connections (e.g. Gulf of Mexico, North Sea, Niger delta); but the richest accumulations in these areas are commonly offshore. In the late 1980s, total potential hydrocarbon resources offshore were estimated to amount to more than half the total on land. Geophysical surveys define the overall structure of hydrocarbon-bearing basins and identify potential oil and gas traps. Drilling for further exploration and subsequent production is fraught with much greater difficulties than on land because of the need to contend with often adverse sea conditions—and in polar regions there is the added complication of ice cover (see Chapter 4).

Salt deposits commonly occur at or near the base of the sediment of the continental shelf, where they were precipitated from seawater in enclosed basins. Under pressure, the deposits are plastically deformed and forced upwards as plug-like columns, doming and then punching through denser overlying sediments (Figure 1.12). The structures associated with them commonly act as traps for oil and gas, so the identification of salt domes is important in petroleum exploration. Some salt domes contain

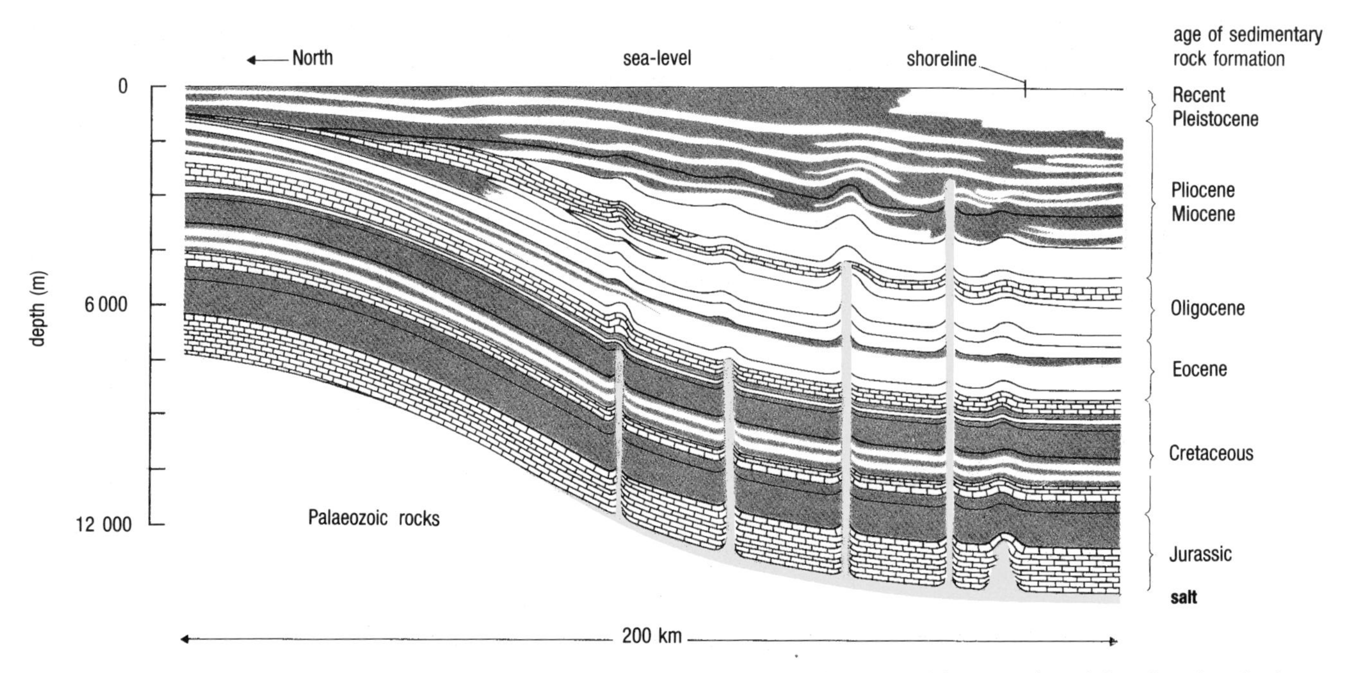

Figure 1.12 Salt domes beneath eastern Louisiana (Gulf of Mexico). Some of the domes have risen through more than 10 000 m of sediment from salt deposits at the base of the sequence (note that the vertical scale is highly exaggerated). This situation is typical of that found in many continental shelf regions.

elemental sulphur in commercial quantities, produced by the chemical reduction of the mineral anhydrite ($CaSO_4$)—one of the salts precipitated from seawater—by hydrocarbons.

1.3.2 THE DEEP OCEAN

Deep-sea sediments consist mainly of very fine-grained clays and calcareous or siliceous microfossils (the skeletal remains of dead **plankton** from the surface layers) in varying proportions. They accumulate at maximum rates of only a few tens of metres per million years. Deep-sea clays are mostly reddish in colour because they are well oxidized; and so they contain little organic matter and hydrocarbons are unlikely to form in them. Thicker accumulations of more varied sediment are brought down onto deep ocean floors by **turbidity currents** from the continental shelves, and sometimes display surface features resembling river systems on land (Figure 1.13). Such sediments could contain hydrocarbons, especially in small ocean basins, where they might have accumulated rapidly to considerable thicknesses—but any hydrocarbons found in them would be difficult and expensive to exploit. However, deep-ocean sediments do contain resources of a quite different kind: manganese nodules and fine-grained metal-rich sediments.

Manganese nodules consist predominantly of manganese and iron oxides and hydroxides arranged in concentric layers round a nucleus (Figure 1.14(a)). Their importance lies not in their content of manganese (*c.*8–40%) and iron (*c.*9–27%), but in their ability to 'scavenge' and incorporate other metals, chiefly copper, nickel and, to a lesser extent, cobalt, which can reach combined concentrations of up to about 3%.

The nodules range in shape from spheroidal to flat; some are only a few microns in diameter, others weigh several tonnes. They are unevenly distributed, but in places may cover more than 50% of the ocean floor (Figure 1.14(b)). They are confined to the topmost layers of sediment, but even so the visible tonnage on the floor of the Pacific Ocean has been estimated at 10^{11} tonnes.

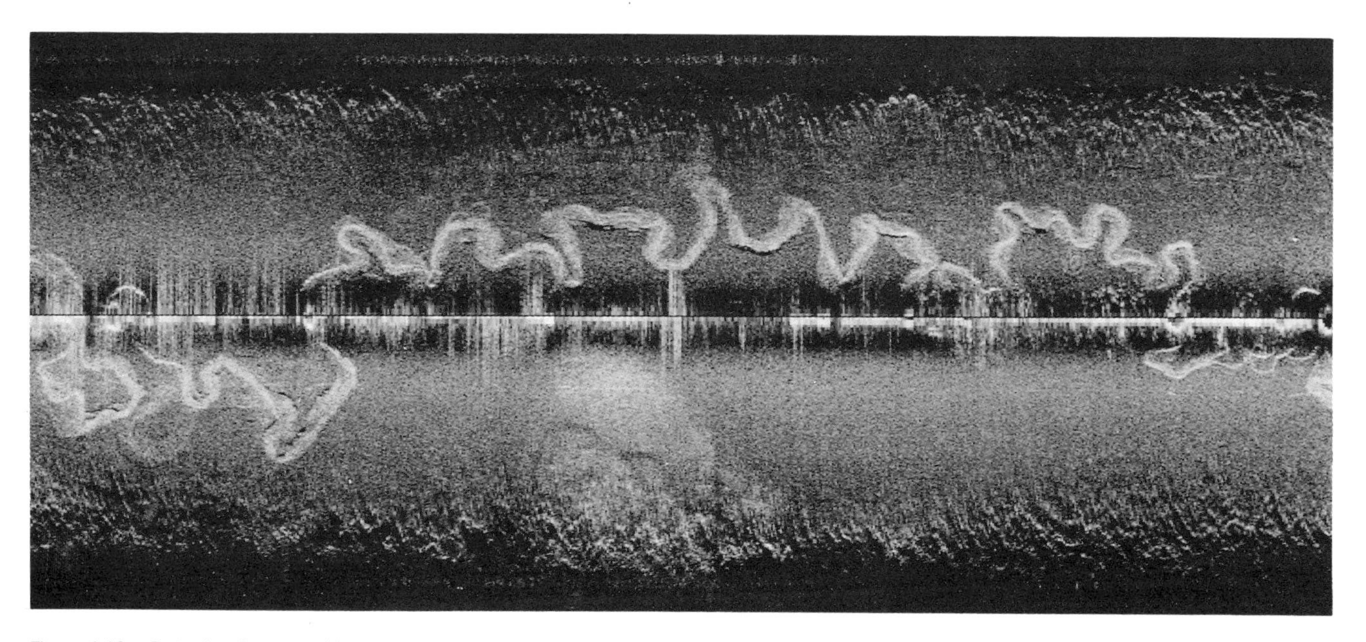

Figure 1.13 Part of a deep-sea side-scan (GLORIA) record, showing a meandering channel on the Indus fan, formed by slowly flowing turbidity currents.

(a)

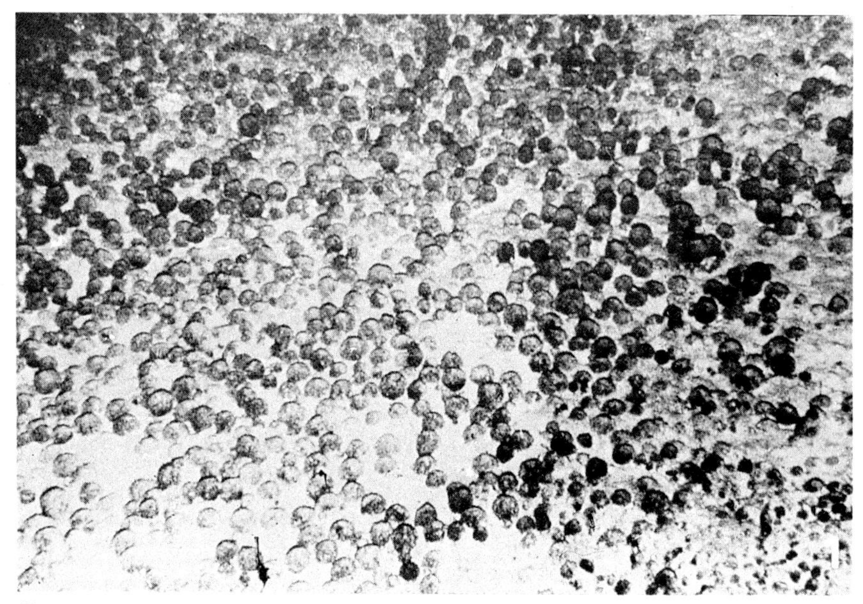

(b)

Figure 1.14 (a) Manganese nodules from 5 400 m depth in the eastern North Atlantic at 31° 25′ N 25° 15′ W. The largest nodule (*c.* 3 cm diameter) has been sectioned and polished, to reveal the concentric structure and the nucleus of altered volcanic rock (basalt) about which it has grown. (b) Typical appearance of a rich manganese nodule field on the deep ocean floor.

A scheme for the recovery of nodules on a commercial scale is illustrated in Figure 1.15. We are dealing here with resources on the sea-bed in international waters, and the problems in reaching agreement on who should benefit from their extraction are considered in Chapter 3. Important scientific questions also remain unanswered, not least about how their exploitation will affect marine ecosystems.

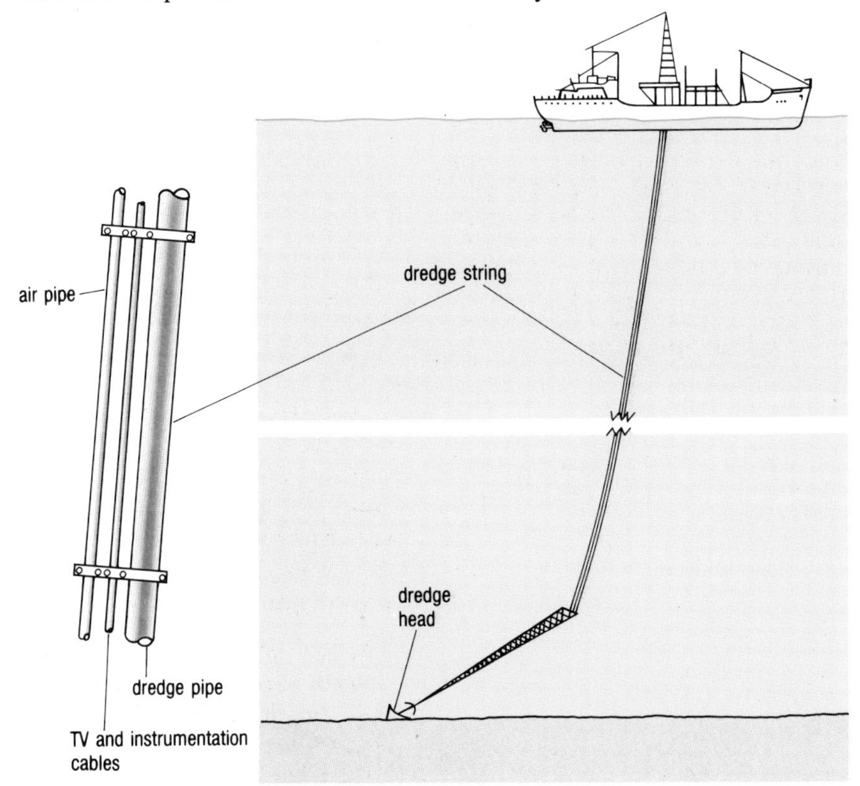

Figure 1.15 The suction method for recovering manganese nodules from the ocean floor.

Pockets of metal-rich sediments form locally in many places along active **spreading axes** of the mid-ocean ridge system, and contain a wider range of metals than those in manganese nodules. The sediments consist mainly of metal sulphides and oxides that have been precipitated from high-temperature **hydrothermal** solutions, generated as seawater circulates through hot oceanic crust and leaches metals and other elements from the basaltic rocks. Precipitation occurs most spectacularly from '**black smokers**' (Figure 1.16), as plumes of the superheated hydrothermal solutions at 300–350 °C emerge from the vents and mix with cold seawater.

Underlying the vents are large and extensive stockworks of metal sulphides and other minerals, precipitated in cracks and fissures in upper parts of the oceanic crust. Each stockwork is a potential resource of millions to tens of millions of tonnes, much greater than the metalliferous sediments that surround the vents at the sea-bed. However, economic exploitation is some way off, because these are sub-surface deposits within oceanic crust which lies 2–3 km below sea-level in mid-ocean settings; and exploitation is also likely to be accompanied by considerable environmental disruption.

The existence of these stockwork deposits was recognized only in the late 1970s but scientists have known since the 1960s that large quantities of metal-rich sediments are accumulating in axial deeps of the Red Sea

Figure 1.16 A 'black smoker' in the axial zone of the East Pacific Rise. Heated and chemically changed seawater emerges from the sea-bed as a clear fluid at about 350 °C, and immediately precipitates metal sulphide particles on contact with cold bottom water, building the vent chimney and forming the dense plume of black 'smoke'. The vent is about 20 cm across.

(Figure 1.17). This is a young ocean basin, in which the sediments of the continental shelf are underlain by thick salt deposits (*cf.* Section 1.3.1). The metals have been precipitated from highly concentrated hot brines, which are hydrothermal solutions heated as a result of high thermal gradients at the spreading axis and circulating through both the salt and the underlying basaltic crust, which respectively increase the salinity and provide the metal content.

The Red Sea metalliferous sediments represent a more accessible resource than the deep-sea deposits. There are over 50 million tonnes of them in the Atlantis II Deep alone (Figure 1.17). Preliminary trials demonstrated that they can be extracted economically, but full-scale exploitation has been delayed because of the massive impact that this major industrial activity would have on the unique Red Sea environment.

Table 1.3 summarizes some data on the physical resources of the oceans which have been discussed so far.

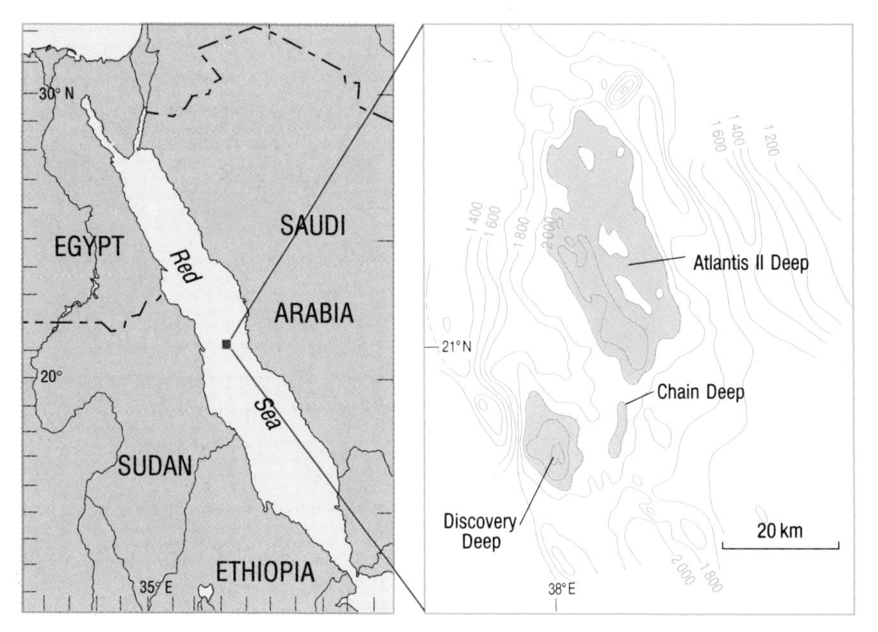

Figure 1.17 Bathymetric details of some major deeps in the Red Sea. Hot, metal-rich brines are found in them, and metalliferous muds are being deposited there. Depths are given in metres.

Table 1.3 Sea-bed materials in world perspective.

Sea-bed deposits	Material commodity	Sea-bed production (10³ t)	World mine production (10³ t)	Estimated average price (US$ per t)	Sea-bed revenues (US$m)*	World revenues (US$m)†	Sea-bed share of world revenues‡ (%)	Sea-bed reported potential resources (10³ t)	World onshore resources (10³ t)	Sea-bed comparison to world resources (%)§	'Resource life'-index (years)‖	Projected onshore depletion by year 2030 (%)¶
Hydrocarbons #	Crude oil	788 834	2 788 913	70	55 218	195 224	28	> 61 429 000	181 857 000	34	65	185
	Natural gas	246 670	1 296 405	95	23 434	123 158	19	> 60 000 000	228 214 000	26	176	45
Sand and gravel	Sand and gravel	112 300	7 620 480	3	334	22 861	1	665 778 000	Very large	Small	Long	
	Industrial sand		181 440	14		2 540		Large	Very large	Small	Long	
Shell	Calcium carbonate	16 667	1 666 667	6	100	10 000	1	90 000 000	Very large	Small	Long	
Sulphur	Sulphur	381	54 000	105	40	5 670	< 1	27 125**	5 000 000	< 1	93	120
Barite	Barite		5 652	31		175		2 087**	453 600	< 1	80	
Phosphorite	Phosphate rock		159 000	24		3 816		7 939 000	129 500 000	6	814	12
Mineral placers	Tin	28	201	6 614	185	1 329	14	2 500	34 500	7	172	105
	Rutile		356	364		130		13 060	181 440	7	510	
	Ilmenite		4 187	49		205		230 500	907 200	25	217	
	Titanium‡		90	12 236		1 101						40
	Zirconium		709	182		129		29 040	54 432	53	77	
	Hafnium		≪1	231 483		17		290	544	53	7 452	
	Yttrium		< 1	35 020		14			172		430	
	Thorium		2	35 850		72		{3 450**}	5 168		2 584	
	Chromite		9 616	42		404		30 158**	32 659 200	< 1	3 396	1
	Gold		1	10 600 000		10 600		< 1**	72	< 1	72	443
	Silver		12	206 667		2 480			743		62	295
	Platinum		≪1	9 000 000		1 980		≪1**	99	≪1	446	13
Nodules and crusts	Platinum§§							2–3		2–3		
	Cobalt		32	25 353		811		6 000–24 000	10 886	55–220	340	
	Nickel		745	5 026		3 744		35 000—131 000	129 730	27–101	174	77
	Manganese		23 406	141		3 300		706 000–2 600 000	10 886 400	6–24	465	17
	Copper		7 805	1 475		11 512		29 000–108 000	1 600 000	2–7	205	86
Massive sulphides	Copper‖‖							5 000–216 000		< 1–14		
	Zinc		6 560	893		5 858		11 000–518 000	1 800 000	< 1–29	274	47
	Lead		3 350	419		1 404			1 400 000		418	46

*Sea-bed production times estimated average price.
†World mine production times estimated average price.
‡Sea-bed revenues times 100, divided by world revenues.
§Sea-bed reported potential resources times 100, divided by world onshore resources.
‖World onshore resources divided by world mine production.
¶Based on low growth case for developing economies.
Hydrocarbons in metric tonnes of oil equivalent.
**Sea-bed estimate for the US only; the number in braces (3 450) is for US sea-bed monazite deposits containing yttrium and thorium.
††Titanium resources are included in rutile and ilmenite resources.
§§See numbers directly above in mineral placers for platinum.
‖‖See numbers directly above in nodules and crusts for copper.

1.4 ENERGY FROM THE OCEANS

The oceans receive energy from solar radiation, from the gravity fields of the Sun, Moon and the Earth itself, from the Earth's rotation and from its internal heat. There are two ways in which the abundant energy of the oceans can be harnessed: we can tap either the energy of water motions (waves, tides, currents), or the energy inherent in vertical temperature, density and salinity gradients.

Tidal motions have so far attracted the greatest attention, mainly because the principle is simply that of hydroelectric power generation. Power is generated by holding the outgoing tide behind a dam, using the head of water so produced to drive turbines for electricity generation. A tidal range of 5 m or more is necessary for electricity generation to be economic, and there are relatively few sites where such a range is found (Figure 1.18).

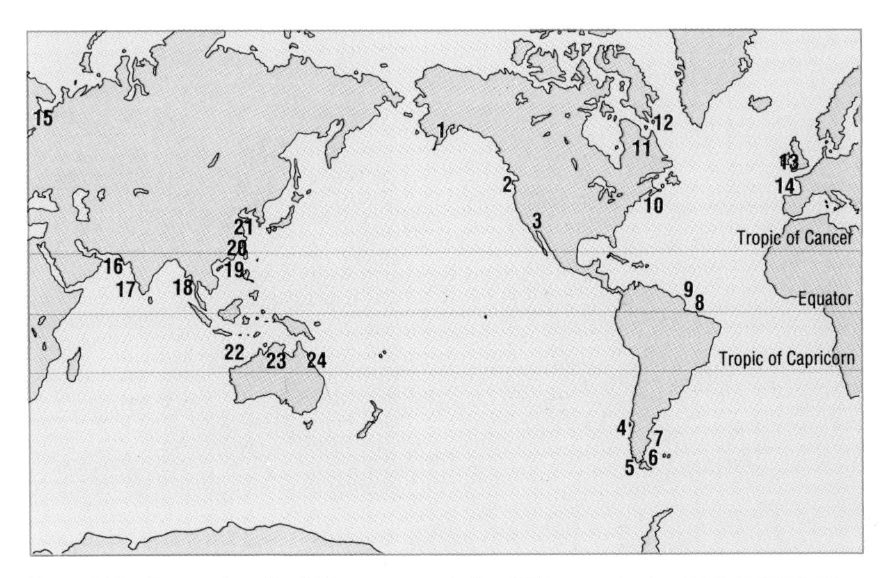

Figure 1.18 Places where the tidal range exceeds 5 m. Tidal power is at present exploited only in north-west France (La Rance installation, Fig. 1.19) and the White Sea. Other schemes are under consideration for the Bay of Fundy, the north-west coast of Australia and the west coast of England (Severn estuary). 1 Cook Inlet; 2 coast of British Columbia; 3 Colorado River estuary; 4 Chonos Archipelago; 5 Magellan Straits; 6 Gallegos and Santa Cruz; 7 Golfo San José; 8 Maranhão; 9 Araguaia River; 10 Bay of Fundy; 11 Ungava Bay; 12 Frobisher Bay; 13 west coast of England; 14 north-west coast of France; 15 White Sea; 16 Gulf of Kutch; 17 Gulf of Cambay; 18 Rangoon; 19 Amoy; 20 Shanghai; 21 Inchon; 22 north-west coast of Australia; 23 Darwin; 24 Broad Sound.

Localities where a sufficient tidal range can be utilized are limited to those sites where suitable dams may be built. One such site is the Rance estuary in France (Figure 1.19) where a tidal power station is operating. Much larger schemes for tidal power in Britain's Severn estuary and Canada's Bay of Fundy have been proposed and discussed many times. Dam construction will of course affect the pattern of currents and so influence the movement of sediments. Ecological disturbance is inevitable, and its extent cannot be reliably predicted.

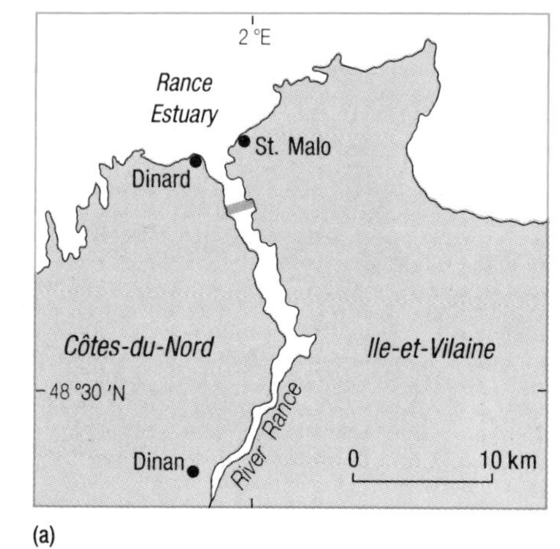

(a)

(b)

Figure 1.19 (a) The location of La Rance tidal power station (location 14 on Figure 1.18). It has been producing about 550×10^6 kW h annually since 1966.
(b) An aerial view of La Rance.

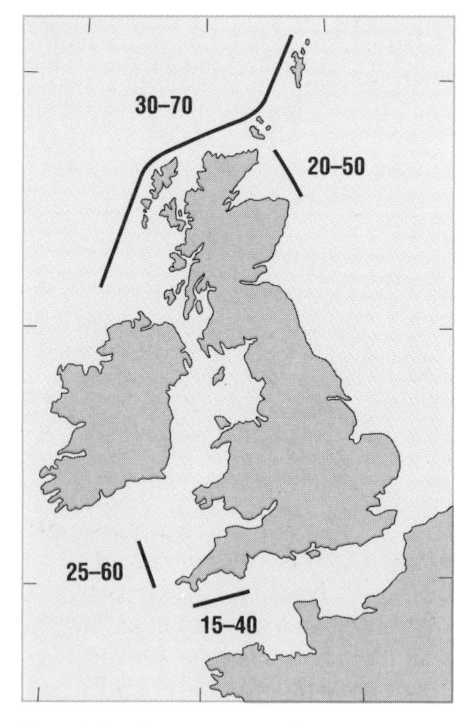

Figure 1.20 Optimum locations for wave energy conversion devices around the UK coastline. Numbers show the estimated power availability in kilowatts per metre of wave crest length averaged over 12 months.

The large-scale harnessing of wave energy is controlled by two basic criteria. First, the prevailing sea conditions must ensure a consistent supply of waves with amplitudes sufficient to make conversion worthwhile (Figure 1.20); secondly, the installations must not be a hazard to navigation or to marine ecosystems. The nature of wave energy is such that rows of converters (Figure 1.21) many kilometres in length would be needed to generate amounts of electricity comparable with conventional power stations. Such rows of converters would form offshore barrages, which might interfere with shipping, although sea conditions would be calmer on the shoreward side. In the calmer conditions, however, water circulation and sediment transport would decrease, and the growth of quiet-water plants and animals would increase—and pollutants would be less easily flushed out. Problems of maintenance would also be formidable, and large-scale conversion of wave power into electricity seems unlikely to be feasible, at least in the short term.

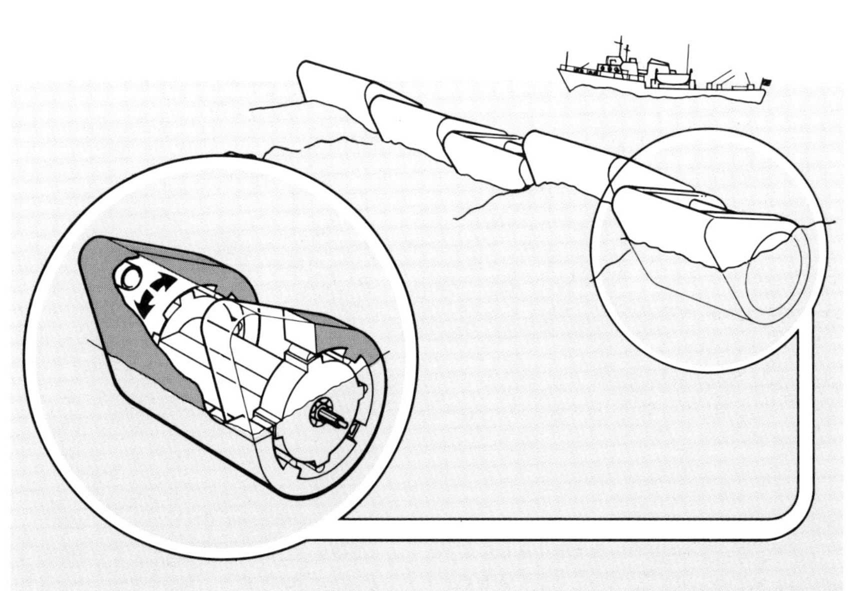

(a)

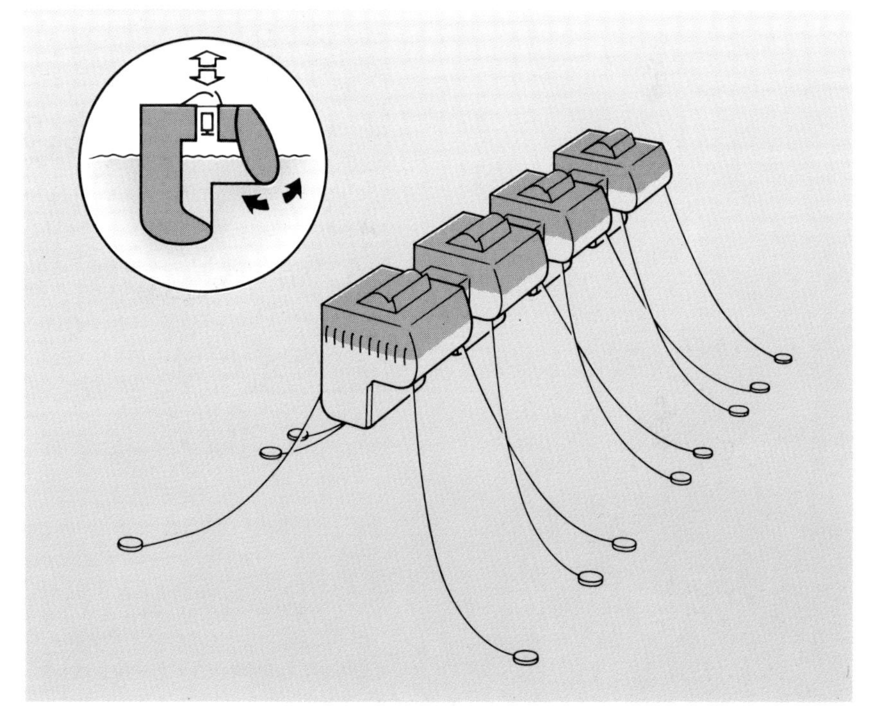

(b)

Figure 1.21 Wave-energy converters.
(a) The Salter duck. This device consists of an oscillating vane within a float that rocks up and down on a central spine, and can be used to drive generators.
(b) An oscillating water column based on the 'inverted-can' principle: incoming waves set up oscillations of the air columns trapped in the inverted vessels; these oscillations can be used to drive turbines or a high-pressure fluid power system. Smaller devices using this principle are used for lighting systems on navigation buoys.

Nonetheless, wave energy *has* been successfully harnessed, but on a scale of hundreds of kilowatts rather than the hundreds or thousands of megawatts of conventional power stations. In Norway, for example, small installations have been built where steep rocky coasts face deep water, with narrow bays and inlets that 'funnel' landwards. The crest length of incoming waves is progressively decreased and their height correspondingly increased. The wave energy can be used either directly to drive converters that work on principles similar to that illustrated in Figure 1.21(b); or indirectly to fill small dams: the resulting head of water is then used to drive turbines, as in any small hydroelectric plant, and the waves act merely to keep the dam filled.

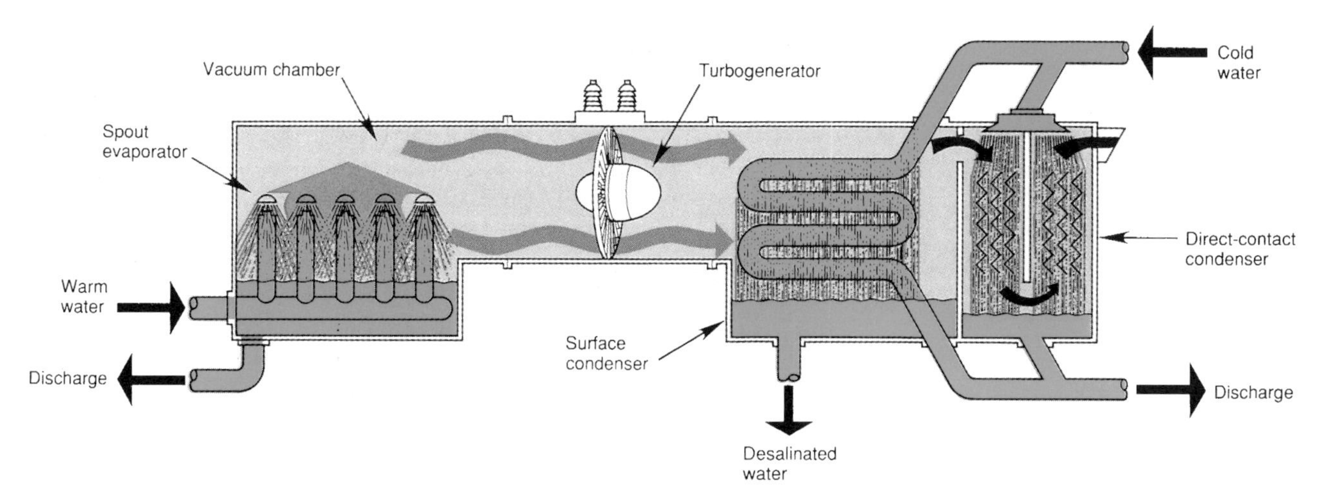

Figure 1.22 Schematic diagram of an OTEC plant on Hawaii. Warm seawater is pumped into a vacuum chamber, where it boils and evaporates after passing through a specially designed spout evaporator, producing steam. The steam is used to drive a turbo-generator to generate electrical power. The steam then passes to a condenser chamber, where cold seawater is used to condense the steam back into water. The cold seawater (which is nutrient-rich) can then be used for mariculture, while the desalinated water (condensed steam) can be used for freshwater and/or irrigation.

The problems of tapping energy from vertical temperature gradients in ocean waters—**Ocean Thermal Energy Conversion (OTEC)**—are mainly those of scale. The basic principle is the same as that used in refrigerators, air conditioners and heat pumps. The original concept was to pump warm surface water at about 25°C into heat exchangers to vaporize ammonia or Freon which would expand to drive turbines to generate electricity. At the same time, cold water at about 4°C from deeper levels in the ocean (below the **thermocline** at several hundred metres depth) would be pumped up to condense the vapour in separate heat exchangers, allowing the cycle to start again. In modern plants, such as the one under development at the Natural Energy Laboratory of Hawaii (Figure 1.22), the warm seawater itself is vaporized under a vacuum and the resulting steam used to drive turbines. The steam is condensed using cool sub-thermocline water which is nutrient-rich and can then be pumped to mariculture ponds, while the condensed steam (which is desalinated seawater) can be used for drinking, washing, irrigation, etc. It has become clear that for the foreseeable future OTEC technology is only viable commercially if electricity generation is treated as a by-product of mariculture.

Such plants are best installed in lower latitudes, where the thermal contrast between surface and deep water is greatest and seasonal changes are least. The Japanese and Americans have advanced furthest with this technology and have built some small plants generating 50 to 100 kW.

For larger power stations (hundreds of megawatts or more), huge installations would be required, comparable in size to oil production platforms on continental shelves. That is because the temperature differential between surface and deep water is at best only about 25–30 °C, so the 'energy density' of the vapour driving the turbines is low compared with that in steam turbines, for example, where the temperature differential is more than 100 °C. What is more, around two-thirds of the power output is needed to drive the pumps and the overall efficiency of OTEC plants is unlikely to be better than 5–10%.

QUESTION 1.4 Look again at Figure 1.20. Taking the average power available as 50 kW per metre of wave crest, how long a 'string' of converters such as those in Figure 1.20 would be needed to generate 500 MW (500×10^3 kW), assuming an energy conversion efficiency of 10%?

1.5 MARINE RESEARCH

Investigating the oceans nowadays involves much more than charting the courses of surface currents and recording the rise and fall of the tides, important though these activities undoubtedly are. A great variety of automatic data-collecting instruments is now available: moored buoys collect a host of oceanographic and meteorological data (Figure 1.23); instrument packages towed behind ships survey the sea-bed or monitor changes of temperature and salinity and other properties in the water column and transponder-equipped neutrally buoyant floats are left to drift passively, so that the speed and direction of currents at various depths can be charted.

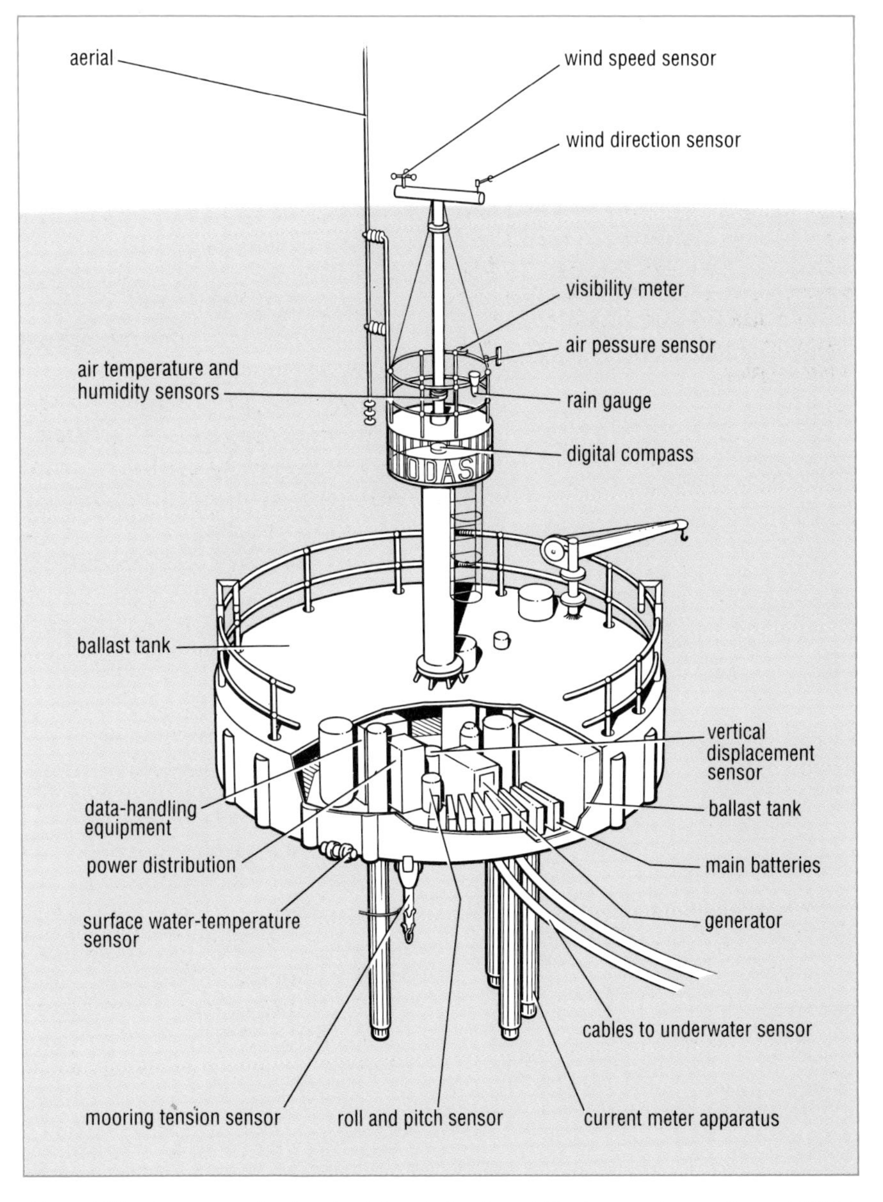

Figure 1.23 An unmanned automatic data-collecting buoy. This diagram shows a buoy of 1970s vintage and displays the variety of equipment well. Later generations of buoys are smaller and more compact, their sensors less exposed to potential wave and storm damage.

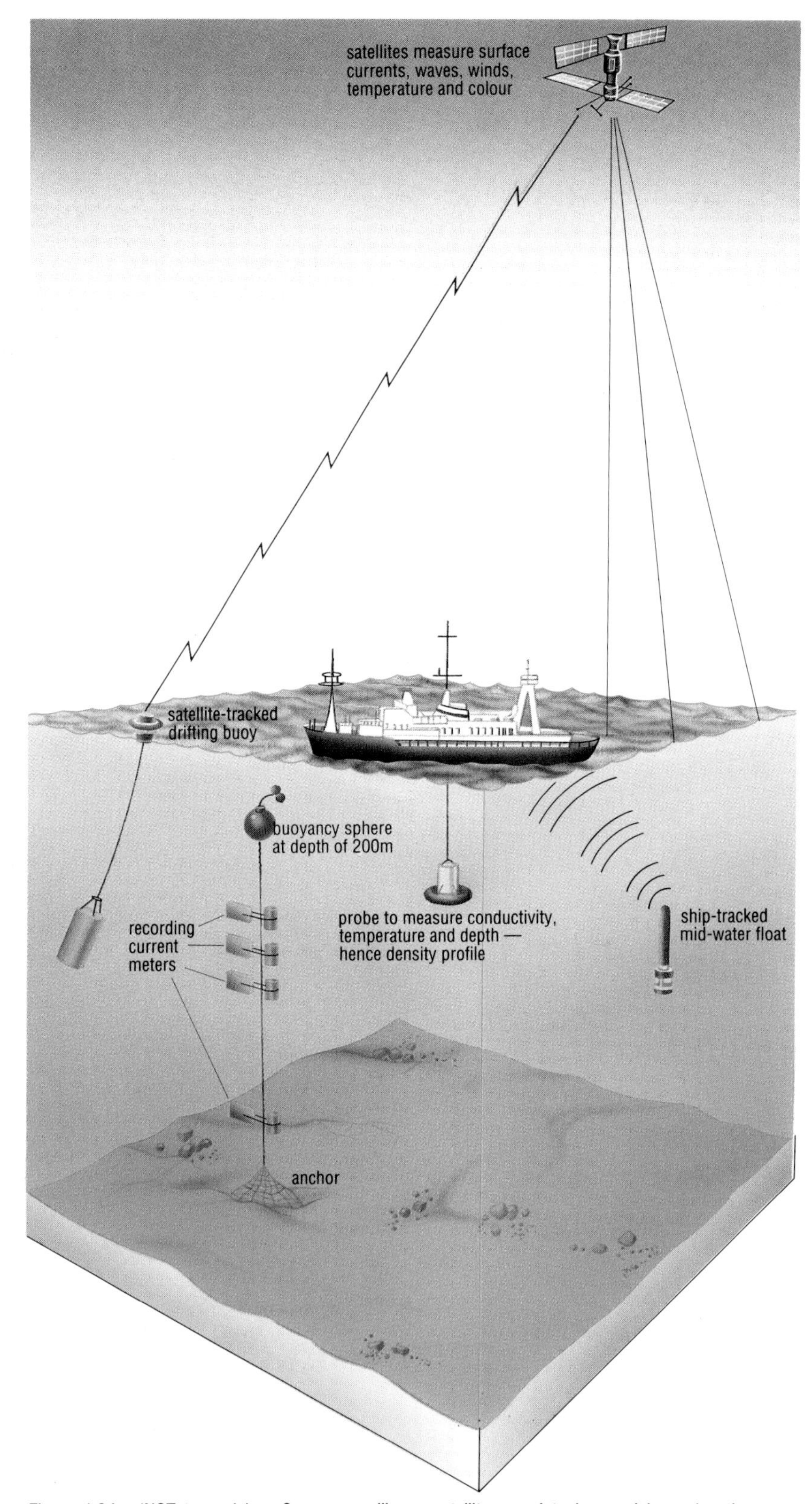

Figure 1.24 (NOT to scale) Ocean surveillance satellites need to be used in conjunction with ships, etc., which provide 'sea truth' for calibration, as well as information about processes occurring within the volume of the ocean. Likewise, research vessels require satellites to provide the large-scale synoptic view. It is important to emphasize that the ship-based equipment shown here uses techniques that have been part of oceanographic research for many decades. The satellite is a relative newcomer to the scene.

Figure 1.24 shows how satellites are important for oceanographic research, supplementing the traditional surface-based methods of data acquisition. They have the great advantage of near-simultaneous global coverage, and can record large-scale seasonal or other transient phenomena that might otherwise go unobserved.

Data collection needs to be done on an international basis, partly because no nation on its own can afford to mount major research programmes, and partly because the effects of changes in the ocean–atmosphere system take no account of national boundaries. Two examples of international multidisciplinary experiments of the 1990s are the TOGA (Tropical Ocean Global Atmosphere) project, and **WOCE (World Ocean Circulation Experiment)**, both involving scientists, ships and equipment from many countries working in close co-operation.

As previous Sections have outlined, study of the oceans and ocean floors is now more important than ever before. The harvesting of food from the sea, the stabilization of fish stocks, the exploitation of ocean-floor minerals and the mitigation of pollution are just a few of the marine activities that cry out for more knowledge through research. But even disinterested 'pure' research has become more difficult.

Up to the 1940s there were few, if any, restrictions on marine research. Scientists wishing to work in the territorial sea of a friendly nation would request permission informally and be granted it with little or no delay; the rights of unfriendly nations or nations considered insignificant would generally be ignored, especially if they had no navy to back up possible protests. In the post-war period, however, attitudes began to change. Governments became more conscious of the economic, military and security aspects of their coastal waters, more concerned to control events and more suspicious of foreign researchers. There were moves to increase the width of the territorial sea (see Chapter 2), sometimes up to 200 miles. The number of independent countries from which permission must be sought has grown rapidly with the progress of decolonization. Most nations now have a greater awareness of the importance of their resource potentials and are suspicious of requests by richer countries for unrestricted permission to carry out 'pure' research. In many countries, bureaucracy has developed to the point where it threatens to prevent the carrying out of the very functions it was designed to serve. There is no doubt that disinterested scientific research in the oceans has been under severe threat, the more so because it has received low priority in international debate. The situation is likely to improve however, with growing international awareness of the oceans' central role in controlling the global environment.

1.6 THE BASIS OF INTERNATIONAL REGULATION

Marine 'resources' can be roughly divided into two groups: *tangible resources* (living, mineral and energy), and *activities* (communications and transport, recreation, waste disposal and military use). It might be thought that this distinction would be fundamental to their exploitation in that the tangible resources of a particular area could be owned outright, whereas activities, which can potentially be undertaken by anyone anywhere in the ocean, would require a different form of international control. The division into things that people do in the marine environment and things they extract from it may be superficially attractive, but it is not particularly useful in practice. This is because outright ownership of parts of the marine environment is the exception rather than the rule, typical only of a limited zone adjacent to the coastline. The position with regard to tangible resources in the greater part of the ocean and its bed is one of rights of exploitation rather than ownership. Many of these resources could not be owned outright anyway, because they are mobile and cannot be identified with certainty. In the case of fisheries and wave power for example, the most that can be 'owned' is the right to extract fish from the sea or energy from the waves, respectively. Realizing the potential of a resource is just another form of activity—extraction—and it is activities, real or potential, that need to be regulated. Essentially the same problems can arise whether it is a question of the ownership of offshore oil or the right to navigate through a particular stretch of water. The impact of an activity is not restricted to one locality nor to one resource: exploitation of migratory fish stocks in one area, for example, reduces the stocks elsewhere, and exclusion zones for other activities have to be set up around offshore structures and pipelines. Thus, not only *can* a single regulatory system in principle be applied to all human involvement with the sea—it *must* be.

Although maritime activities are global in extent, they inevitably become more intense towards the coast, where much of the world's population lives. Most of the tangible resources are found within a hundred miles of the shore and activities such as waste disposal are concentrated near centres of population and industry. But there is little uniformity in the distribution of these features among nations: the accidents of political geography which favour one with a natural harbour or rich fishing grounds may deprive its neighbour of trade and an important food supply. The potential for dispute between neighbouring coastal states is considerable, as we saw in Section 1.2.3 for example. Activities far out to sea are unaffected by these differences and are therefore more amenable to uniform regulation.

Among the resources themselves, those which are fundamentally biological are also renewable—the living resources and the sea's capacity to act as a sewage treatment works. But this renewability has limits. Stocks can be overfished beyond the point of recovery. Excessive pollution by organic wastes can locally exhaust the oxygen supply so that oxidizing bacteria are replaced by bacteria that metabolize anaerobically and produce sulphide which kills other bottom-living (benthic) organisms. Persistent components of industrial wastes, such as mercury and PCBs, *accumulate* in the marine environment, so as a receptacle for these pollutants the sea is a finite or non-renewable resource, at least on a time-scale of decades, and the need to regulate its use is greater than ever.

1.7 A RECIPE FOR CHAOS?

Territory above sea-level has been contested since history began. Competition between societies for areas more favoured by position, climate, soil or other natural resources is the theme underlying most of the archaeological record. Disputes were sometimes settled by cultural assimilation, but often led to open warfare.

In the case of marine territory, the analogous process was delayed to later stages of cultural development. Throughout prehistory the same rules that applied to the land probably applied to estuaries, bays and the immediate nearshore zone, but conflict further offshore would have been only a projection of conflicts generated on land. Claims to exclusive rights of navigation or resource ownership on a larger scale only became credible in the context of relatively advanced civilizations. The prerequisites were the accumulation of sufficient knowledge to recognize the potential of a resource and the ability to exploit it (see Chapter 4). Without these, the idea of maritime rights remained hypothetical. Once a nation acquired the knowledge and the technology, the scene was set for 'continuing use' and the equivalent of squatters' rights or traditional rights of way. Even so, there was no need to claim a right until exercising it could be perceived by competitors as being to their disadvantage.

Where competing claims arose, if matters did not immediately proceed to violence, whatever law came to hand was adapted and pressed into service to challenge or defend such claims. This could lead to a peaceful settlement of the dispute, and, perhaps, to agreements about how future disputes might be settled along equitable lines. Such were the first steps toward a Law of the Sea.

With the idea of maritime rights came the potential for conflicting uses, and another area of application for the emerging Law of the Sea. Conflicts of use are becoming increasingly common in the coastal seas as the range of marine resources being exploited grows; for example, oil rigs and pipelines versus trawlers; dumping and effluents versus fisheries. Nations currently developing their marine resources are naturally suspicious of oceanographic researchers from other countries and sometimes seek to restrict their activities (Section 1.5, see also Section 2.3). In addition, geographically disadvantaged and land-locked states are prompted to seek a 'slice of the action'. There is obviously a need for agreed rules for the conduct of these manifold and often potentially conflicting activities. Such rules have been introduced by a range of authorities for application within, and sometimes beyond, their accepted jurisdictions. In the next two Chapters we shall concentrate on *public international law*, that is to say the law operating between nations, as it applies to their use of the seas. This is to be distinguished from private maritime law, which governs the commercial transport of goods by sea and will not be considered in this Volume.

1.8 SUMMARY OF CHAPTER 1

1 The possession or lack of marine 'space' has done much to condition national attitudes and aspirations and has been a major influence on the political geography of the world. This is chiefly because throughout human history the most important use of the sea has been for transport, trade, warfare and communications as well as fishing.

2 Humanity's oldest marine activity is fishing. About three-quarters of the global catch is taken in coastal and continental shelf waters. Fish are caught in different ways depending upon whether they are demersal or pelagic, shallow or deep water species. Improved technology has

increased the range and size of fishing vessels and the tonnage of the catch; and overfishing is a growing threat to the marine harvest. Agreements to control fishing are not easy to reach, because different countries have different degrees of dependence on their fishing industries.

3 Commercial fish and the ecosystems of which they are part are also under growing threat from pollution of many kinds, which has shown an almost exponential increase in amount and variety during the 20th century, with great influxes of heavy metals, organochemicals, radionuclides, hydrocarbons, and so on. Many artificial substances have natural counterparts, but the rates at which they are added to the oceans and how they react with organisms in the marine environment are the main factors which determine their behaviour as pollutants.

4 Seawater itself is an important physical resource, not only for the freshwater obtained by desalination, but also for the dissolved substances in it, notably common salt, magnesium, bromine and other elements. The extraction of valuable trace elements (e.g. gold) would require huge volumes of water to be processed, and is not economically feasible.

5 The physical resources of continental shelves comprise both surface and sub-surface accumulations. Surface accumulations include sands and gravels, placer deposits of heavy minerals and chemically precipitated limestone and phosphorite. The principal sub-surface resources are hydrocarbons in the form of offshore oil and gas fields.

6 The deep sea has not so far been tapped for its sea-bed resources, but these are enormous. Best known are metal-rich manganese nodules, but there are also sulphide stockworks within oceanic crust, formed beneath hydrothermal vents along ocean ridges; and metal-rich muds in axial deeps of the Red Sea. Exploitation of any of these deposits would cause considerable environmental disruption.

7 Extracting useful energy from the oceans presents difficulties, and only relatively small amounts of the enormous quantities potentially available have so far been tapped, chiefly in tidal power projects. The main difficulty is that the oceans represent a relatively 'dilute' energy source compared with (for example) hydrocarbons, and large-scale power generation would require correspondingly large-scale installations.

8 Marine scientific research is now more important than ever before, and it is also more international both in scope (with the help of satellite technology) and organization (TOGA, WOCE, etc.). However, as national attitudes and vested interests of coastal states lead to the imposition of stricter controls on navigation almost everywhere except in the open oceans, the logistics of shipborne marine research have become progressively more difficult.

9 While it may be convenient to recognize a dichotomy between maritime activities on the one hand (transport and communications) and tangible marine resources on the other (living and physical resources, energy), the distinction is less useful when it comes to regulation, because resources are of no use unless they can be exploited, and that is also an activity. The uneven distribution of both living and physical resources in relation to political geography, however, makes regulation difficult. It may also be important to distinguish those resources which are renewable (mainly biological) from those which are not (mainly 'physical'). Any international system of regulation—a Law of the Sea—must deal adequately with all these variables.

10 Analogies can be drawn between territory on land and 'maritime' territory. Nations have sought to conquer and exploit distant seas in the same way as they conquered and exploited distant lands, although it was often only necessary to claim seas not already 'occupied' or claimed by others. But it is more difficult to defend maritime 'territory' against appropriation by another power than it is to defend land areas; and there is abundant scope for argument and even open warfare between adjacent coastal states. As the range of activities and resources has expanded in the marine environment during the 20th century, the need for an international Law of the Sea has become overwhelming.

Now try the following questions to consolidate your understanding of this Chapter.

QUESTION 1.5 Three-quarters of the global fish catch is taken in coastal and continental shelf waters. According to Figure 1.4, where does most of the rest come from?

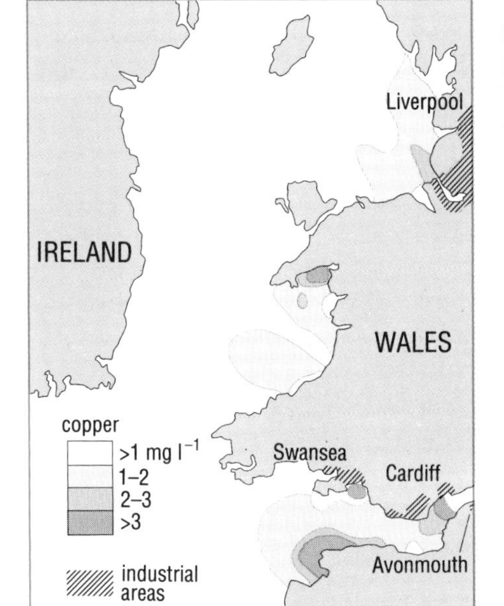

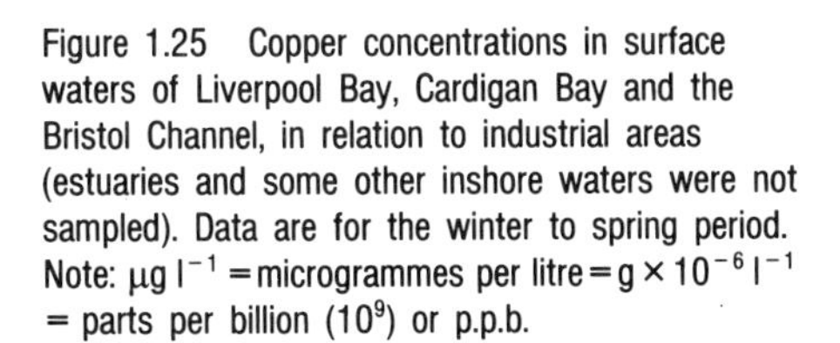

Figure 1.25 Copper concentrations in surface waters of Liverpool Bay, Cardigan Bay and the Bristol Channel, in relation to industrial areas (estuaries and some other inshore waters were not sampled). Data are for the winter to spring period. Note: $\mu g\ l^{-1}$ = microgrammes per litre = $g \times 10^{-6}\ l^{-1}$ = parts per billion (10^9) or p.p.b.

QUESTION 1.6 A conspicuous absentee from the discussion of pollution in Section 1.2.4 was carbon dioxide (CO_2) and the greenhouse effect, chiefly because this topic has already received so much publicity. Explain (a) why CO_2 is an example of a 'pollutant' for which it will be difficult to reduce emissions; and (b) why the increase in atmospheric CO_2 concentrations is now a major focus of oceanographic research.

QUESTION 1.7 Examine Figure 1.25.

(a) Would you say that all the high copper concentrations in coastal waters off western England and Wales could be attributed to industrial effluents? If not, can you suggest another cause?

(b) If you were concerned with the effects of copper on marine life, what other information would you seek to amplify the data in Figure 1.25?

QUESTION 1.8 Examine Table 1.3.

(a) Which is the sea-bed resource with the greatest potential tonnage; in what parts of the sea-bed would you expect to find the largest deposits; and why might you expect to find them mostly in higher northern latitudes?

(b) What is the second largest potential sea-bed resource, and how does its distribution compare with the one you identified in (a)?

QUESTION 1.9 Why is it not particularly useful to distinguish marine resources from marine activities as a basis for international regulation?

CHAPTER 2 THE LEGAL BACKGROUND

'Yet still his claim the injured ocean laid,
And oft at leap-frog o'er their steeples played,
As if on purpose it on land had come
To show them what's their *mare liberum*'

From Marvell's 'The Character of Holland'.

International law is an essential part of the modern world—but it is not a unified code of statutes just waiting to be applied. Its sources are many and its means of enforcement weak. Whatever the formal position may be, international law in general and the Law of the Sea in particular rest ultimately on the good faith of nations. This means that it is not always obeyed either in the spirit or in the letter, especially (but far from exclusively) during time of war, declared or otherwise. Moreover, good faith notwithstanding, the Law of the Sea, like any other creation of the intellect, is open to alternative interpretations which form the basis of disputes fuelled by economic and political ambitions.

2.1 EVOLUTION OF BASIC CONCEPTS

In November 1718 the head of Edward Teach was hung from the bowsprit of the British naval patrol vessel HMS *Pearl.* Teach, better known to Victorian romantics and their successors as Blackbeard the Pirate, had finally met his end at the hands of a Lieutenant Maynard, ostensibly because of a proclivity to break the law of the high seas. But what law? The fact is that there was no law of the high seas—at least no universal law recognized by international agreement.

It should not be concluded from this, however, that activities on the high seas occurred entirely in a conceptual vacuum, for the theoretical basis of a possible Law of the Sea had been debated on and off since Roman times. No doubt few practical seafarers were familiar with such discussions; and those who were could perhaps have been forgiven for concluding that legal argument could be adduced to support any point of view. The fact remains, however, that the Law of the Sea had already been an issue among scholars and lawyers for about 2 000 years even as Blackbeard's head was swinging in the wind.

According to traditional Roman law, for example, seas and seashores were common to all and could be used by anyone wishing to do so—a philosophy of freedom that was extended even to rivers and ports. In his *Digest*, Ulpian wrote that the Sun, air and waves 'are by nature things open to the use of all', because they were produced by nature in the first place, and 'have never yet come under the sovereignty of any one'. In *Metamorphoses*, Ovid made a similar claim: 'Why', he asked, 'do you deny me water? Its use is free to all. Nature has made neither Sun nor air nor waves private property; they are public gifts'. Cicero also enjoined that the waters be denied to no one, an injunction that Virgil extended to shores. In short, the classic Roman view was that rivers, seas and oceans should be open to all. This concept, known as ***res***

communis, assigned the seas to the community in its widest sense and carried with it the implication that they be preserved as community assets for all time.

However, this is not the only way of viewing the oceans, for there is also a rather different concept, ***res nullius***, which takes as its starting point the view that the seas are not common to all, but in their original state belong to no one and are thus available for appropriation by the first party to establish a claim. One of the first states to take this more restrictive view was Venice, which during the late 13th century claimed sovereignty over the whole Adriatic, although she was not in possession of much of the opposing shores. Moreover, by this time the Venetian state was affluent and pre-eminent in commerce and navigation, so it was in a position to enforce its claim, levying tariffs on all ships using the Adriatic and, when it chose, even preventing certain vessels from entering that sea. This was mastery through strength rather than moral argument; and for a time other European states, including the Vatican, had no choice but to comply with Venice's demands. As a particular practical application of *res nullius*, it was not to last, coming to an end as Venetian power waned. But the example was firmly set and was later copied by Genoa in the Ligurian Sea and by the Scandinavians in the Baltic.

Needless to say, the application of the *res nullius* concept was a prime recipe for conflict and resulted in numerous multilateral treaties and agreements covering navigation, fishing and trading. These were not destined to be permanent but to change, die out or be cancelled as historical circumstances changed. The Venetian initiative and the discussion that followed it did, however, give rise to one long-term consequence of supreme importance, namely that most nations have ever since regarded the doctrine of *res nullius* as entirely appropriate to a narrow band of sea immediately adjacent to their coastlines. Thus became universal the concept of what is now known as the **territorial sea**, or **territorial waters**, over which the coastal state claims virtually total sovereignty.

Outside the territorial sea lie the **high seas**, or **international waters**, and here attempts to apply *res nullius* have been much less successful, at least in the long term. In 1493, for example, Pope Alexander VI succumbed to pressure from Spain and Portugal by agreeing to divide between them all the world's undiscovered areas with a line from the North to the South Pole passing to the west of the Azores and the Cape Verde Islands. A year later, Spain and Portugal agreed by the *Treaty of Tordesillas* to move the line further to the west, but the principle of division remained unchanged. Thus, Spain claimed sovereignty over, and was given Papal authority to control, a zone including the western Atlantic and the Pacific; and Portugal likewise took the Indian Ocean and the eastern Atlantic south of Morocco. This meant that in theory the two countries could prohibit or license non-local navigation and trading throughout the world, and hence control the development of all international maritime activity. Attempts to put theory into practice were carried out with a blatant ruthlessness made possible by naval superiority; it often succeeded, but not always. When Drake sailed to the Pacific in 1580, the Spanish Ambassador to England complained bitterly to Queen Elizabeth I, only to be told firmly that in her view the sea was open to all nations and, by nature, subject to no national sovereignty.

Frontispiece to Grotius's book.

The most serious challenge to Spanish–Portuguese hegemony came, however, from the Dutch, who by the 17th century had come to resent the attempted abolition of what they held to be their right to trade with the Indies. In 1609 Hugo Grotius, under commission from the Dutch East India Company, published a short book entitled *Mare Liberum*, in which he sought to justify the Dutch position and hence destroy Portugal's claim to exclusivity. The essence of his argument was a more explicit, detailed and up-to-date statement of the concept of *res communis* in respect of the open ocean. Thus, he appealed to 'the civilized world for the complete freedom of the high seas for the innocent use and mutual benefit of all', an aim which if necessary should be achieved by force. However, he specifically excluded coastal seas from consideration on the grounds that the principle and implications of *res nullius* were already widely accepted as being applicable to such zones.

Another participant in this debate was an Englishman, John Selden. In his *Mare Clausum*, a direct reply to *Mare Liberum* published 25 years later, Selden backed the principle of *res nullius* in support of Charles I's pretensions to 'Dominion of the British seas' which were very generously defined. By quoting numerous precedents, he sought to show that appropriation of the seas was not only justifiable but had in many cases actually been carried out. Having established to his satisfaction the validity of *res nullius* in respect of coastal seas, he would by logical extension apply it to the open ocean. The debate did not end there, but the philosophy expressed in *Mare Liberum* ultimately triumphed so far as the high seas are concerned.

The 18th and 19th centuries saw a trend away from 'natural law', based on reason and sustained by appeal to scripture and the authors of antiquity, towards the 'positivist' view. In international matters, this is more concerned with what nations *actually do* than with what it might be thought they *should do*. Its sources are custom and treaties, rather than 'universal principles' which may turn out to be a matter of national opinion. By the dawn of the 20th century, international law was generally considered to be the product of the voluntary subscription of nations to legal controls—what is commonly termed *customary law*, a term we shall use again later. Thus, the present position regarding the concepts discussed above and set out in the various international conventions reviewed in succeeding pages, may be summarized as follows, in very broad terms:

1 The concept of *res communis* or *mare liberum* is held to apply to the seas in general and to the high seas in particular. The high seas may not be appropriated by any single power or group of powers and must remain open to all nations for all purposes.

2 Notwithstanding this general rule, the concept of *res nullius* or *mare clausum* is held to apply to territorial seas (although, in fact, states do not need to acquire or claim a territorial sea: it is traditionally regarded as part of the sovereign area by virtue of its being adjacent to the land territory). Within its territorial sea a coastal state may exercise complete jurisdiction, with the sole exception that it must not interfere with vessels of other states going about their peaceful business—the right of **innocent passage**.

A cautionary digression is appropriate at this point. It is generally assumed (and the foregoing has done nothing to contradict such an assumption) that modern Law of the Sea originated in Europe in the 17th century as a result of interactions among European states and as a consequence of influential publications, particularly Grotius's *Mare Liberum*. There is good evidence in their writings, however, that Grotius and other classical marine jurists adduced arguments based on long-established maritime practices in South-East Asia:

> 'For at least 2000 years, deltaic, coastal and archipelagic empires have distinguished the Southeast Asian region as a zone of maritime transit and transaction. Innumerable explorers, emissaries, traders, missionaries, raiders and refugees have for centuries traced and retraced a dense pattern of maritime traffic and flourishing trade in this important region of the world . . . the region has always been an important centre of maritime commerce.'
>
> R.P. Anand (1983) *Origin and Development of the Law of the Sea*, Martinus Nijhoff.

As we continue with our somewhat Western-oriented view of maritime legal history, it is well to bear in mind that it does not represent the whole story.

2.2 20TH-CENTURY DEVELOPMENTS

When the concepts of *res communis* and *res nullius* were first formulated, the seas were used only for navigation and fishing and by a very small world population—a situation that obtained more or less up to the middle of the 19th century. Today, the position is quite different. From what has been said already it should be clear that the uses to which the oceans may be put are so varied and the views of the world's nations so diverse that it would be extremely difficult, perhaps even impossible, to draft a set of rules that would satisfy everyone and be fair to all. Nevertheless, that is precisely what international jurists have repeatedly sought to achieve this century, in response to the accelerating tempo of maritime activity.

During the previous three centuries, the Law of the Sea had developed in a fragmented way, largely dictated by the relative strengths and shifting spheres of influence of the major maritime powers. New concepts were introduced unilaterally, prompted by national interest. For example, in 1811 the British excluded foreign fishermen from pearl beds beyond the territorial sea around Ceylon (now Sri Lanka); in 1839, the British and French agreed that French oyster fishermen should have exclusive use of certain parts of Granville Bay which lies between Jersey and the coast of Normandy; and in 1904 Tunisia took control over all sponge fishing between its territorial sea and the 50 metre isobath. It became apparent that the various contradictory national regimes which had grown up over the years needed to be harmonized.

The first official attempt to codify the Law of the Sea was initiated by the League of Nations in 1924. After extensive preliminary studies, a conference was convened at The Hague in 1930. Its central topic was the territorial sea, but it was unable to reach agreement on the width of territorial waters and its draft articles on other aspects of the territorial sea were simply referred to Governments for their consideration. These efforts were not wasted, however, because they provided a major input to preparations for the First United Nations Conference on the Law of the Sea, held at Geneva in 1958. This was the first intergovernmental conference to consider the Law of the Sea as a whole. Eighty-six nations participated, many more than were represented at The Hague Conference. It adopted four Conventions:

1 Convention on the Territorial Sea and the Contiguous Zone.

2 Convention on the Continental Shelf.

3 Convention on the High Seas.

4 Convention on Fishing and Conservation of the Living Resources of the High Seas.

Note: A convention is a document in which the points agreed at a conference or conference session are set down. It is an agreement in writing between states and/or international organizations, and it is governed by international law. When the representatives of the states involved in the negotiations have reported back to their respective Governments, these Governments decide whether or not to ratify the terms of the convention (see also Chapter 3).

The four Conventions adopted in 1958 were based largely on existing customary law (see Chapter 3) and emerging practice. The first three Conventions duly became the generally accepted legal framework in their areas, though the vexed question of the width of the territorial sea still remained unanswered. The fourth Convention also entered into force, but it failed to attract sufficient support to become fully effective (see Section 2.2.3). A Second United Nations Conference on the Law of the Sea which met in 1960 to reconsider this problem and the associated issue of fishery limits failed to reach agreement by one vote (see Section 2.2.4).

As the 1960s progressed, it became increasingly obvious that the rapid development of maritime activities demanded a thorough revision of the legal regime governing them—a task that would be complicated by the

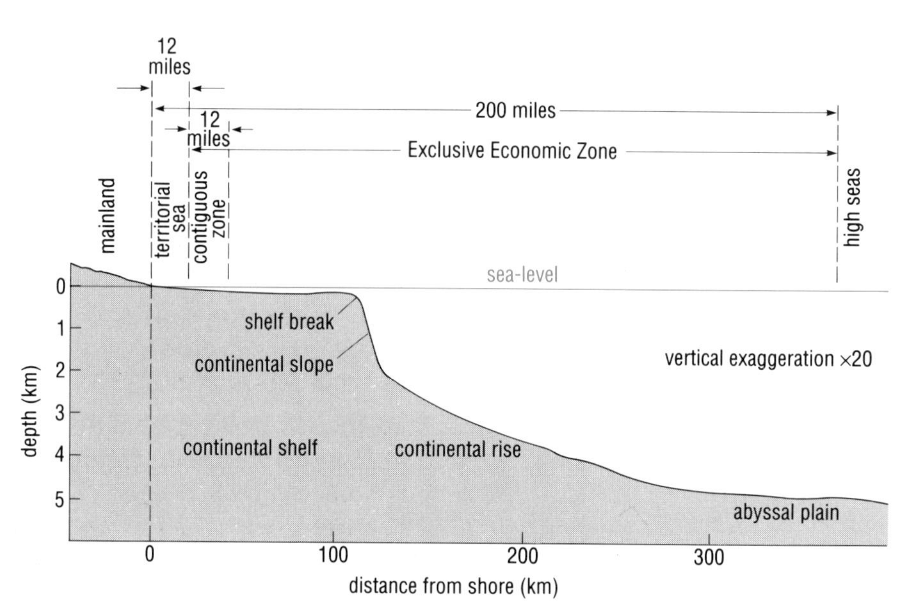

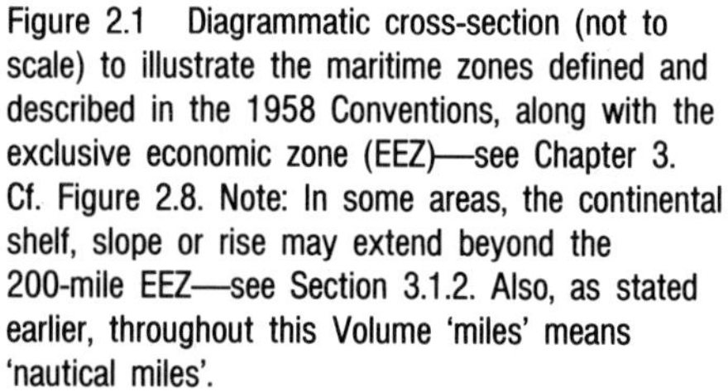
Figure 2.1 Diagrammatic cross-section (not to scale) to illustrate the maritime zones defined and described in the 1958 Conventions, along with the exclusive economic zone (EEZ)—see Chapter 3. Cf. Figure 2.8. Note: In some areas, the continental shelf, slope or rise may extend beyond the 200-mile EEZ—see Section 3.1.2. Also, as stated earlier, throughout this Volume 'miles' means 'nautical miles'.

growing number of independent nations concerned. The Third United Nations Conference on the Law of the Sea was convened in 1973. Negotiations continued for nine years, culminating in the opening for signature of the **United Nations Convention on the Law of the Sea (UNCLOS)** in 1982. Although unlikely to come into force before the early 1990s (and perhaps much later), it will be examined in Chapter 3, because it provides the framework within which current developments are taking place. But first we must look at the earlier United Nations Conventions, as a general introduction to the problems of definition and measurement, and because most of the underlying principles have been carried over into UNCLOS. Figure 2.1 shows the arrangement of the various maritime zones described in the earlier Conventions.

2.2.1 CONVENTION ON THE TERRITORIAL SEA AND THE CONTIGUOUS ZONE

Although territorial waters were established as an international principle by the 17th century, the Convention on the Territorial Sea and the Contiguous Zone was the first clear statement of the rights and duties of states within them. In its own words, 'The sovereignty of a State extends, beyond its land territory and its internal waters, to a belt of sea adjacent to its coast, described as the territorial sea, diminished only by the right of innocent passage. This sovereignty extends to the air space above the territorial sea and to the sea-bed beneath'.

This Convention also introduced the concept of the **contiguous zone**, which lies just outside the territorial sea. The contiguous zone, though seldom mentioned in the media, is perhaps the most contentious concept in the whole Convention, for it gives the coastal state some control in what is technically a part of international waters. On the other hand, very few states have declared contiguous zones. Within its contiguous zone a state may only exercise the control necessary to prevent or punish infringements of its customs, fiscal, immigration or sanitary regulations committed within its territory or territorial sea, i.e. not within the contiguous zone itself.

Baselines

A fundamental difficulty in establishing a territorial sea, whatever its width, lies in defining the coastal **baseline** from which it is to be measured. Up to 1930, individual nations designed their own baselines for the territorial sea, and nations settled their own demarcation disputes, peacefully or otherwise.

Incidentally, an interesting example of an early application of straight maritime boundaries was in 17th century England (Figure 2.2). Straight lines were drawn linking 27 headlands traditionally used in coastal navigation. These lines were the outer limits of neutral zones in which the Crown claimed the right to prevent hostilities between third parties.

The Hague Conference of 1930 made some attempt to frame universal baseline rules which would preclude international disagreement. Although it failed, the meeting cleared the air and led to further study and discussion, which culminated in the Convention on the Territorial Sea and a set of baseline rules that, though imperfect, are much better than no rules at all.

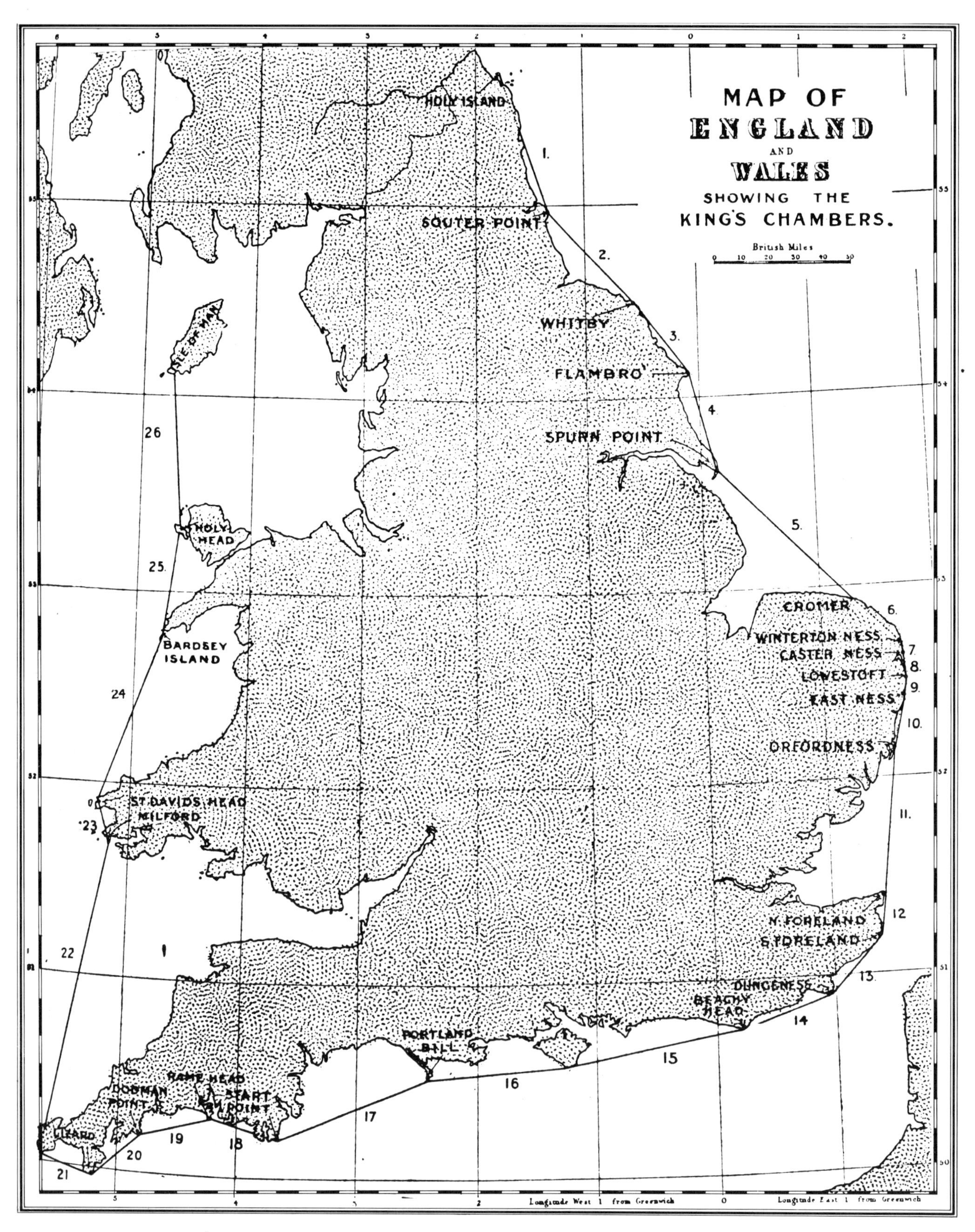

Figure 2.2 A system of straight maritime boundaries adopted for England and Wales in the 17th century. 'King's Chambers' refers to the areas enclosed within each segment.

In general, the territorial sea is to be measured from the coastline, defined as the low-water line. Unfortunately, however, the term 'low-water line' was not itself defined, so there is room for dispute even here. Is 'low water' to be taken as average low water, the lowest low water ever recorded, or something in between? The difference could be several miles, especially where the offshore gradient is low. It is true that the Convention refers to 'the low-water line along the coast as marked on large-scale charts officially recognized by the coastal State' but that could mean almost anything. The only other clue comes from the recommendation of the 1930 Hague Conference that the low-water line should not 'depart appreciably' from mean low water of spring tides, a recommendation that responsible states would endorse. Marine waters lying between such a baseline and the coast are just as much part of a state's **internal waters** as any freshwater lake or river.

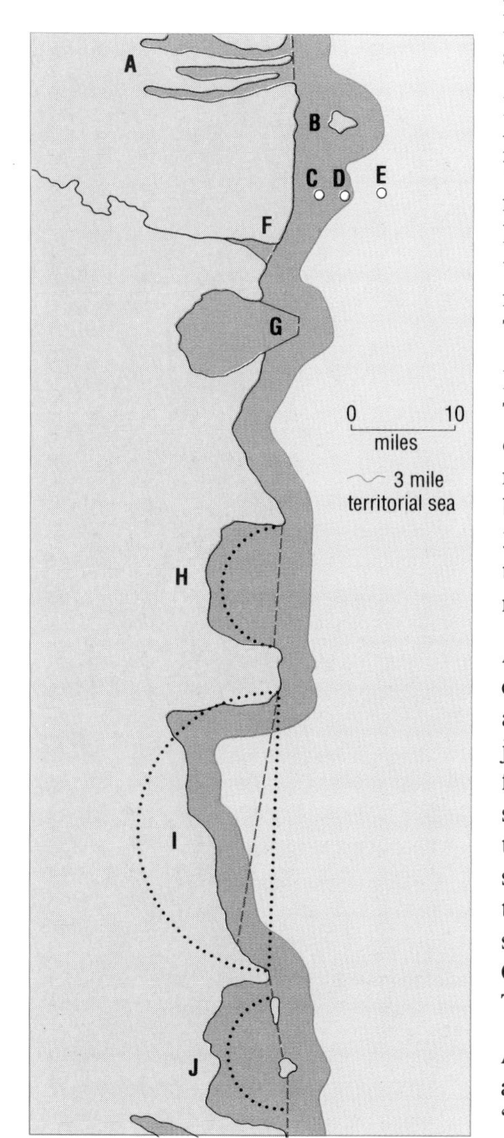

Figure 2.3 The rules for establishing territorial sea baselines, according to the Convention on the Territorial Sea (1958). Letters A to J on the map are referred to at various points in the following text, and you will need to come back to it later. See text for explanations of dashed and dotted lines.

If all coastlines comprised gentle curves, there would be no further problems. Unfortunately, nature is seldom like that. Real coasts have bays, gulfs, estuaries, promontories, peninsulas, fjords, offshore islands, man-made earthworks, reefs, and 'islands' that are only above water at low tide—so-called **low-tide elevations**. When it comes to deciding where the baseline for measuring the territorial sea should be, all such features must be examined closely and rules laid down to accommodate them. This is not a simple matter.

It is important to stress that the low-water line is described in the Territorial Sea Convention as 'the normal baseline'. However, the variety of geographical circumstances for which special provisions are laid down make it doubtful whether in practice the low-water line *is* the normal baseline for most states. (The Law of the Sea Convention—see Section 3.1.2—appears to recognize this situation, for in Article 14 it provides that '. . . the coastal State may determine baselines . . . by any of the methods provided for . . . to suit different conditions.')

According to the Territorial Sea Convention: 'In localities where the coastline is deeply indented and cut into, or if there is a fringe of islands along the coast in its immediate vicinity, the method of straight baselines joining appropriate points may be employed in drawing the baseline from which the breadth of the territorial sea is measured'. However, such baselines must not depart appreciably from the general direction of the coastline; they should take no account of low-tide elevations unless such elevations have installations always above sea-level (e.g. lighthouses); they must not cut off from the high seas the territorial sea of another state; and they must be clearly marked on charts. Unfortunately, the Convention did not specify the maximum length of straight baselines. They are therefore open to abuse, and have been abused in some cases.

As with low-water baselines, the waters landward of a straight baseline are regarded as internal waters. The Convention notes, however, that 'Where the establishment of a straight baseline. . .has the effect of enclosing as internal waters areas which previously had been considered as part of the territorial sea or of the high seas, a right of innocent passage. . .shall exist in those waters'. An example of this situation is described for the Canadian Arctic Archipelago in Chapter 4.

Figure 2.3 is a diagram of a hypothetical coastline with a variety of features and shows both low-water baselines and straight baselines (we shall examine different parts of Figure 2.3 later on). For deep indentations such as A in Figure 2.3, a straight baseline (dashed) is clearly appropriate—the most obvious real examples are fjords such as those of Norway and Chile.

Figures 2.4 and 2.5 illustrate contrasting examples of how different states have interpreted the Convention on the Territorial Sea, with particular reference to the specification of baselines. Figure 2.4 shows the baselines proclaimed by the Irish Government in 1959. They consist of six stretches of natural coastline and about 40 lengths of straight baseline.

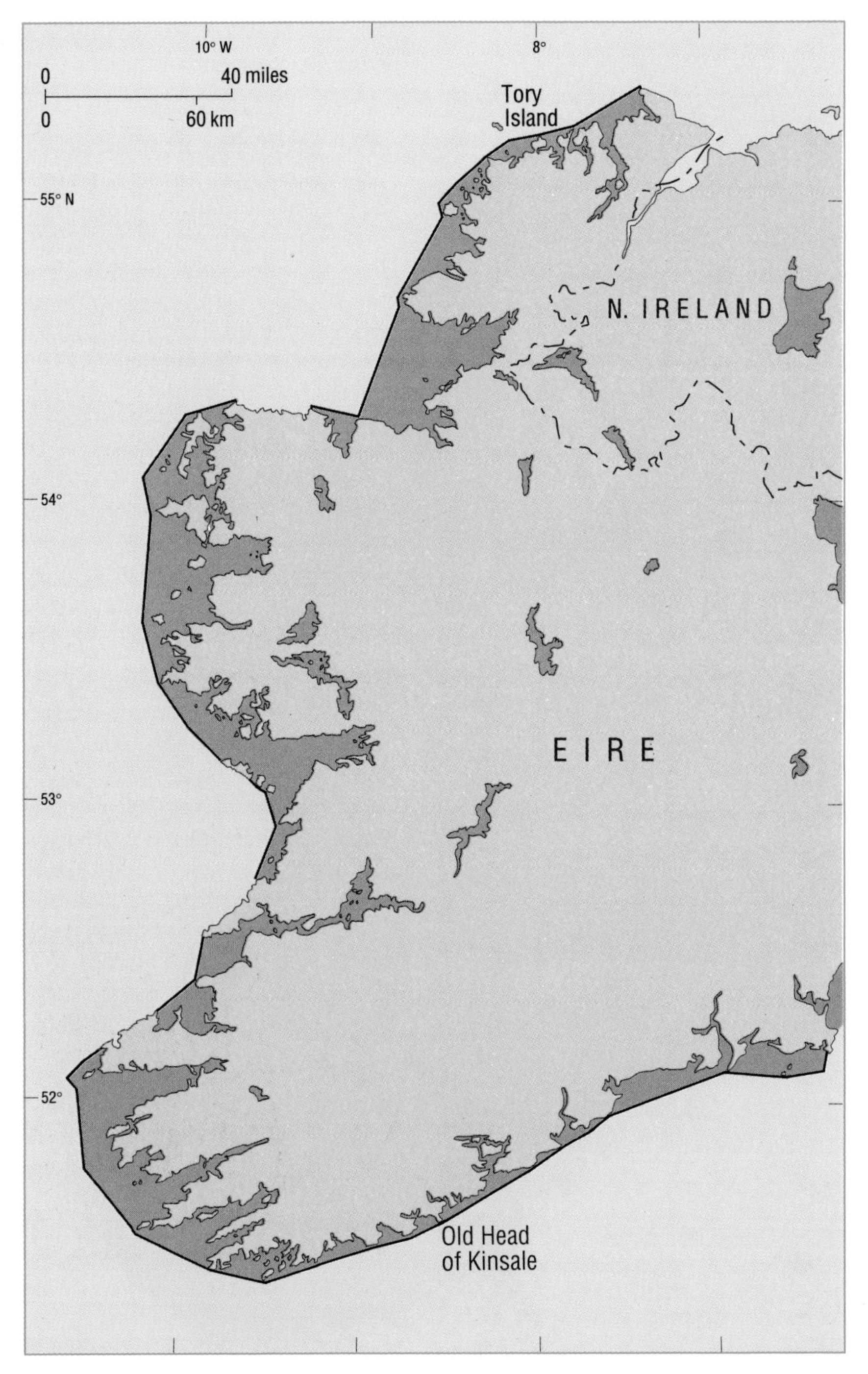

Figure 2.4 The straight baselines for western and southern Ireland, proclaimed in 1959.

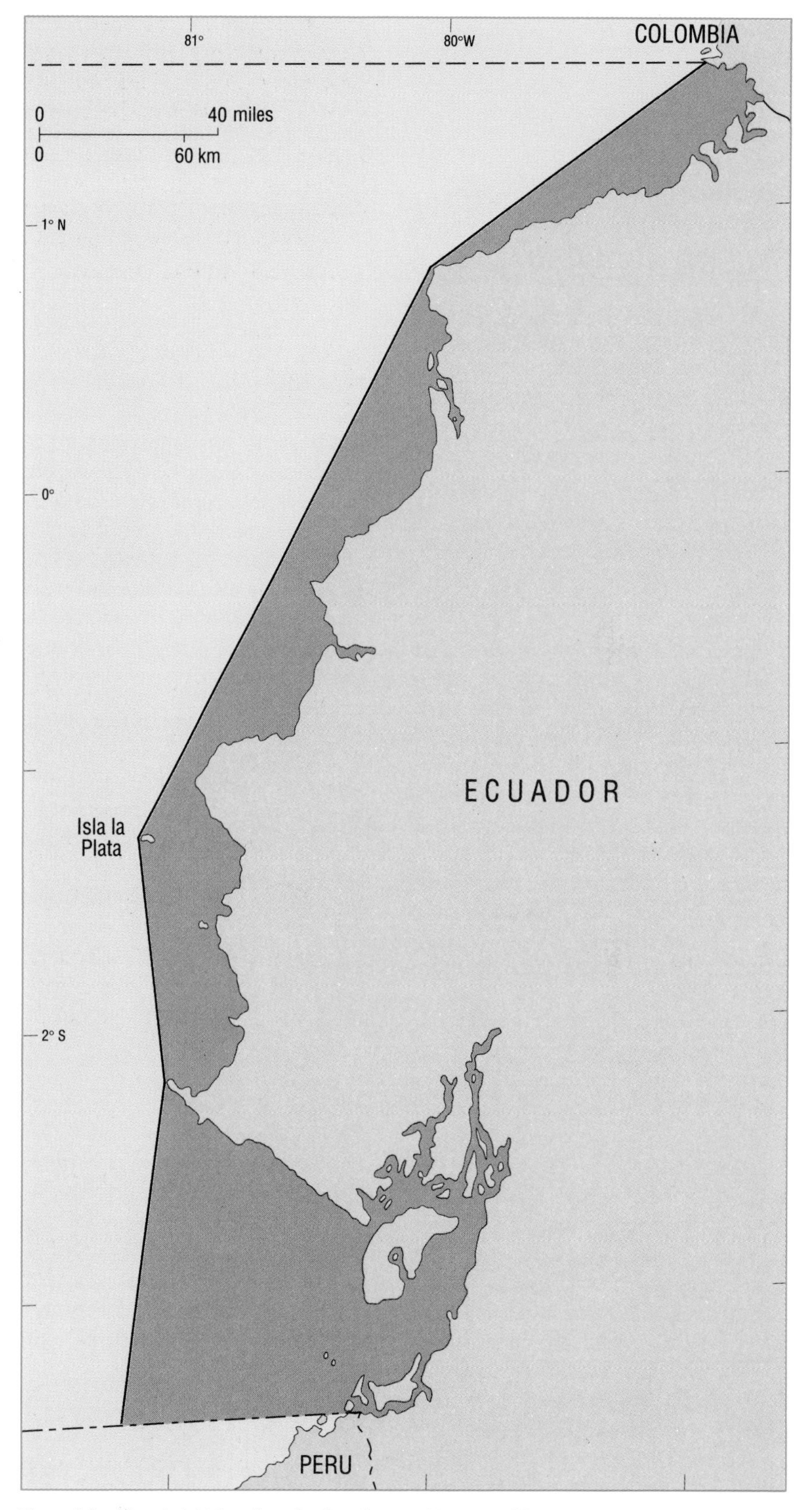

Figure 2.5 The straight baselines for Ecuador, proclaimed in 1971.

It is arguable that straight baselines are inappropriate east of the Old Head of Kinsale, and that it would have been preferable to adopt the normal low-water line procedure there. On the other hand, the scheme is conservative in that it took no account of islands such as Tory Island, which now lie outside the proclaimed baselines. These minor points apart, the Irish baselines conform with the Convention in both letter and spirit.

The same cannot really be said of the straight baselines encompassing its entire mainland coast that were declared by Ecuador in 1971 (Figure 2.5).

There are four ways in which Ecuador's scheme is at variance with the Convention. Can you identify what they are?

First, the coast of Ecuador is neither deeply indented nor fringed with islands, so straight baselines are inappropriate anyway. Secondly, the lines do not conform with the general direction of the coast, the southernmost leg being particularly errant in this respect. Thirdly, one of the reference points used is Isla La Plata, an isolated island 14 miles* from the coast (i.e. not strictly a 'fringing island'); and finally, the southernmost leg actually ends at Cape Blanco in Peru, and not in Ecuador at all (the internal waters so enclosed are split between Ecuador and Peru). This is an extreme example, but over 50 states have applied the method of straight baselines to some part of their coasts, with greater or lesser degrees of justification.

We have already seen (e.g. Figure 2.4) how fringing islands can project national baselines seaward, thereby displacing the territorial sea further offshore and bringing a new area of high seas under the legal regime of territorial waters. A question that arises naturally from this is whether baselines drawn round an individual island generate its own territorial sea. The Convention established that they do, regardless of the size of the island. Thus, an island near the coast may effectively extend the territorial waters of the adjacent mainland, as shown by B in Figure 2.3 (far away from shore, of course, it could have its own territorial sea—and contiguous zone—separate from those of any parent land-mass). A low-tide elevation, however, may be used to extend territorial waters only if it falls within the territorial sea that would exist if the elevation were not there. Moreover, a low-tide elevation 'wholly situated at a distance exceeding the breadth of the territorial sea from the mainland or an island . . . has no territorial sea of its own'. Examples of low-tide elevations in various positions are shown as C, D and E in Figure 2.3. C lies wholly within the territorial sea attributable to the mainland and hence may be used to extend it. Both D and E lie outside the territorial sea attributable to the mainland, and therefore have no effect. Thus, D cannot be used to extend the territorial sea further even though it falls within the extension produced by C. Note, however, that if C had been an island, D could have been used for this purpose.

For completeness, we need also to consider the effects of river mouths (F on Figure 2.3) and harbour (and other coastal) works (G on Figure 2.3). According to the Convention, if a 'river flows directly into the sea, the baseline shall be a straight line across the mouth of the river . . .'. This sounds simple enough—see F on Figure 2.3.

*As stated in the Introduction, throughout this Volume 'miles' means 'nautical miles'.

Estuaries pose a problem, however, which is not addressed by the Convention. Is an estuary part of the river, which can therefore be regarded as flowing directly into the sea, automatically giving a (straight) baseline across the estuary mouth? Or is the estuary a bay, to which other rules apply (see below)? Naturally, there is a tendency for states to use the interpretation most favourable to themselves.

Permanent harbour works are regarded as part of the coast for the purpose of measuring the territorial sea (G, Figure 2.3). Roadsteads (offshore anchorages) 'which are naturally used for the loading, unloading and anchoring of ships' may be regarded as part of the territorial sea where they would otherwise lie partly or wholly outside it.

Archipelagos present a different problem, which the Convention did not address. Many states consisting of groups of mid-ocean islands, such as the Philippines, to which the fringing islands provision of the Convention clearly does not apply, have nevertheless sought to establish straight baselines around their territory rather than applying the normal island regime to each part of it (i.e. drawing baselines round individual islands). A frequent effect of this has been to appropriate large areas of the high seas of importance to international navigation, provoking vigorous objections from maritime powers.

Bays

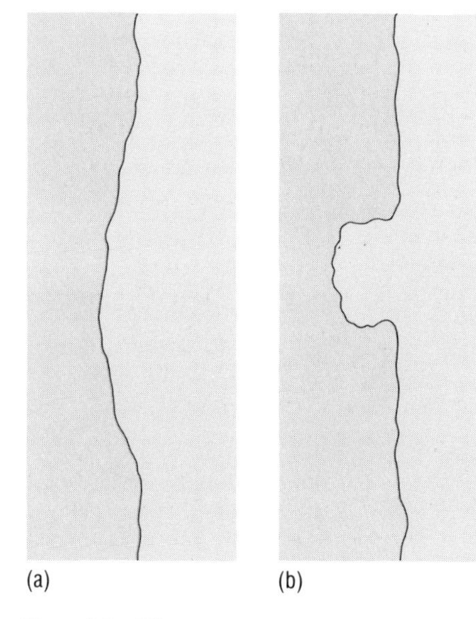

Figure 2.6 When is a bay a legal bay?

It would seem reasonable to regard a bay as part of a state's internal waters and thus completely under the state's jurisdiction. In that case, the mouth of the bay would be closed by a straight baseline and the territorial sea would be measured outward from there. But when is a bay not a bay? Consider, for example, the two hypothetical coastlines shown in Figure 2.6. You may well conclude that the coastline in (b) has a bay whereas that in (a) does not; but if so, you have made a subjective judgement, however reasonable. In international affairs, subjectivity is undesirable; to avoid disputes it is preferable to have more objective criteria. How, then, can the bay in (b) be clearly differentiated from the 'bay' in (a)? The Convention states that 'a bay is a well-marked indentation whose penetration is in such proportion to the width of its mouth as to contain landlocked waters and constitute more than a mere curvature of the coast. An indentation shall not, however, be regarded as a bay unless its area is as large as, or larger than, that of the semicircle whose diameter is a line drawn across the mouth of that indentation'. But where 'because of the presence of islands, an indentation has more than one mouth, the semicircle shall be drawn on a line as long as the sum total of the lengths of the lines across the different mouths'.

Once it has been decided that a bay falls within this definition of a **legal bay**, the position of the baseline from which the territorial sea is to be measured depends upon whether the **closing line** across the mouth of the bay is longer or shorter than 24 miles. If it is shorter, it becomes the baseline. If it is longer, 'a straight baseline of twenty-four miles shall be drawn within the bay in such a manner as to enclose the maximum area of water that is possible with a line of that length'.

These rules are not as complicated as they may at first appear, although it is sometimes difficult to apply them in real situations. Three examples of their application to hypothetical cases are shown in Figure 2.3 (H, I and J), to which you should again refer.

In the case of bay H, the area of the semicircle drawn on the closing line (*dashed line*) is evidently smaller than the area of the bay within the closing line. A is therefore a legal bay and part of the state's internal waters. The closing line is less than 24 miles long and is thus the baseline from which the territorial sea is measured. In the case of bay I, on the other hand, the area of the semicircle drawn on the closing line (*dotted line*) is greater than the area of the bay within the closing line. I is therefore not a legal bay; it is not part of the state's internal waters and the territorial sea is measured from the shore. Note, however, that if the area of the semicircle had been smaller than the area of the bay within the closing line, I would have been a legal bay, but the closing line would not have been the baseline because it is more than 24 miles long. The baseline would have been the line 24 miles in length enclosing the maximum area of water (*dashed line*). Bay J is different again, for it has islands across its mouth. The total length of the closing line (*dashed line*) excluding the islands is 10 miles; the area of the semicircle with that diameter is smaller than the area of the bay within the closing line. J is therefore a legal bay.

Although the Convention on the Territorial Sea provided these rules for the definition of legal bays, it also recognized that there are bays lying outside the strict definition that may nevertheless be regarded as legal bays by virtue of traditional usage. These are known as **historic bays**. No rules were laid down to guide claims to historic bays because of the difficulty involved in defining them. This creates a legal loophole, because a case could probably be made for regarding most of the world's bays as historic on one basis or another. Nevertheless, bays widely accepted as legal for historical reasons include the Bristol Channel (Britain), Chesapeake Bay, Delaware Bay and Long Island Sound (United States), the Zuider Zee (Netherlands), the Gulf of Panama (Panama), Granville Bay (France), Hudson Bay (Canada), the White Sea, the Bay of Riga and Peter the Great Bay (Russia) and about 50 others. The significance of legal bays, however defined, cannot be over-emphasized. The key point is that they are subject to the absolute sovereignty of the coastal states concerned, so there is no right of innocent passage through their waters. The idea of sovereignty being acquired through traditional usage has been extended to waters other than bays; the importance of *historic waters* in determining legal status will be discussed in Chapter 4 in the context of the Arctic.

This discussion of one aspect of maritime legislation may seem technical, but is no less important for that. The truth is that much of the Law of the Sea is precisely about such details. When philosophers have argued about justice to developing countries, when academics have quibbled over principles, and when politicians have rationalized national self-interest into a respectable bargaining posture, someone has to translate agreed general positions into a practical form that can operate under day-to-day conditions. Unless the position is made clear, disputes may arise, wars may be fought and lives may be lost. Or as one writer has put it: 'It is fine to talk in the abstract, but at some point a line may have to be drawn'.

Width

The width of the territorial sea has been perhaps the most contentious issue of all in the history of the Law of the Sea. By the middle of the

19th century, three miles had come to be the figure most widely adopted, but the origins and development of this situation are complex.

The concept of a continuous band of protected waters around the coast really began to develop in Scandinavia four centuries ago to restrict fishing and trade by foreigners. Over much of Europe, the width of such a band came to be determined by the 'cannon-shot-rule': a principle, based on the range of coastal ordnance, mainly to define neutral zones in times of war. The two ideas converged in 1782 when the Italian economist and jurist Galiani, accepting that the waters over which coastal states exercised jurisdiction formed a continuous zone, suggested that its width should be three miles, that being the maximum range of cannon in those days. Thus was born the three-mile limit for territorial waters, which persists in some areas to the present day.

The basic reason for the longevity of the three-mile limit was that the chief maritime powers (Britain, USA, Germany and Japan) had a vested interest in keeping the territorial sea as narrow as possible, thereby maximizing the area of high seas over which they could roam unhindered. They claimed a three-mile territorial sea and refused to recognize the claims of others to wider jurisdiction. However, the first attempt to establish this limit by international agreement, at The Hague Conference in 1930, failed, as we have seen. In Denmark, Norway and Sweden, the four-mile Scandinavian league persisted, Spain wanted six miles and Italy ten. Only twelve participants were satisfied with a three-mile limit.

The Convention on the Territorial Sea was somewhat more successful in this respect, but the only guidance given as to the permissible extent of a territorial sea was that its width combined with that of the contiguous zone should not exceed twelve miles. What the Convention did confirm was that choice of baseline and width of claim are independent issues which together determine the geographical position of the sea areas over which a coastal state can exercise jurisdiction. After that, the pressures for wider territorial waters increased.

You are now in a position to explore some consequences of real-life decisions about baselines and the width of the territorial sea.

QUESTION 2.1 Look at Figure 2.4.

(a) Given that the territorial sea declared by the Republic of Ireland is only three miles wide, and that the passage between Tory Island and the mainland is four miles wide, explain why this passage lies entirely within Irish territorial waters.

(b) If Ireland were to extend its territorial sea limit to, say, ten miles, what effect would Tory Island then have on Irish territorial waters?

(c) Could any of these arguments (relating to a three- or ten-mile wide territorial sea) be applied to Isla La Plata in Figure 2.5?

Straits

Special problems arise in narrow straits connecting two areas of the high seas. The legal status of the waters in the straits is affected by the width of the territorial sea claimed by the adjacent coastal state(s). If the distance between the baselines on opposite shores is greater than twice the width of the territorial sea, then there remains a strip of international waters down the middle of the strait; but if the distance between the baselines is less, the whole width will be occupied by territorial waters. Where such straits are used for international navigation, i.e. they are **international straits**, the outcome is a matter of concern to all maritime nations, because foreign shipping in the strait becomes subject to regulation by the coastal state(s).

Increasing the width of the territorial sea from three to twelve miles has the potential effect of regulating passage through more than 100 straits. Most are of minor importance, but at least 15 are of high strategic value to maritime powers and especially to the USA and USSR. They include the Bering Straits; Malacca Straits; Old Bahamas Channel; Straits of Hormuz; Straits of Gibraltar; and Straits of Dover (Figure 2.7). The list of such strategic straits will vary according to the criteria of selection, but there is clearly a group of straits that virtually no maritime nation would wish to see closed to navigation. The Convention recognized that problem by incorporating a clause forbidding the suspension of the right of innocent passage through straits used for international navigation.

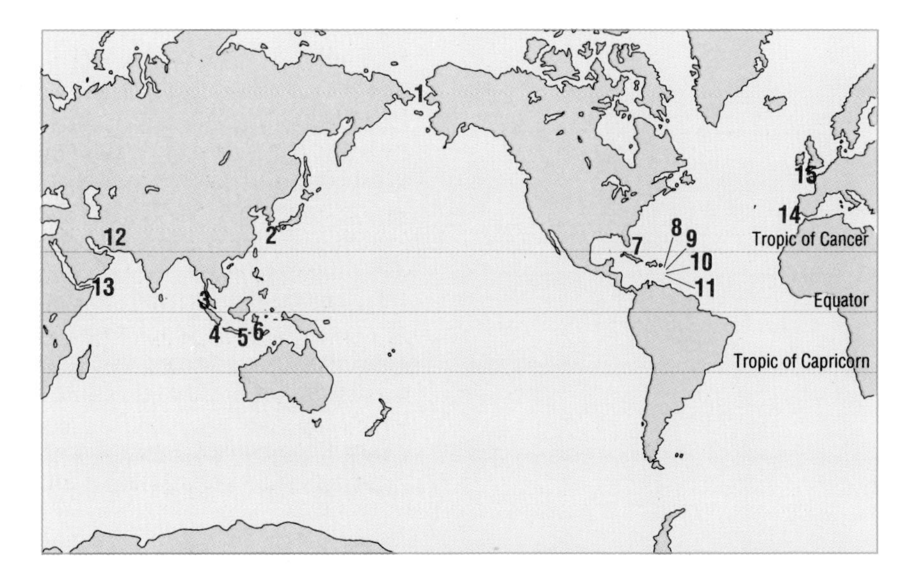

Figure 2.7 Map showing 15 strategic straits that would, in theory, be closed because they are less than 24 miles across (2 × 12-mile territorial sea width). 1 Bering; 2 Western Chosen; 3 Malacca; 4 Sunda; 5 Lombok; 6 Ombrai; 7 Old Bahamas Channel; 8 Dominica; 9 Martinique; 10 St. Lucia; 11 St. Vincent; 12 Hormuz; 13 Barb el Mandeb; 14 Gibraltar; 15 Dover.

2.2.2 CONVENTION ON THE CONTINENTAL SHELF

The right of a coastal state to the sea-bed underlying its territorial sea, which the Convention on the Continental Shelf established, had developed from historical claims based on the exploitation of living sea-bed resources such as oysters. Some of these early claims had been recognized as extending beyond territorial limits, but it was the modern quest for hydrocarbons that made inevitable far broader claims, whose intention was pre-emptive rather than regulatory.

The first step was taken in 1942 when Venezuela and the United Kingdom (acting on behalf of its colony, Trinidad and Tobago) signed a treaty agreeing on a boundary dividing 'the submarine areas of the Gulf of Paria', by which each agreed not to assert any claim to sovereignty or control in the other's sector and to recognize any rights lawfully acquired by the other therein. Although shelf resources were not specifically mentioned, the aim of the treaty was to assist the search for oil beneath the waters of the gulf outside territorial limits, involving about 5 000 km^2 of continental shelf. The next important step was not long in coming; it was the so-called Truman Proclamation of 1945. Through it the USA became the first nation to claim sovereign rights over the natural resources of its continental shelf, without qualification.

This document set the precedent for a rising tide of similar claims which led to the Convention on the Continental Shelf in 1958. By then, about 20 nations had claimed some form of jurisdiction over their shelves. The Convention followed the general line taken by the Truman Proclamation, but is somewhat narrower in scope. Thus, it declares that 'The coastal State exercises over the continental shelf sovereign rights' [only] 'for the purpose of exploring it and exploiting its natural resources'. It does not, however, affect the legal status of the waters overlying the shelf.

But what *is* the continental shelf? Its geological definition is clear enough, but when it comes to fixing its outer limit for legal purposes we enter very murky waters indeed.

Definitions

In the words of this Convention, 'the term "continental shelf" is used as referring (a) to the sea-bed and sub-soil of the submarine areas adjacent to the coast but outside the area of the territorial sea, to a depth of 200 metres or, beyond that limit, to where the depth of the superadjacent waters admits of the exploitation of the natural resources of the said areas; (b) to the sea-bed and sub-soil of similar submarine areas adjacent to the coasts of islands'. This formulation is already some way from the reality of the physical world, so let us briefly return our attention to the bathymetric facts.

Figure 2.8 shows what textbooks usually call a section through a 'typical continental margin'. The continental shelf, a submerged extension of the

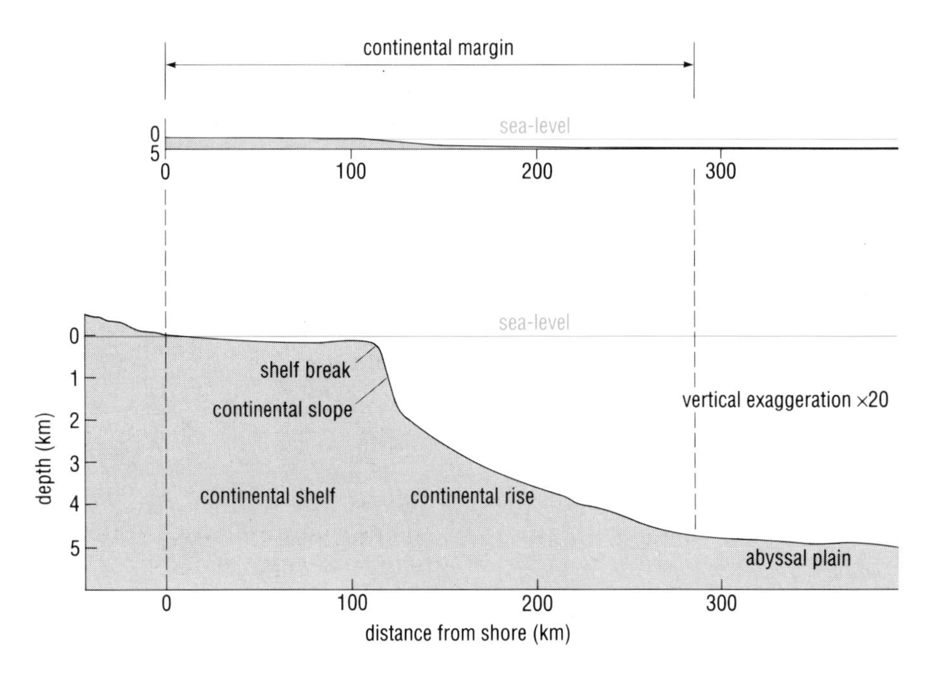

Figure 2.8 A 'typical' continental margin. As outlined in the text, distances, depths and gradients vary greatly from place to place.

land modified by erosion and deposition, gives way at the shelf break to the continental slope, which merges into the continental rise and finally the abyssal plain. The continental shelf has an average width of about 40 miles and descends with a gradient of about 0.1° to an average depth of 130 m at the shelf break. The continental slope has an average gradient of about 4°, merging with the continental rise beneath several kilometres of water. (Along some continental margins, of course, there is no rise, for the slope leads into an oceanic trench.) These averages conceal wide variations, however. In the Barents Sea, the width of the continental shelf goes up to about 700 miles, whereas west of the Niger Delta it can be as little as 20 miles. At the outer edge of the shelf the water depth varies from 20 m to more than 500 m. Off the north-west coast of Australia, the gradient of the continental slope is about 1°, but off the coast of Chile it is 45° in places. The variability shown by continental margins is illustrated in Figure 2.9.

In the face of such variations, any hope that the outer edge of the continental shelf would provide a universal jurisdictional boundary soon disappears; moreover, the shelf break is seldom sharp enough to fulfil that purpose accurately because it commonly extends over several miles. In practice, the attraction of using a natural boundary is far outweighed by problems arising from its natural variability and imprecision. Instead, in an attempt to compensate for these inconvenient properties, the definition of the continental shelf has been distorted for legal purposes resulting in a concept almost as artificial as that of the territorial sea. This is the **legal continental shelf**, a term adopted to distinguish it from geological reality.

The dual definition quoted from the Convention on the Continental Shelf is a case in point. Its depth criterion has the merit of bearing some, albeit loose, relation to real continental shelves, but at the same time it fails to avoid the very problem that an essentially arbitrary limit might have been expected to solve: the unequal distribution of shelf area per unit length of coastline. Replacing the shelf break with a depth limit brings equality no closer, because the width of the legal continental shelf still depends on its gradient. States with gently sloping shelves found themselves with areas of jurisdiction stretching hundreds of miles out to the 200 m isobath; those with steep shelves could see the limit of their jurisdiction from the beach. So, we are left with an uneasy compromise which satisfies neither physical reality nor legal pragmatism.

The second 'exploitability' criterion mitigated this failure to some extent, but only at the cost of further uncertainties. For where does such an elastic concept end? The risk was recognized that the shelf regime could expand, as new technology gave access to ever greater depths, until it covered the entire ocean floor, but it was generally considered that it would be impossible to exploit sea-bed resources beyond the 200 m isobath in the foreseeable future. This view was mistaken. Although the 200 m limit was foreshadowed in a press release accompanying the 1945 Truman Proclamation, by 1958 the US had granted oil exploration leases for areas in water depths greater than 400 m. The capability to recover polymetallic nodules (Section 1.3.2) from the abyssal plains on an industrial scale has existed since the 1970s, although it has yet to be applied.

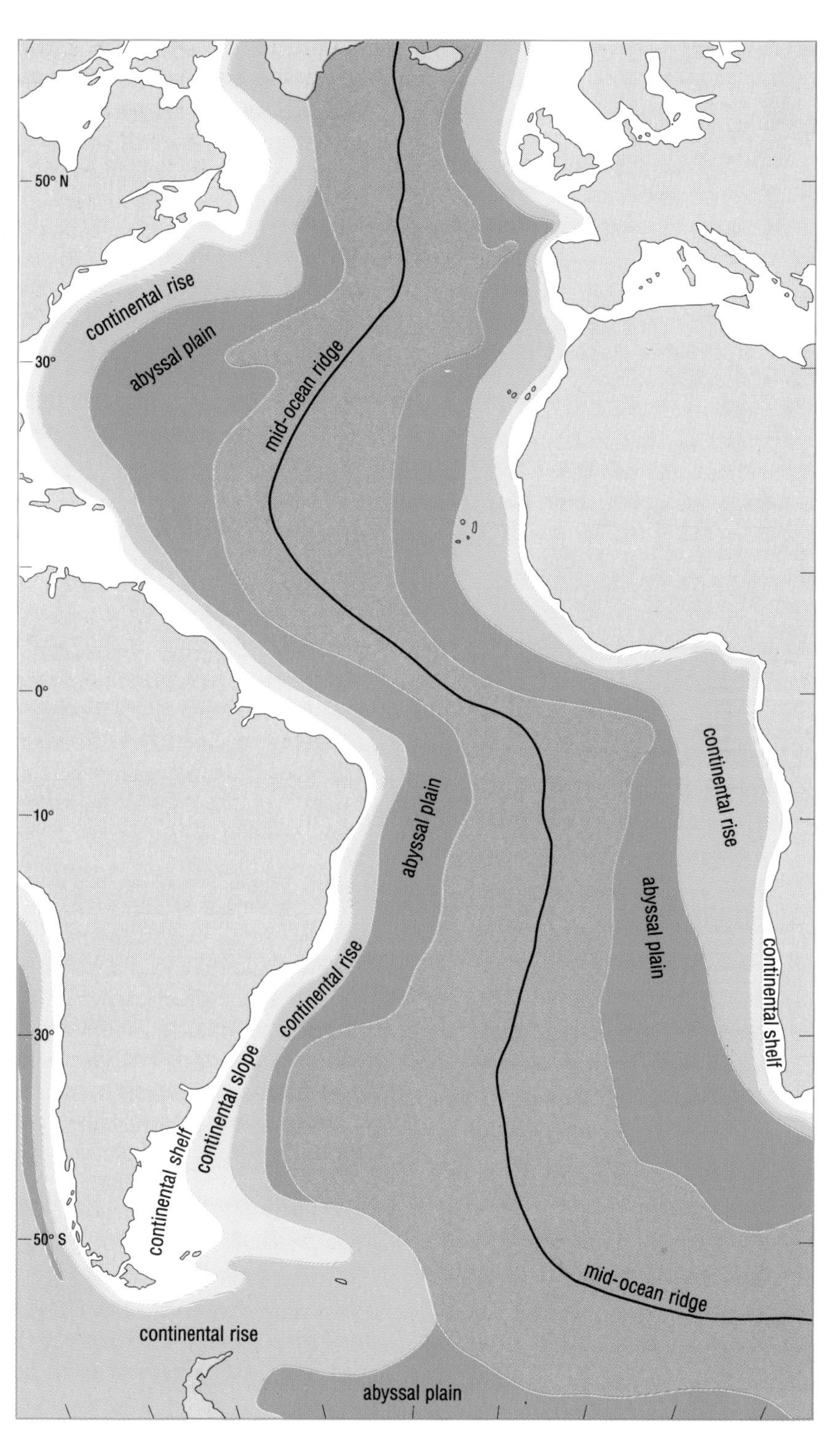

Figure 2.9 Simplified map showing the extent of variability of continental margins in the Atlantic alone. Variability between different oceans is even greater.

Under this Convention, the rights of a coastal state over its legal continental shelf exist independently of any action on its part, so the seaward boundary would expand automatically. However, there are other difficulties. Even the phrase 'exploitation of the natural resources' is ambiguous. Does it imply national control over commercial operations, or would simply collecting research samples be sufficient to qualify? And who is envisaged as undertaking this hypothetical activity? Would there be a universal seaward limit, defined by the capabilities of the most advanced technology currently available anywhere in the world, or would each state's boundary be determined by the technology available to its own organizations? If the latter, the less developed countries would obviously be at a disadvantage. If the former, it would only be a matter of time before claims could be extended over the whole area of the world's ocean.

The introduction of this hybrid definition of the continental shelf, consisting of a fixed numerical limit with an open-ended rider, could only doom the subject to inconclusive argument and, eventually, a further round of international negotiations, as we shall see in Chapter 3. However, it is worth noting that ambiguities are often deliberately left in proposals, so that, even though conventions are agreed and signed, the momentum towards proper regulation is still maintained.

2.2.3 CONVENTION ON THE HIGH SEAS

This is the simplest of the 1958 Conventions as regards the definition of maritime zones. It applies to 'all parts of the sea that are not included in the territorial sea or in the internal waters of a State', and declares that 'The high seas being open to all nations, no State may validly purport to subject any part of them to its sovereignty'. Thus, the traditional position on 'freedom of the seas' (**freedom of navigation**) was preserved throughout most of the ocean, including the contiguous zone and waters overlying the legal continental shelf. Much of the Convention on the High Seas is concerned with the regulation of shipping, covering such matters as nationality and registration of vessels and jurisdiction over them; collisions at sea; piracy and 'hot pursuit' (of suspected law-breakers); and, to a limited extent, pollution.

The other major human activity on the high seas—fishing—was dealt with in the ill-fated Convention on Fishing and Conservation of the Living Resources of the High Seas, already mentioned at the start of Section 2.2. Perhaps because it was more innovative than the others, this attracted insufficient support to become really effective, and it will not be considered further here. However, we still have to consider one important group of maritime zones which has developed *outside* the international context of the 1958 Conventions in recent decades: exclusive fishing zones and similar unilateral claims.

2.2.4 UNILATERAL CLAIMS OUTSIDE THE 1958 CONVENTIONS

In the years following the 1958 Conventions, continued restriction of the width of territorial seas led to unilateral declarations of broader zones of sovereignty by frustrated maritime nations seeking exclusive access to the real or presumed resources of their offshore areas. This process began not long after World War II, but the 1958 Conventions probably gave it additional impetus. Cases of a nation actually *reducing* its claim are virtually unknown: it is much more difficult for a state to moderate an excessive claim than to make a more modest one in the first place and subsequently extend it.

Whatever the purpose or motive of these unilateral declarations, their effect was to encroach upon the high seas. In some cases, the claims amounted to extending full-blown territorial seas, but mostly they were 'diluted' extensions of territorial waters for the regulation of fisheries and other activities, such as sea-bed prospecting.

Exclusive fishing zones

For several millennia, fishermen contented themselves largely with fishing in their own near-coastal waters, but during the past few centuries improving technology has enabled them to go further afield (*cf.* Section 1.2.2). By the mid-1500s, the Portuguese were catching cod on the Grand Banks off Newfoundland, and by the 18th century American whalers were dispersed over much of the globe. In recent decades this trend has accelerated, the Russians and Japanese in particular being well known for their large distant-water fishing fleets with sophisticated fish-detection equipment and giant factory ships for on-the-spot processing of fish.

So, it is not surprising that retaining control of fisheries has for some time provided the motivation for many claims to territorial waters. The pearl fisheries of Sri Lanka and the oyster beds of Granville Bay, France (Section 2.2), are two examples that go back over 150 years.

All this has led to the concept of the exclusive fishing zone—a zone beyond the territorial sea as defined in the Convention on the Territorial Sea, within which the coastal state has complete control over the exploitation of fish stocks. The operative word here is 'control'. A nation declaring an exclusive fishing zone does not necessarily reserve the fish in the zone entirely to itself. Others may fish there, but only under licence and usually within a quota system.

In Section 2.2.2, we mentioned the Truman Proclamation covering natural resources of the sea-bed and sub-soil of the continental shelf. At the same time as this document was issued (1945), there was another Truman Proclamation on conservation of 'coastal fisheries in certain areas of the high seas'. This declaration was as influential as the other, and opened the way to claims by other nations, some of which went even further. For instance, in 1946, Panama claimed the waters superadjacent to the continental shelf 'for purposes of fisheries in general' and in 1951, Honduras declared complete sovereignty over its epicontinental sea.

Despite much discussion at the First (1958) Law of the Sea Conference, it was not possible to reach agreement about exclusive fishing zones: the Convention on the Territorial Sea had little effect on existing fisheries regimes, while the Convention on the Continental Shelf affected only 'sedentary species' (growing corals, oysters, sponges, seaweeds, etc., and possibly certain crustaceans: crabs, lobsters, shrimps, etc.).

There were further discussions at the Second (1960) Conference, when the USA and Canada made a strong case for the so-called $6 + 6$ formula—a proposal that there should be a six-mile territorial sea plus a six-mile exclusive fishing zone (a $3 + 9$ formula was also possible for nations wishing to keep a three-mile territorial sea). This proposal was defeated by only one vote and the great majority of nations still had fishing limits of twelve miles or less as recently as 1972 (Appendix 1, Table I). After that, however, the situation changed rapidly.

Iceland, which had increased its fishing limit to twelve miles in 1958, increased it again to 50 miles in 1972, and to 200 miles in 1975—moves which led to the 'cod wars' with Britain. Norway increased its limits to 200 miles in 1976; in the same year the USA announced that it would follow suit in 1977; and on 1 January 1977, the European Economic

Community (in which negotiating rights for all member countries were vested in 1976) introduced its own 200-mile limit.

As for fisheries on the high seas, these remained without any international regulation because of the failure of the ill-fated fourth Convention mentioned at the start of Section 2.2.

Wider unilateral claims

Many claims went beyond fisheries. Substantial maritime zones were claimed by various Latin American nations in the 1950s and 1960s. These claims were diverse and often vague, ambiguous or even internally inconsistent. Most were designed to gain control over the continental shelf and its superadjacent waters without necessarily restricting the traditional rights of freedom of navigation and innocent passage on the high seas. They were often based explicitly on the philosophy that coastal communities have an ancestral interest, and hence ancient rights, in their offshore zones with which they form a natural interdependence, especially in respect of the living resources—this is the concept of the *patrimonial sea*, which is not unrelated to the concept of historic waters (Section 2.2.1). There are some parallels with the Convention on the Continental Shelf (Section 2.2.2), which these early claims pre-dated, and the patrimonial seas can be seen as putative or embryonic exclusive economic zones (see Chapter 3).

Whatever the motives, claims for wider marine territorial limits multiplied in the years after 1958. The most popular figure was twelve miles (see above), in line with the provisions of the Convention on the Territorial Sea (Section 2.2.1). Nine countries, mostly in South America (but including Iceland, see above) had even raised their limits to 200 miles. For example, Ecuador (by a decree of 1966), Panama (1967) and Brazil (1970) were evidently claiming a 200-mile territorial sea *sensu stricto*. Between these extremes, Costa Rica, for example, while claiming only a three-mile territorial sea in 1972, nevertheless established 'state protection' (whatever that means) over 200 miles. On the other hand, Argentina, Chile, El Salvador, Peru and Uruguay recognized within their 200-mile limits the freedom of navigation (as on the high seas) and not simply the right of innocent passage (as in the formal territorial sea).

Detailed study of the relevant proclamations is often required to discover just what individual claims imply, although ambiguity, vagueness and inconsistency often thwart this aim. The general point to bear in mind, however, is that many countries no longer regarded themselves as having a single all-embracing territorial sea, but a series of two or more maritime zones designed for different purposes.

QUESTION 2.2 The earliest and largest claims for increased marine limits came from the west coast of Latin America. Why do you think this might be?

This Section in particular has disclosed some of the ragged edges of the international legal regime, as it stood (or purported to stand) until the early 1980s. The position was obviously chaotic, with different limits for different purposes, a collection of *faits accomplis* which were to be regularized by UNCLOS in 1982, as we shall see in the next Chapter.

First, though, we look at an example of just how desirable it is to have a coherent Law of the Sea. We have chosen the Aegean, partly because of its long history (in keeping with the historical sweep of this Chapter) and partly to illustrate why the Law of the Sea is relevant to oceanographers.

2.3 THE AEGEAN

Who owns what in the Aegean has been a cause of conflict for at least 4 000 years. As shown in Figure 2.10, the Aegean is bordered by both Greece and Turkey; nevertheless, almost all the many Aegean islands are Greek, even those very close to the Turkish mainland—Samos and Rhodes, for example. An exception is the island of Imrozo, to the west of the Dardanelles; but because most of the islands are Greek, the international boundary in the Aegean lies very close to Turkey.

In such circumstances, the delimitation of marine boundaries may prove difficult. Consider, for example, the two bays on the Turkish side with headlands indicated by A and B, and C and D, respectively (Figure 2.10).

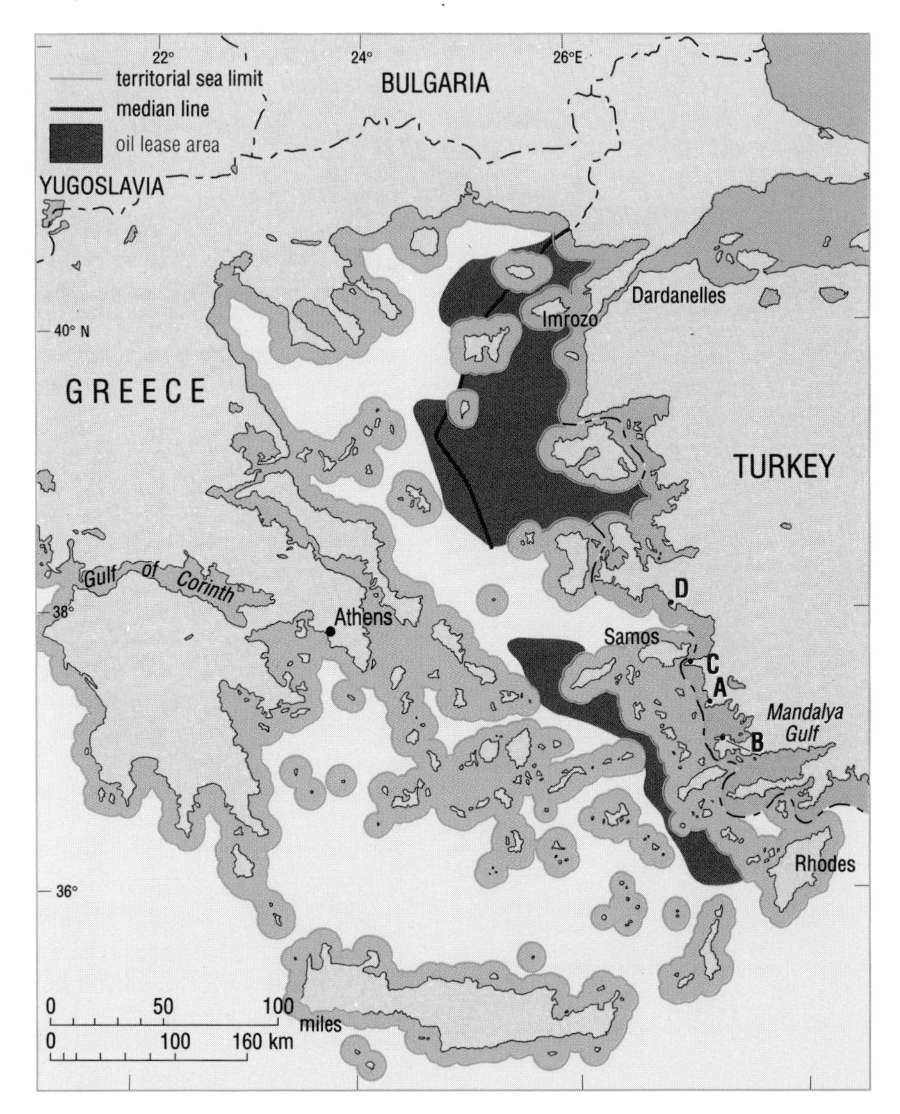

Figure 2.10 Map of the Aegean area, showing the limit of (12-mile) Greek and Turkish territoral seas. (For an explanation of the 'median line' and the coloured areas, see text; for letters A to D see Question 2.3.) Broken lines show international boundaries.

QUESTION 2.3

(a) Is Mandalya Gulf (between headlands A and B) a legal bay?

(b) Is the bay with headlands C and D a legal bay, according to geometrical criteria?

(c) Assuming that both *are* legal bays, where would their 'closing lines' be drawn?

(d) What is the significance of the position of closing lines as far as the coastal state (in this case Turkey) is concerned?

As you know from Section 2.2.1, the territorial sea is normally measured from the coastline (strictly, the low-water mark); Figure 2.10 shows the boundaries of twelve-mile territorial seas for both Turkey and Greece. In disputed areas like the Aegean, even the delimitation of territorial seas can cause problems. While attempting Question 2.3, you will no doubt have noticed that the Greek island of Samos—which has its own territorial sea—is extremely close to the Turkish mainland. As the island is less than 2 × 12 miles from the coastline, it is not possible for both Turkey and Greece to claim a full-width territorial sea. The intervening waters have been divided by a negotiated line which is, in effect, the international boundary between Turkey and Greece. If you look along the Turkish coastline in Figure 2.10, you will see that a similar situation obtains along much of its length.

Not surprisingly, the fact that the Greek and Turkish territorial seas are contiguous is a cause of confusion and bad feeling. In 1974, the British research vessel *Shackleton* was arrested by the Greek Government for working in Greek territorial waters without permission. Those working on the vessel thought they were operating in Turkish territorial waters and therefore had obtained permission from the Turkish Government—which agreed with them. As a consequence of this misunderstanding, the researchers on the *Shackleton* lost all the data they had collected—their drill cores, records and maps were all confiscated by the Greeks.

Why in particular might Greece and Turkey be concerned about the presence of a foreign research vessel in 'their' waters?

As we have seen, coastal states feel a need to defend their marine resources, whether above or below the sea-bed. In the Aegean, there are thick accumulations of sediment—potential reservoirs of oil and gas. Unfortunately, if it is difficult to define Greek and Turkish territorial waters clearly, it is even trickier to determine how the sea-bed of the Aegean should be apportioned between the two states.

How much of the Aegean sea-floor would count as continental shelf (Greek *or* Turkish), according to the Convention on the Continental Shelf?

According to the Convention, the term 'continental shelf' refers to the 'sea-bed and the sub-soil of submarine areas adjacent to the coast but outside the area of the territorial sea, to a depth of 200 m or, beyond that limit, *to where the depth of the superadjacent waters admits of the exploitation of the natural resources of the said areas*'. Only about one-quarter of the Aegean sea-bed is less than 200 m deep; however, the more advanced the technology available—whether to Greece and/or

Turkey or to a more wealthy nation is not clear—the more of the sea-bed that can be claimed as 'exploitable' and therefore also as 'continental shelf'.

If it is assumed that all of the Aegean sea-floor *is* exploitable (most of it is less than 500 m deep), then it all becomes legal continental shelf, and must be divided between the two countries. How this might be done to the satisfaction of both Greece and Turkey is not obvious, especially given the large numbers of Greek islands in the eastern Aegean. According to the 1958 Conventions, islands have not only their own territorial seas but also their own legal continental shelves, with the proviso that if islands are situated where they impinge on the continental shelf of another state, the position of the dividing line should be negotiated separately.

Assume for the moment that the Greek islands do *not* have their own continental shelves (or that the islands in the eastern Aegean are not Greek but Turkish). How would the line dividing the Aegean sea-floor between Greece and Turkey be drawn, according to the Convention on the Continental Shelf?

A median line could be drawn in the Aegean. According to one of the articles of that Convention, such a **median line** would be 'equidistant from the nearest points of the baselines from which the breadth of the territorial sea of each State is measured'. The northernmost section of this line is shown in Figure 2.10. This median line can have no legal force, because the assumptions made in drawing it are not valid (i.e. most islands in the eastern Aegean are Greek). In general, a median line is required only if no agreement has been reached between the parties, and there are no special circumstances; but the States concerned should be parties to the Convention on the Continental Shelf.

In November 1973 and July 1974, the Turkish Government leased part of the Aegean to the Turkish Petroleum Company, for exploration and exploitation. The area involved is shaded in maroon on Figure 2.10; as you can see, most of it lies on the Turkish side of the median line. Not surprisingly, Greece protested, arguing that, as the Greek islands *do* have their own legal continental shelves, the Greece–Turkey dividing line should come to the east of the Greek islands close to the Turkish coastline. Essentially, they claimed that the area leased by the Turkish Government was not theirs *to* lease.

The extent of disagreement and confusion about the delimitation of the Greek and Turkish continental shelves is well illustrated by the exchange of letters in *The Guardian*, reproduced below. These letters, written in 1979, demonstrate not only the importance of the 'Law of the Sea' for economic issues in the Aegean, but also (according to interpretations put upon it by the respective states) its strategic importance.

Let us return briefly to the issue of scientific research and look at how the Convention on the Continental Shelf affects rights to participate in research in the Aegean. The Convention states that the consent of the coastal state shall be obtained in respect of any research concerning the continental shelf and undertaken there '. . . .and that the coastal State shall have the right . . . to participate or to be represented in the research . . .'. As much of the Aegean is claimed by Greece *and* Turkey, it is necessary not only for researchers to obtain permission from *both* countries, but also for provision to be made to accommodate representatives of *both* countries on board the research vessel. With space at a premium on such vessels, this can be a severe problem. Even in situations where there is no conflict of 'ownership', the need to obtain permission from coastal states is an important consideration in the planning of research cruises, not least because the amount of time needed to obtain such permission may often be as long as two years.

Given that the dividing line between pure scientific research and applied research is extremely blurred (especially where study of the sea-bed is concerned), it is perhaps not surprising that coastal states take such a proprietorial interest in 'their' continental shelves. The Aegean is unlikely to be free of disputes for some time to come. As recently as 1987, arguments erupted between Greece and Turkey over oil exploration in the northern Aegean. If contemporary press accounts are to be believed, the issue was sufficiently provocative to bring both nations to a condition approaching war fever. It is perhaps worth noting that—as implied in the first letter reproduced below—many oil experts believe that there is not enough oil in the northern Aegean to be worth arguing about, let alone going to war over.

A Greek sea of troubles (6 February 1979)

Sir,
In your Special Report on Turkey (January 31) Michael Simmons contributed an interesting article on Ankara's foreign policy headlined 'Oil on troubled waters'. Despite the title this piece dealt only cursorily with the substance of the Aegean problems, in a single paragraph that was almost wholly inaccurate. The facts are as follows:

First, the Turkish Prime Minister, Mr. Ecevit, is on record as saying on February 24, 1975: 'Even if we know now that there is no oil in the Aegean, we have to pay great attention to the continental shelf in the Aegean.'

Secondly, the sea-bed problem originated with the Turkish grant of exploration licences in areas of the Aegean continental shelf which, according to international law, would appertain to Greece.

Thirdly, in the light of the 1958 Geneva Convention, Greece has always held that her islands have the right to a continental shelf.

Fourthly, Greece has declared that the right to a 12-mile territorial sea stems from international law and therefore cannot be renounced.

Finally, regarding the topic of Aegean air space, the civilian air-corridors were closed in 1974 because of the danger to air traffic posed by Turkey's violation of ICAO arrangements which had functioned smoothly for some 20 years.

Peter Thompson
Press Officer
Greek Embassy,
1a Holland Park,
London W11 3TP.

Divisions over the Aegean (14 February 1979)

Sir,
I would like to clarify some of the points raised by Mr. Peter Thompson of the Greek Embassy (Letters, February 6) in connection with the Aegean sea and air space. The concept of continental shelf applies not only to the oil or gas in the sea-bed, but it also encompasses vital rights in the sea-bed, sea column and surface and the air space over it.

Contrary to Mr. Thompson's claim, the Aegean continental shelf problem was not born out of the granting of exploration licences by the Turkish authorities, but of the Greek refusal to consider the very special character of the Aegean Sea, and its insistence not to recognize Turkey's right to have a continental shelf beyond her existing territorial waters.

It is important to remember that the Turkish and Greek continental shelves have never been delimited, although the contrary was implied in Mr. Thompson's letter. Had there been any delimitation, there would have been no need for the experts from both countries to meet, as they have done over the last three years, or for Greece to apply to the International Court of Justice.

Furthermore, Turkey has never been a party to the Convention to which Mr. Thompson referred. If the continental shelf pattern of the Aegean is to be determined according to Greek views and claims, then the continental shelf of mainland Turkey would hardly extend beyond the limits of its present territorial waters. An extension of the 12-mile limit would be tantamount to converting 63 per cent of the Aegean Sea into Greek territorial waters.

Ercan Citlioglu
Press Counsellor
Turkish Embassy,
43 Belgrave Square,
London SW1X 8PA.

Sea Law (7 March 1979)

Sir,
It is odd that nobody has come forward to rebut the astonishing assertion by the Press Counsellor of the Turkish Embassy that 'the concept of continental shelf applies not only to the oil or gas in the sea-bed, but it also encompasses vital rights in the sea-bed, sea column, surface and the air space over it' (Letters, February 14).

This argument is, of course, untenable in international law. The continental shelf is an exclusively resource-oriented concept. Sea-bed rights imply no rights whatsoever in the sea column (fishing), sea surface (navigation), or air space.

Peter Thompson
Press Officer
Greek Embassy,
1a Holland Park,
London W11 3TP.

2.4 SUMMARY OF CHAPTER 2

1 Until the 19th century, it was largely the ability to enforce maritime 'territorial' claims militarily that determined the outcome, as exemplified by the cannon-shot rule. In general, the concept of *res nullius* (*mare clausum*) was effective only where it could be maintained by force. Where it could not, *res communis* (*mare liberum*) prevailed. However, there was a long and fairly well-established distinction between the (typically) three-mile wide territorial sea, over which coastal states could claim near-total control (subject only to rights of innocent passage), and the high seas, where there was virtually complete freedom of navigation.

2 During the 19th and (especially) the first half of the 20th century, there was some progress from this basic situation to one in which it was proposed that a uniform rule of law should govern the majority of peacetime activities at sea. The first UN Conference on the Law of the Sea, in 1958, adopted four Conventions: on the Territorial Sea and Contiguous Zone; on the Continental Shelf; on the High Seas; and (unsuccessfully) on Fishing and Conservation. Two other Conferences were convened, in 1960 and 1973, and the UN Convention on the Law of the Sea (UNCLOS) was opened for signature in 1982.

3 Convention on the Territorial Sea: The width of the territorial sea generally ranges from three to (typically) twelve miles, though in some cases 200-mile limits have been claimed. One difficulty of establishing the distance of the territorial sea boundary from shore lies in defining the baseline from which to measure it. Internal waters (where the state's jurisdiction is complete) lie within the baseline; the territorial sea (where there is right of innocent passage) begins outside it. Rules were established for dealing with bays including legal bays and historic bays; with islands; with river mouths, low-tide elevations, harbour works and so on. The rules have been interpreted liberally by some states, leading to inordinately large areas claimed for internal waters and the territorial sea. Straits present particular problems where they are narrower than the combined widths of adjacent territorial seas and may be designated international straits. A contiguous zone (declared by a minority of states) lies adjacent to the territorial sea; in this zone, the coastal state has partial jurisdiction.

4 Convention on the Continental Shelf: This Convention covers the sea-bed and sub-soil of the continental shelf, but not the superadjacent waters (*cf.* the exchange of letters at the end of Section 2.3). Problems of definition are even more formidable than with the Convention on the Territorial Sea. The legal continental shelf is not the same as the 'geological' continental shelf. The legal definition has to contend not only with natural variability (e.g. variations in the distance to the shelf break and water depth), but also with the technological ability of states to exploit shelf resources.

5 Convention on the High Seas: The high seas are the simplest maritime zone to define. The Convention deals mainly with limitations to complete freedom of navigation, with rules governing nationality and registration of vessels, piracy, 'hot pursuit', etc.

6 Many unilateral claims outside the UN Conventions were made in the decades following World War II, mostly concerned with fishing. 200-mile exclusive fishing zones were claimed by many nations. Other

nations included shelf resources and research activities by foreign nationals, as well as fishing, in the scope of their 200-mile exclusive zones. These claims were effectively the precursors of the exclusive economic zones to be discussed in Chapter 3.

Now try the following questions to consolidate your understanding of this Chapter.

QUESTION 2.4 Look back at Figure 2.10. Would you describe the Dardanelles as an international strait of high strategic value?

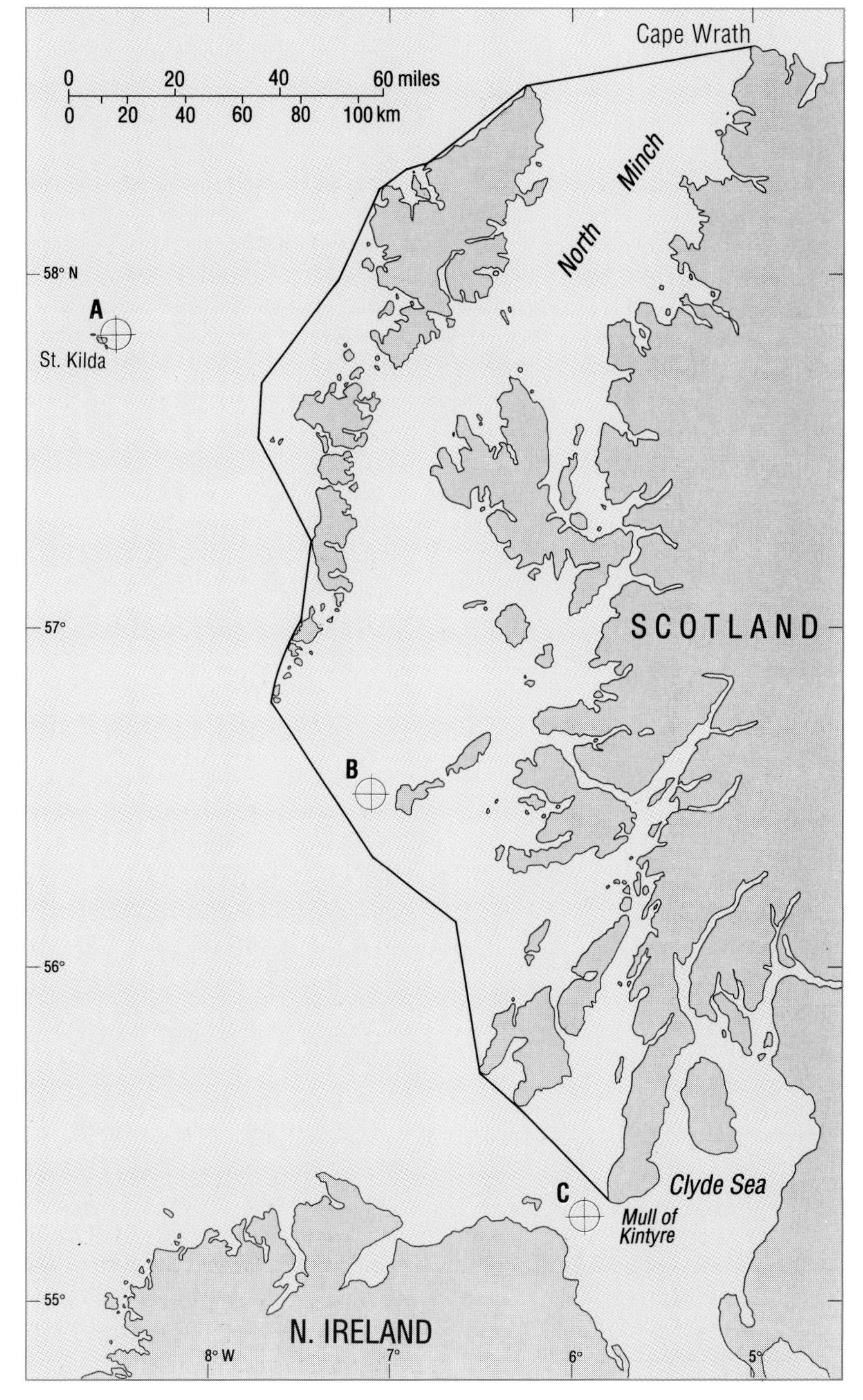

Figure 2.11 Map of western Scottish waters, for use with Question 2.5 overleaf. (Note that the ends of straight baseline segments apparently out at sea are in fact on small islands, with lighthouses on them, but too small to show at this scale.)

QUESTION 2.5 (a) The coastline of the west of Scotland is so convoluted that the straight baseline method has been adopted to determine the territorial sea between Cape Wrath and the Mull of Kintyre. Figure 2.11 shows this coast and its associated straight baselines. What is the legal status of waters at the three positions listed below (give your reasons)? Assume a three-mile territorial sea, and a contiguous zone of nine miles.

(i) A (57°50′ N,8° 30′ W)

(ii) B (56°30′N,7°00′ W)

(iii) (55° 15′ N,5° 55′ W)

Also, summarize the scope of UK jurisdiction in each case.

(b) The Clyde Sea (Figure 2.11) is a legal bay with a closing line some 27 miles in length. Why does it qualify as a legal bay and how is its seaward limit defined? (It is *not* an 'historic bay'.)

CHAPTER 3 THE PRESENT INTERNATIONAL LEGAL REGIME

> 'The further off from England, the nearer it is to France.'
>
> Lewis Carroll, *Alice in Wonderland.*

Before discussing the present regime, it would be useful to remind ourselves of the principal maritime interests of nations, and how these relate to their geographical, political and economic circumstances, topics that were touched on at the start of Chapter 1. The situation of an island state in the middle of the Pacific is clearly very different from that of a European coastal state, but how do such differences affect attitudes towards the Law of the Sea?

From a purely geographical point of view, there are three obvious categories of states: island, coastal and land-locked. It may be asked whether a land-locked state has any legitimate interest in maritime issues; but every nation in the modern world depends to some degree on the transport of goods by sea—and land-locked states, like all other states, have a right to sail ships on the seas under their own flag. Coastal and island states are directly involved in seaborne trade, and they have access to the resources of their own maritime zones; but many other factors play a part in determining even the *potential* importance of the sea in a nation's life. Amongst these are land area, length of coastline, area of continental shelf and, more particularly, the ratios between these indicators. Some of these factors are listed by nation in Appendix 1, Tables II and III.

Within these somewhat esoteric parameters are many more specific factors such as the geology of the shelf, the productivity of coastal waters, and the presence of harbours. Given this geographical context, the *actual* importance of the sea to a particular state is largely a matter of socio-economic factors such as population and degree of development. These in turn are compounded with political considerations to define a Government's policy. The outcome is that each state has a unique attitude to international maritime law.

Fortunately, three main roles can be identified in all this complexity: strategic, mercantile and resource-orientated. But these do not necessarily coincide with identifiable groups of nations. Thus, strategic considerations influence not only the superpowers, but also, through their membership of defence pacts such as NATO or involvement in regional disputes, other nations whose primary maritime interest may be mercantile or resource-orientated. Broadly speaking, every coastal or island state has an interest in the exploitation of its own marine resources; some of them also have important mercantile interests, and a minority are involved in strategic issues.

Not only do these roles co-exist at a national level, but the balance between them changes with time. Several developing countries have become major maritime trading nations in recent years and the rapid growth of their registered merchant fleets has been at the expense of those of the traditional maritime powers. The word 'registered' is important here. The fleet size is less a matter of national *ownership* of the vessels than of whether the vessels are registered under the state's flag. As it has become more expensive to operate under the strict requirements of flags of the major developed states, vessels have been transferred to those of perhaps less demanding states. In recent decades, for example, the South Korean and Taiwanese-registered merchant fleets have grown to become among the largest in the world, whereas Britain's has declined from its previous pre-eminent position. As regards the strategic role, the Royal Navy no longer maintains a global presence, whereas the American and Russian navies have expanded.

The relative importance of these three roles will affect a state's maritime policy and hence its attitude to the Law of the Sea. A Pacific island state is probably very sensitive about the exploitation of fisheries around its shores by foreign vessels and keen to extend its jurisdiction seawards, while a European state will favour unrestricted passage through international sea lanes and protection of its offshore oil interests. The international legal regime is an expression of the interactions between these differing attitudes, albeit delayed and distorted by the process of formulation.

So, what is the present regime of the Law of the Sea? The four United Nations Conventions negotiated in 1958 (Chapter 2) entered into force during the period 1962–66. They marked a great advance in obtaining international agreement and spelled out this agreement in some detail. Yet they contain legal loopholes, are open to considerable variation of interpretation, and are therefore liable to abuse. Moreover, most of the present large number of developing nations, who had not achieved their independence in 1958, played no role in the negotiation and thus regarded much of the regime as inappropriate to their interests. Formally, the 1958 Conventions remain the basis of the present legal regime, but are being superseded by a new international agreement—the UN Convention on the Law of the Sea (**UNCLOS**)—which overcomes most of their deficiencies. Although not yet in force, it is already the major influence on developments in maritime law.

3.1 UN CONVENTION ON THE LAW OF THE SEA (UNCLOS)

The essential features of an international legal regime are that it is a multilateral agreement (in this case negotiated under the aegis of the UN) and that it has global application. In the case of Law of the Sea, this means that its provisions cover the whole ocean, right up to national baselines, i.e. with the exception of internal waters. The 1958 Conventions form the core of UNCLOS, but other sources contribute, as outlined in Section 2.2.4 and further described in the following Sections.

To a considerable degree, the UN Conventions represent the codification of **customary law**. Regardless of the formal status of UNCLOS (i.e. despite the fact that it has not yet been ratified as a whole by enough states to become a legally binding treaty), with growing application and acceptance its various provisions become part of customary law. Just as the 1958 Conventions codified and reinforced existing customary law, and the application of their provisions generated new customary law binding on non-adherents (a form of positive feedback), so UNCLOS is influencing the development of the legal regime even before it enters into force and becomes a legally binding treaty.

A note on ratification

States are not generally bound to implement a treaty or convention by signing it (unless they specifically accept to be so bound). Deposit of a formal written 'instrument of acceptance to be bound' is required. Generally, states can ratify a treaty or convention, i.e. become bound to implement it *in toto* when it comes into force, only if their representative at the relevant conference has signed the document, either at the conference itself or within a specified time limit thereafter. Ratification can be effected later and generally is, being usually delayed by the need for reference to and approval by the national legislature. Non-signatories may subsequently *accede* to the treaty or convention, and this has the same force as ratification. Signature requires that the signatory state does nothing to undermine the objectives of the convention, pending ratification by it; but signature does not bind the state eventually to ratify the convention or to apply any of its provisions (though some states consider they are under an obligation to do so).

3.1.1 THE CONFERENCE

As we saw in Chapter 2, maritime law was still in a state of flux at the end of the 1970s. The Third United Nations Conference on the Law of the Sea, begun at Caracas in 1973 (Section 2.2), remained unconcluded, while discussions continued at the UN and at national and regional levels. After nine years of negotiations, the Conference was finally concluded in April 1982 and the Convention adopted. In December 1982, at Montego Bay, Jamaica, the United Nations Convention on the Law of the Sea (UNCLOS) was finally opened for signature.

By the end of 1990, 159 states had signed the Convention and 45 had ratified it (Table 3.1). The *signatories* included most Western European states, most Eastern Bloc states, and ten of the twelve EC (European Community) states, as well as the EC itself (adherence of the Community being specially provided for in UNCLOS). The USA is among the few that have not yet signed, having found the regime proposed for the exploitation of the international sea-bed area (see later) unacceptable. It is around this topic that most of the arguments centre, as we shall see.

Table 3.1 The 45 states which had ratified UNCLOS up to the end of 1990, with their respective dates of ratification. It will enter into force one year after the deposit of the 60th ratification.

State	Date	State	Date	State	Date
1 Fiji	December 10, 1982	17 Togo	April 16, 1985	33 Democratic Yemen	July 21, 1987
2 Zambia	March 7, 1983	18 Tunisia	April 24, 1985	34 Cape Verde	August 10, 1987
3 Mexico	March 18, 1983	19 Bahrain	May 30, 1985	35 Sao Tome and Principe	November 3, 1987
4 Jamaica	March 21, 1989	20 Iceland	June 21, 1985	36 Cyprus	December 12, 1988
5 Council for Namibia	April 18, 1983	21 Mali	July 16, 1985	37 Brazil	December 22, 1988
6 Ghana	June 7, 1983	22 Iraq	July 30, 1985	38 Antigua and Barbuda	February 2, 1989
7 Bahamas	July 29, 1983	23 Guinea	September 6, 1985	39 Kenya	February 3, 1989
8 Belize	August 13, 1983	24 Tanzania	September 30, 1985	40 Zaire	February 17, 1989
9 Egypt	August 26, 1983	25 Cameroon	November 19, 1985	41 Somalia	July 24, 1989
10 Côte d'Ivoire	March 26, 1984	26 Indonesia	February 3, 1986	42 Oman	August 17, 1989
11 Philippines	May 8, 1984	27 Trinidad and Tobago	April 25, 1986	43 Angola	December 5, 1990
12 Gambia	May 22, 1984	28 Kuwait	May 2, 1986	44 Botswana	May 2, 1990
13 Cuba	August 15, 1984	29 Yugoslavia	May 5, 1986	45 Uganda	November 9, 1990
14 Senegal	October 25, 1984	30 Nigeria	August 14, 1986		
15 Sudan	January 23, 1985	31 Guinea-Bissau	August 25, 1986		
16 Saint Lucia	March 27, 1985	32 Paraguay	September 26, 1986		

On the face of it, a few abstentions from UNCLOS might appear not to matter, in view of the (much) more than two-thirds majority in favour. However, the Convention's actual or potential effectiveness is somewhat vitiated because the minority include some of the major maritime nations. In such circumstances, a majority-consent treaty can dangerously polarize the world community into the majority, which have acted 'responsibly' (at least in their own eyes), and the minority, which includes states with the power to enforce their will. It is small comfort to the smaller maritime nations to find themselves among the majority who have adopted UNCLOS, if wealthy and powerful nations like Germany, the UK and the USA are among the minority who have not. But this might be to overdramatize the situation: signing the Convention is not the same as ratifying it—and the number of nations which have made this final commitment remains small.

3.1.2 THE CONVENTION

UNCLOS comprehensively updates the four separate Conventions of 1958 (Chapter 2) into a single package. It is intended to establish a unified international legal regime for all purposes, symbolizing the unity of the marine environment and the need for international standards in the management of fish stocks, pollution, and so on. There is further recognition not only of functional divisions, such as the territorial sea and the exclusive economic zone (EEZ, see below); but also of the political dimension, especially the interests of developing countries.

The Convention can also be seen as a sort of 'umbrella', setting out general principles (whose technical details had previously been formulated elsewhere), and encouraging the development of other specialized agreements in the future by providing standards and a consistent framework. Thus, the UNCLOS equivalent of Article 6 of the Convention on the High Seas forms the basis of international shipping law, but the regulations governing ships' safety and the control of pollution from ships are negotiated within the International Maritime Organization (IMO), a specialized agency of the UN. The provisions of the Convention, which has yet to come into force, are summarized in Appendix 2. Some of the more important ones are described below.

Table 3.2 Statistical summary of claims to territorial waters (1972).

Width of territorial sea (miles)	Number of states
3	35
4	4
6	11
10	1
12	52
18	1
25	1
30	1
50	1
130	1
200	9
Other*	5

*No single-figure claim: Korea, Lebanon, Maldive Islands, Philippines, Portugal.

The territorial sea

Included under this heading are the contiguous zone, bays, islands, archipelagos, and straits. Rules for establishing the baselines of maritime zones under the Convention are very similar to those described in the Convention on the Territorial Sea (Section 2.2.1). Neither the 1958 nor the 1960 Conference solved the breadth problem (Sections 2.2.1 and 2.2.4), and the range of territorial sea claims was considerable, as Table 3.2 illustrates. UNCLOS finally succeeded in reaching agreement on this question, with a twelve-mile territorial sea, plus a twelve-mile contiguous zone for control over infringements of the coastal state's customs, fiscal, immigration or sanitary regulations (see Figure 2.1).

The *rights* of coastal states within their territorial seas also differ little from those available under the Convention on the Territorial Sea, as described in Section 2.2.1: ' . . . sovereignty . . . mitigated only by the right of innocent passage. . .', and extending '. . . to the air space above the territorial sea and to the sea-bed beneath'. By custom (recognized in this Convention), the right of innocent passage through international straits cannot be temporarily suspended for national security reasons, as it can elsewhere in the territorial sea. Indeed, the principle of **transit passage** was introduced to cope with the problem of international straits (Section 2.2.1), striking a balance between the innocent passage of territorial seas (Section 2.1), and the freedom of navigation of the high seas (Section 2.2.3). It permits 'the exercise ... of the freedom of navigation and overflight solely for the purpose of continuous and expeditious transit of the strait between one area of the high seas or an exclusive economic zone [see below] and another. . . .', but it does not preclude '. . . entering, leaving or returning from a State bordering the strait, subject to the conditions of entry to that State'.

It should be remembered that the other side of every right is a *duty*, and that includes responsibility for policing the territorial sea: administering various regulations in time of peace (including those to ensure safe passage of ships and protection of the marine environment); maintaining neutrality in wartime; and so on.

Archipelagos

The long-standing problem of archipelagos (Section 2.2.1) was resolved by giving archipelagic states sovereignty over the sea area enclosed by straight baselines joining the outermost points of islands, irrespective of whether the islands are separated by more than twice the territorial sea width (24 miles) or not. However, ships of all other states enjoy the right of passage through these waters, using sea lanes designated by the archipelagic state(s). The territorial sea and other maritime zones begin where this area ends. UNCLOS thus provided a new right, called *archipelagic sea-lanes* passage, which is not to be confused with the innocent passage in both territorial sea and so-called archipelagic waters. This clause achieved a balance of interests, providing for uninterrupted passage while recognizing the new concept of archipelagic waters and their enclosing straight baselines.

The exclusive economic zone

Of the changes introduced by UNCLOS, the most far-reaching apply to waters further offshore and the sea-bed underlying them. As you read in Section 2.2.4, in the years following the signature of the 1958 Conventions, and even before that in some cases, coastal states were seeking to extend their jurisdiction over marine resources by claiming wider territorial seas or declaring exclusive fishing zones up to 200 miles in width. Such claims were resisted by traditional maritime states but they were unable to reverse this trend. UNCLOS provides for these developments by adopting the concept of an **exclusive economic zone (EEZ)**.

The EEZ is a maritime zone extending 188 miles beyond the limits of the territorial sea (*cf.* Figure 2.1) within which the coastal state is accorded 'sovereign rights to conserve, manage, explore and exploit' all living and non-living resources in the water, and on and under the sea-bed; and it has exclusive jurisdiction over all activities relating to those resources. It combines the concept of exclusive fishing zones with that of access to continental shelf sea-bed resources; and adds new jurisdictional controls for purposes of protection of the environment and scientific research. It is thus a single multipurpose zone that falls short of being a traditional territorial sea: a compromise between rigid territorial claims on the one hand and complete freedom for all comers on the other. In this latter respect, the EEZ is the clearest symbol of changes in the 'freedom of the seas' philosophy which had dominated for so long: it balances the coastal state's control over its marine resources with residual freedom of navigation.

Figure 3.1 shows how much—or rather how little—of the world's ocean is now classified as 'high seas'. Only about 60% of the ocean is unclaimed. The rest is claimed by individual states as exclusive economic zones, continental shelf zones or exclusive fishing zones.

The continental shelf

At the end of Section 2.2.2, we left the continental shelf as a maritime region doomed to 'a further round of international negotiations' because of the hybrid definition that was inherent in the Convention on the Continental Shelf. It is evident that this Convention posed as many problems as it solved, for by redefining the continental shelf in arbitrary terms the advantages of having a real oceanographic feature to deal with were lost and yet the acknowledged disadvantages remained. So, would it be better to return to a true oceanographic standard? In 1972, Hollis D. Hedberg, an oceanographer (i.e. not a lawyer, sociologist, politician, or economist), stated his belief that it would. He based his proposal on the premise that there should actually be a limit to the extent of national jurisdiction over the ocean floor—an idea that the Convention on the Continental Shelf approved in principle but failed to put into practice.

Hedberg believed that as knowledge of the distribution of natural resources on the ocean floor increased, international agreement would be 'made infinitely more difficult by the contentions of self interests for specific local economic prospects', if the matter was not resolved. An important advantage of a natural as opposed to an 'invented' boundary is that it would be less amenable to arbitrary political change either by individual nations or by the world community as a whole. It would also be fairer to land-locked states which would be satisfied that coastal nations could not reduce the communal wealth of the high seas by gerrymandering the national–international boundary.

The boundary proposed by Hedberg is the base of the continental slope, which he argued is the nearest one can get to the natural division between continental and oceanic crust. Hedberg's case was a scientific one, not a practical one. He did not suggest that the base of the continental slope is any less or more difficult to determine than the edge of the continental shelf. But, given that any ocean-floor boundary will be difficult to measure accurately, the choice might as well go to one that has geological logic (as opposed to being just a topographic feature, such as the shelf break). The practical difficulties of measurement mean, however, that the slope base cannot be used directly as the national–international boundary, but must act as a guide.

Hedberg proposed that:

1 The slope base should be used to define a *boundary zone* at least 100 km wide to accommodate irregularities and uncertainties in the position of the slope base, and possibly up to 300 km wide; the precise width to be decided by international agreement and applied uniformly throughout the world.

2 The landward limit of the boundary zone should be the best scientifically determined position of the slope base, and its seaward limit would then be fixed by the zone-width previously agreed.

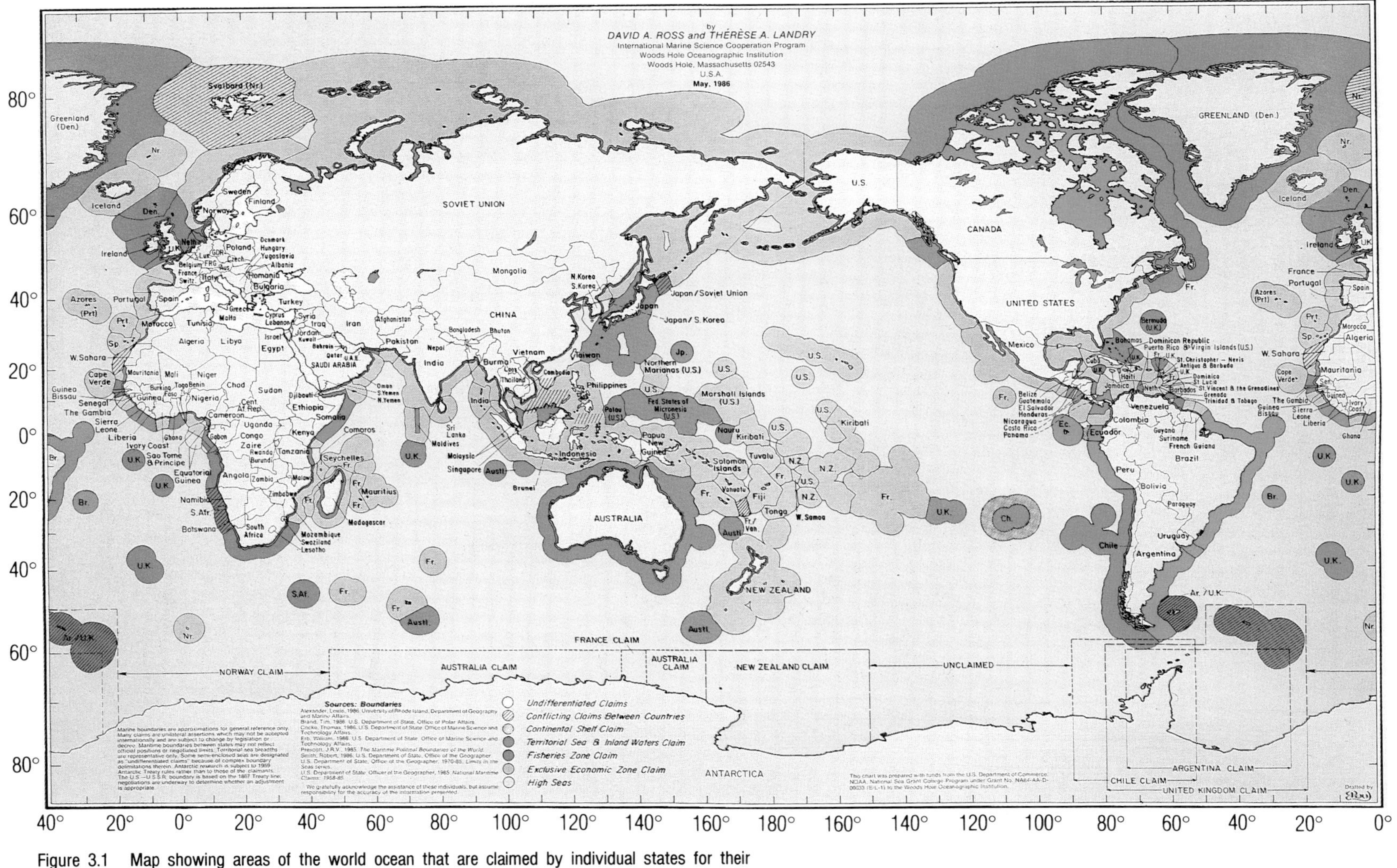

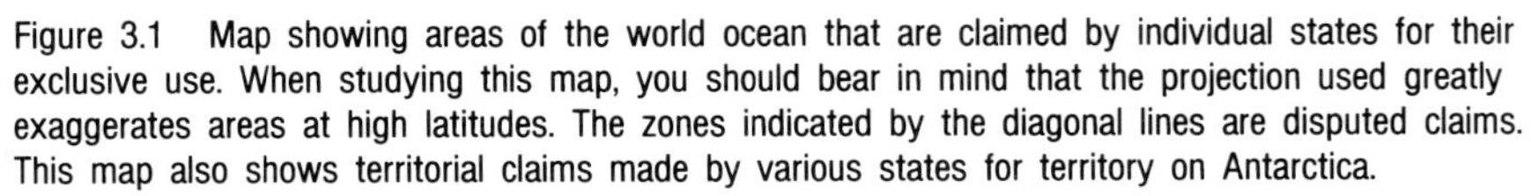

Figure 3.1 Map showing areas of the world ocean that are claimed by individual states for their exclusive use. When studying this map, you should bear in mind that the projection used greatly exaggerates areas at high latitudes. The zones indicated by the diagonal lines are disputed claims. This map also shows territorial claims made by various states for territory on Antarctica.

3 The coastal state should be free to decide on the precise position of the political national–international boundary within that zone. For practical reasons, that boundary should consist of straight lines or arcs joining points fixed by latitude and longitude.

Figure 3.2 illustrates the Hedberg proposal, which must have influenced the final definition of the legal continental shelf laid down in UNCLOS. The UNCLOS definition of the legal continental shelf is expressed in terms of both distance and natural (geological) prolongation. All coastal states can therefore claim continental shelves up to a distance of 200 miles from the coastal baseline (where distance from any opposite state permits), without the need to establish natural (geological) prolongation. Where there is a natural prolongation of the shelf beyond 200 miles, the coastal state can claim shelf rights to 'the outer edge of the continental margin' up to 350 miles or even beyond (depending on the nature and configuration of the continental margin), *or* 100 miles beyond the 2 500 m isobath. Coastal states have a free choice between these alternatives. In this context, 'the outer edge of the continental margin' is defined in terms of the maximum rate of change in gradient at the base of the continental slope. The advantage of this definition is that the legal continental shelves of all coastal states have a guaranteed minimum width and take in *at least* the whole geological continental margin.

QUESTION 3.1 (a) To what extent does the UNCLOS definition follow Hedberg's proposal as illustrated in Figure 3.2?

(b) 'The UNCLOS definition has the advantage over the Convention on the Continental Shelf in that it provides equal areas of "shelf" per unit length of coastline.' To what extent is that statement valid?

(c) Why is it necessary to distinguish a legal continental shelf, if a coastal state has already declared an EEZ?

It is important to bear in mind that the rights of the coastal state extend also to the waters overlying its legal continental shelf out to the 200-mile limit, because these waters lie within its EEZ. If the legal shelf projects

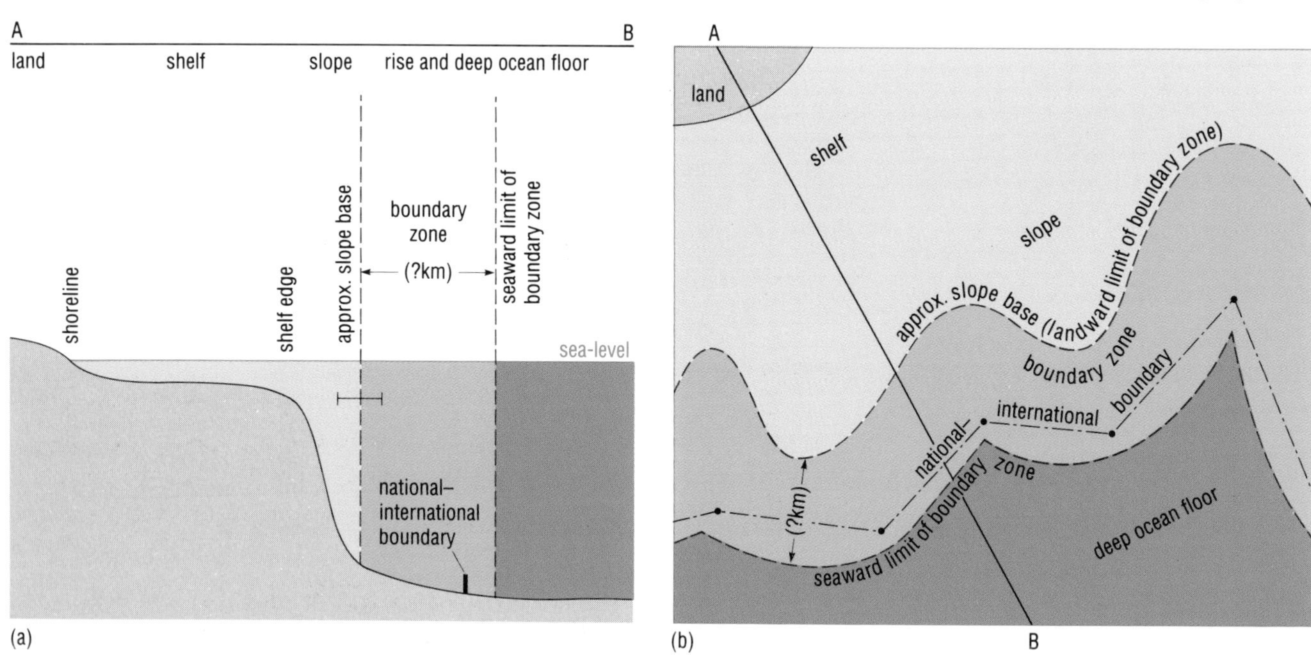

Figure 3.2 Hedberg's 1972 proposal for the definition of a political national–international ocean-floor boundary within a finite boundary zone related to the base of the continental slope. (a) Profile (vertical scale highly exaggerated); (b) map.

beyond that limit, the state's rights do not extend to the overlying waters, except insofar as the coastal state also has the duty to ensure that sea-bed operations do not damage either the marine environment in general or the interests of other actual or potential users of those waters. *Beyond* the 200-mile limit the coastal state is required to pay a proportion of the revenue from exploitation of shelf resources to the International Sea-Bed Authority (see below).

Specifically, the coastal state has the right to exploit the natural resources of its legal continental shelf, which are defined (as they are in the Convention on the Continental Shelf) as: 'the mineral and other non-living resources of the sea-bed and sub-soil together with living organisms belonging to sedentary species, that is to say organisms which, at the harvestable stage, either are immobile on or under the sea-bed or are unable to move except in constant physical contact with the sea-bed or the sub-soil.'

In legal cases heard during the late 1980s concerning delimitation of continental shelf boundaries between states, the International Court of Justice determined that geology is now irrelevant as a factor to be considered in relation to shelves that lie within 200-mile EEZs. It remains a consideration, however, in determining the outer limits if states do have shelves that 'naturally prolong' beyond 200 miles.

The high seas

The high seas are currently used chiefly for transport, both of people and cargo, and commercial fishing. Other uses are relatively unimportant economically, though the economic (and political/environmental) impact of commercial exploitation of metals on the ocean floor (manganese nodules and hydrothermal sulphide deposits, Chapter 1) would be enormous. On the other hand, cables on the ocean floor are likely to become less and less important as satellite communication technology improves. In the meantime, the high seas and/or the underlying sea-bed are also used for political and security purposes (especially by the USA, the USSR and, to a declining extent, France and the United Kingdom), as routes for pipelines and power cables, for scientific research, for recreation (fishing and cruising) and for waste disposal.

All of these activities (both on the high seas themselves and—to a minimal extent—on the underlying sea-bed and sub-soil) were governed by the Convention on the High Seas, the general tone of which was one of freedom, albeit freedom modified by the need to safeguard the rights of other nations to exercise *their* freedoms (Section 2.2.3). Under UNCLOS, however, activities *on* the high seas are now distinguished from activities relating to the underlying sea-bed and sub-soil (see below). Moreover, as noted earlier, with the introduction of the EEZ, the area to which the provisions on the high seas apply has been considerably diminished (Figure 3.1).

Note also that some of the high seas activities mentioned above impinge on EEZs, e.g. deep sea pipelines and cables must of necessity cross them to reach their destination. This they are permitted to do (see Appendix 2, item 5).

The International Sea-Bed Area

The lure of deep-sea minerals beyond the legal continental shelf led to the inclusion in UNCLOS of a new legal framework for their exploitation, in an attempt to forestall a dangerous scramble under the high seas. This was not covered in the 1958 Conventions, perhaps because such a scramble was not anticipated (this was in the days before the scientific recognition of continental drift and plate tectonics, when very little was known about deep-sea resources); the legal framework developed as a consequence of fears about the extension of the shelf regime in favour of industrialized nations.

The Convention defines the **International Sea-Bed Area** as 'the sea-bed and ocean floor and sub-soil thereof, beyond the limits of national jurisdiction'. Most of the reservations about the Convention were held by developed countries and concerned the regime for deep-sea mining in this region. It proposed a parallel system in which two mine sites are offered by each applicant; one of these two mine sites is to be worked by the successful applicant under contract with a new International Sea-Bed Authority, and the other one reserved, either for the Enterprise (the Authority's commercial arm) or for developing countries. At a late stage of the Conference, this framework was substantially modified in an attempt to satisfy objections from those who had already invested heavily in exploration for deep-sea minerals.

Eight such 'pioneer investors' would be accepted, one each from France, India, Japan and the USSR, and up to four others with members from one or more of the following eight nations: Belgium, Canada, the Federal Republic of Germany, Italy, Japan, the Netherlands, the UK and the USA. They could apply for one site each, where they could engage in certain preliminary activities, but they would not be permitted to commence production until the Convention entered into force, and then only if all the states involved were parties to the Convention, i.e. had ratified it or acceded to it. After that they would still be subject to production limits sets by the International Sea-Bed Authority and might be required to transfer technology to the Enterprise, to assist it to develop one or two sites on its own.

These changes were incorporated as two Resolutions appended to the Conference's Final Act, the formal record of its proceedings. One Resolution set out the provisions summarized above, and the other established a Preparatory Commission to lay the groundwork for the two main institutions to be established under the Convention. These are the International Sea-Bed Authority and the International Tribunal for the Law of the Sea, whose seats will be in Jamaica and Hamburg, respectively. (The Preparatory Commission held several meetings and by 1990 had registered some pioneer investors; many applications have been received from member states.)

Despite these last-minute changes, the compromise eventually reached by the Conference failed to satisfy objectors to the proposed regime for deep-sea mining. The USA was the most outspoken of these critics, claiming that adherence to the Convention would restrict its access to strategic minerals on the sea-bed by making them commercially unattractive. The two EC members (UK and Germany) that have not signed the Convention were less categorical. The UK position can be considered fairly representative: while supporting the principle of a comprehensive treaty acceptable to the international community as a whole, and finding many of its provisions helpful, the UK Government would not sign the Convention as it stands for a number of reasons, primarily related to the proposed sea-bed regime. These are the cost and complexity of the Authority, the fees and production limits to be imposed on commercial operators, their obligation under the parallel system to share exploration data and technological information with the Enterprise, and the terms of reference of a review conference to be held 15 years after mining begins.

The Final Act was signed by 150 delegations. Signature of the Final Act does not imply a commitment to accept the Convention, and non-signatories are entitled to observer status in the Preparatory Commission, where they will be able to take part in its deliberations, but without the right to vote. It is through this mechanism that some states intend to lobby for changes in the sea-bed regime. It is hard to see how such fundamental amendments as those sought by the USA and its allies could be incorporated by the Preparatory Commission, especially when so many states have accepted the Convention in its present form—the more so when discussions are taking place in the absence of the principal objector (the USA), which has chosen not to be represented in the Commission.

In 1982–83, the USA made strenuous efforts to get its allies to sign a 'mini-treaty' setting up a rival regime. However, the other potential sea-bed mining states were prepared only to join in 'Reciprocal Agreements', under which they agreed merely to respect each others' potential mining sites. No machinery for joint management was established, nor were any terms and conditions of mining laid down. Most of the potential mining states have, however, unilaterally enacted national legislation laying down the terms of mining by companies or consortia registered with them. This 'picking and choosing' is regarded by many as contrary to the spirit of the Convention, which was designed as a balanced package of rights and responsibilities. The legal wrangling over deep-sea resources is reminiscent of the ancient dichotomy between the concepts of *res communis* and *mare liberum* on the one hand, and *res nullius* and *mare clausum* on the other.

During 1989–90, however, there were indications in the UN that some developing countries were willing to enter into a 'dialogue' about the possibility of modifying the regime in such a way as to attract more states to ratify the Convention. The UN General Assembly's annual resolution on the Law of the Sea in 1989 was more moderate in tone in relation to non-signatories; but it remained unclear about what modifications could be made and how they might be brought about. Nonetheless, 'dialogue' became something of a 'buzzword'.

Indeed, the number of ratifications is steadily rising towards the minimum of 60 required to bring the Convention into force as a Treaty (one year after the 60th ratification, Table 3.1). But the last few will be the hardest to get, perhaps chiefly because of the financial implications of bringing the International Sea-Bed Authority into existence.

Other matters (research, conservation, pollution control, etc.)

Of particular interest to oceanographers are the Convention's implications for marine scientific research.

In the *territorial sea*, the position is little changed from what it was in 1958: coastal states 'have the sovereign right to conduct and regulate marine scientific research in their territorial sea.'

In the *EEZ*, research must only 'be conducted with the consent of the coastal State' but the coastal state must not withhold its consent unless (i) the project 'bears substantially upon the exploration and exploitation of the living or non-living resources'; (ii) it 'involves drilling or the use of explosives'; (iii) it 'unduly interferes with economic activities performed by the coastal State'; or (iv) it involves the construction or use of artificial islands. In return, the research state must give the coastal state six months' advance warning of a project, provide extensive information about it, give the coastal state the opportunity to participate, provide the coastal state with the data obtained, and secure the coastal state's permission to publish data relating to the resources of the EEZ.

If within three months a coastal state fails to respond to an approach from a research state, including refusing consent outright or demanding further information, the research state may go ahead with its project when the six months' notice expires. Even if both parties fulfil their obligations as best they can, however, the coastal state would still seem to have ample opportunity to cause extensive, if not indefinite, delay. It can, for example, refuse consent if apparently inaccurate information is supplied by the researching state, or if it requires still more relevant information or considers that not all obligations under a previous project have been fulfilled. Disputes arising from a coastal state's regulation of research in the EEZ are among the few types of dispute that are not subject to a compulsory and binding settlement process under the Convention.

The *continental shelf* is treated in the same way as the EEZ, for as we have seen, most of the legal continental shelf is the floor of the EEZ (the waters above that part of the shelf extending beyond the EEZ remain of course the high seas). Restrictions on research in these zones is the biggest non-scientific problem now facing oceanographers, especially as in recent years much research has been focused on processes at and near the shelf edge, both in the water column and at the sea-bed: internal wave dynamics, shelf-edge eddies, mid-water fauna, sediment movements, and so on. Another glance at Figure 3.1 shows how essential it will be for such research to be conducted in a spirit of international cooperation in future, anywhere outside the 'developed' world.

On the *high seas*, one of the explicit freedoms is (as it was in 1958) 'freedom of scientific research'. States and 'competent international organizations shall have the right . . . to conduct marine scientific research in the water column beyond the limits of the economic zone.' Will complete freedom of research apply to the sea-bed beneath the waters of the high seas?

The Convention does not specifically rule out research by interested parties, despite efforts by some at the negotiating stage to enable the Authority to control *all* matters relating to exploration and exploitation of the resources of the deep sea-bed (including research directed towards those ends). Much depends on how the Authority interprets its role and powers. It 'may' carry out research itself; it is also required to promote research, though this does have to be 'for peaceful purposes' and 'for the benefit of mankind as a whole'.

The umbrella function of the Convention, mentioned at the start of this Section, is relevant to other activities under this general heading (conservation, pollution control, and so on): ' . . . setting out general principles . . . and encouraging the development of . . . specialized agreements . . . by providing standards and a consistent framework'. Thus, coastal states also have some jurisdiction over conservation/environmental preservation within their EEZs. There is also a general requirement that states which are parties to UNCLOS should promote the transfer of relevant technology where appropriate.

International boundaries

The Aegean (Section 2.3) exemplifies the dictum 'Definition is not delimitation', and illustrates some of the difficulties that can arise where there are overlapping claims, shared bays, international straits in territorial waters, and where median lines may have to be drawn. In the simplest case, to define the limits of jurisdiction of adjacent coastal states over their territorial sea, EEZ, continental shelf and so on (e.g. between Canada and the USA over Georges Bank), it would seem necessary only to draw the boundary as nearly perpendicular as possible to the shoreline or to a mutually agreed straight baseline. Less simple solutions are required in regions of complicated physical and political geography such as the Aegean and parts of the western Pacific.

On these issues, the UNCLOS formula merely provides for delimitation 'by agreement on the basis of international law in order to achieve an equitable solution', which leaves the outcome somewhat more subjective and uncertain than it might have been under the 1958 Conventions (Convention on the Territorial Sea and Convention on the Continental Shelf, Sections 2.2.1 and 2.2.2).

3.1.3 COMPARISON WITH 1958

In many respects UNCLOS is a consolidation and clarification of the earlier UN Conventions, incorporating some additional concepts that had developed outside them, such as exclusive fishery zones. But it does introduce several ideas new to the field, notably the principle that the resources of the deep sea-bed are the common heritage of mankind, as well as several innovatory provisions on protection of the marine environment (e.g. vulnerable areas in the EEZ, ice-covered areas, rare and fragile ecosystems).

Major differences reflect the changing 'balance of power', with introduction of the concepts of the EEZ and the International Sea-Bed Area. As we have seen, the general adoption of a twelve-mile territorial sea plus a 188-mile exclusive economic zone brings some 37 750 000 square miles of sea within national control, leaving 67 517 000 square miles (about 60% of the total) as international waters (Figure 3.1). This clearly represents a huge increase in areas under national jurisdiction, compared with the situation under the previous UN regime and it may seem that the concept of *res nullius* is being revived and is replacing that of *res communis*. However, it would be more correct to say that claims eroding the spirit of *res communis* are now being controlled—to some extent, at least—by a recognized international legal regime. Furthermore, when the Law of the Sea Convention comes into force, much of the principle of *res communis* will, for the first time, be given legal muscle.

3.2 OTHER CONTRIBUTIONS TO MARITIME CUSTOMARY LAW

As a result of the constant interaction between customary and treaty law, the legal situation is continually evolving and in a permanent state of flux. To be sure, some parts of international maritime law are widely accepted as being more or less 'frozen' by general acceptance, or through judgements of the International Court of Justice, or through binding arbitration in specific cases. Otherwise, however, deduction of the rules of international law must be from existing state practice and/or general principles. The states which made excessive claims before 1958 did not retract them when the Conventions on the Territorial Sea and Continental Shelf entered into force; indeed, the trend continued in the years leading up to UNCLOS (Section 2.2.4).

The influence of political and economic considerations is fundamentally a matter of self-interest, generally overlaid with a fragile concern for international harmony. Thus, several groups were formed at UNCLOS to strengthen negotiation of special interests, such as land-locked states, island states, archipelagic states, and so-called margineers (states with large continental shelves). These formed in addition to more conventional regional and economic groupings (e.g. the Group of 77, an informal coalition that represented developing countries). Many groups cut across normal alliances, ideologies and political differences. For example, the 'land-locked group' included Afghanistan, Austria, Hungary, Lesotho, Nepal, Switzerland and Uganda.

However, the Conventions of 1958 and UNCLOS are not the sole inputs to the international legal regime. Some others are listed below.

(a) International agreements in specialized areas, e.g. the IMO Convention for regulating control of pollution from ships (Section 3.1.2); and the London Dumping Convention (a resolution which in 1989 recommended that the British Government should not implement plans either to dump worn-out nuclear submarines on the sea-bed, or to drill holes from oil platforms and drop cut-up parts into specially prepared 'graves'). An informal moratorium on dumping low-level nuclear waste at sea has been in force since 1983, and waste incineration at sea should cease by the mid-1990s. As mentioned at the start of Section 3.1.2, the subject of maritime transport is largely outside the scope of the UN Conventions. It is covered in general terms by UNCLOS and, before it, by the Convention on the High Seas. Whaling and the IWC have already been mentioned in Section 1.2.3.

(b) Regional and/or multilateral agreements on particular issues (e.g. the UN Environmental Programme, UNEP), and the continuation of earlier *ad hoc* diplomacy or bilateral/multilateral treaties (e.g. the North-East Atlantic Fisheries Convention and the 23-nation Convention for the Conservation of Antarctic Marine Living Resources, set up in 1982), frequently under the aegis of intergovernmental organizations.

(c) Multilateral or bilateral agreements, self-imposed restrictions on activities in internal waters, overlapping territorial seas and contiguous zones, and the high seas, for mutual benefit which often extends beyond the zone(s) directly involved (e.g. hydrocarbon exploration and exploitation, fish quotas, pollution control, traffic separation zones).

(d) National management, i.e. development and conservation, in the coastal zone (internal waters and the territorial sea).

3.3 DISPUTES AND THEIR RESOLUTION

"Let's show these guys *our* version of the Law of the Sea."

Given that a valid regime exists, how is compliance of states achieved? Probably by various combinations of moral, political and (in the final resort) military pressure. There are practical advantages in uniformity of behaviour and obvious disadvantages in 'package deals' or mini-treaties, and in 'picking and choosing' those parts of the legal regime which suit and rejecting those which do not. The most effective thrust towards implementation of existing regimes is reciprocal—states cannot expect to benefit from the exercise of their rights and duties unless they concede the same to others.

Outrageously illegal behaviour will be tolerated (but not ignored) until it affects or threatens to affect another state's interests. Typical disputes arise between individual states over issues that are either of immediate practical importance or are perceived as likely to become so in the future. They can be settled by negotiation, (binding) arbitration, the International Court of Justice, and by special procedures laid down in the Conventions. Enforcement in the event of non-compliance could go so far as to involve international sanctions—but if they fail, what alternative remains? Nonetheless, if states do not protest about illegal behaviour, they may not be able to oppose any change in the law that it might bring about. Acquiescence may be dangerous in this respect, but disputes must be settled by peaceful means. In any case, full international sanctions are only possible if the UN Security Council declares the situation a threat to peace and world order; they are then organized by the Security Council itself (though states often resort to unilateral sanctions).

Disputes can of course be over all sorts of different issues, such as freedom of navigation (straits, bays and so on), boundary disputes relating especially to resources, and—best known of all—fishery disputes. You can no doubt think of examples of such disputes, for they are often given considerable coverage in the press and on radio and TV, and they may well concern more than one issue. Thus:

The *Icelandic cod wars* (Chapters 1 and 2) were obviously about fishing limits and quotas; but not exclusively so (see Appendix 3).

In the *Aegean* (Section 2.3), disputes seem to be mainly about resources, but they also concern freedom of navigation.

The *Gulf of Sirte/Sidra dispute* between Libya and the USA in 1985 (which involved the shooting down of Libyan aircraft allegedly threatening American naval vessels) was about freedom of navigation and right of innocent passage in bays—but there were decidedly military overtones.

Following the *Falklands war* of 1982, which was a dispute about territory, resources, and Britain's presence in the South Atlantic, the UK Government first declared a 150-mile Maritime Exclusion Zone (later Total Exclusion Zone) which foreign vessels entered at their peril (the presumption being that they had hostile intent). This zone was not measured from the coastal baselines but from a point in the middle of the main island so that its outer limit formed a perfect circle. The later 200-mile fishing limit was not declared as an EEZ or EFZ (Exclusive Fishing Zone), in order to mollify Argentina, which claims sovereignty over the Falklands (Las Malvinas) and hence claims rights also in any Falklands EEZ or EFZ. Instead, this zone was declared as a Fisheries Conservation Zone, also measured from the central point, within which 'foreign' fleets might operate under licence and within specified quota limits. The zone thus avoided any association with the UNCLOS approach.

The long-running *Beagle Channel dispute* between Chile and Argentina was about access to resources and the placing of maritime boundaries affecting ownership of offshore islands and the zones they 'generated'; it was settled privately in the mid-1970s by papal conciliation—reminiscent of the Treaty of Tordesillas (Section 2.1)—though it is clear from the result that trends in the Law of the Sea were taken into account.

In the next Section, we look at a couple of other examples of potential or actual disputes in more detail.

3.3.1 ROCKS OR ISLANDS?

If you look at any good map of the ocean floor, in the region of the Pacific to the south of Japan, you will see a line of seamounts extending roughly north–south, lying midway between the Mariana–Bonin Trench system and the Ryuku–Philippines Trench system. Many of the seamounts come close to the sea-surface and have been colonized by coral reefs. In some cases, tectonic movements have lifted the corals above the sea-surface, so that they now form coral islands. One of these islands, at about 21° N, has been known as Parece Vela since the 17th century when it first appeared on charts, and would be the southernmost outpost of Japanese land territory if it were not for another smaller coral island about 850 km to the south at about 13° N (Figure 3.3). This island is known to the Japanese as Okinotorishima—Remote Bird Island. At low tide, the island is about 5 km by 2 km, but at high tide only two small rocks are left protruding a metre or so above the sea. These rocks—North Dew Rock and East Dew Rock—are very important because they control the southern limit of Japanese rights to the Pacific. Unfortunately for the Japanese, the coral island is in the cyclone belt and is being rapidly eroded by wind and wave action. Six years ago there were four rocks above high-water mark; now the remaining two—which are mushroom-shaped and are only joined to the underlying reef by fragile columns—may be swept away at any time.

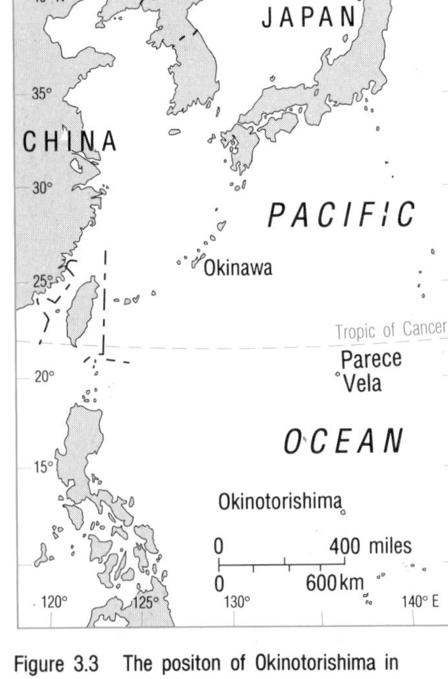

Figure 3.3 The positon of Okinotorishima in relation to Japan. The original island was under the government of Tokyo in 1931 and was reaffirmed as Japanese by a treaty drawn up in 1951.

The Japanese are determined to arrest this erosion. As part of a three-year plan, they have dropped thousands of four-pointed steel anchors on to the reef to form 55-metre diameter circular walls around each of the rocks. Their aim is to fill the space between the anchors and the columns with concrete. So far, the project has cost £31 million; the final cost is likely to be about £135 million.

QUESTION 3.2 Why do you think the Japanese consider this enormous sum a worthwhile investment?

The seas around Okinotorishima are rich in tuna and bonito, and cobalt-rich polymetallic crusts have been found on the sea-floor. Preservation of North Dew and East Dew Rocks would seem to assure the Japanese the right to exploit these resources because, according to Article 121 of UNCLOS, an island is a 'naturally formed area of land, surrounded by water, *which is above water at high tide*'. However, the Convention also says that '*rocks which cannot sustain human habitation or economic life of their own shall have no exclusive economic zone or continental shelf*' (see item 8 in the summary in Appendix 2). North Dew and East Dew Rocks are only a few square metres in area at high tide, and the only use planned for them is as a site for an automatic meteorological observation station. It is therefore difficult to see how the Japanese hope to claim exploration rights over 400 000 km^2 of surrounding sea-bed. They should, however, be able to prevent the southernmost part of Japanese territory abruptly moving about 850 km to the north during a cyclone!

Rockall

The subtle though important difference between 'rocks' and 'islands' was the cause of a loyal British subject living on the inhospitable surface of Rockall (Figure 3.4) for a few days in 1985. The seas around Rockall include rich fishing grounds and, perhaps more importantly, the sea-bed may contain hydrocarbons. A 200-mile exclusive economic zone around Rockall would be of enormous potential value, and the UK Government has already declared a 200-mile fishing zone around it.

Figure 3.4 Rockall on a calm day. The rock is about 22 m high, and is located 370 km (200 nautical miles) west of the Outer Hebrides. The summit of Rockall was removed in 1971 to instal the navigational beacon visible here (*cf.* montage page).

Rockall was annexed on 18 September, 1955. Lt. Cdr. Desmond Scott, in charge of the landing party from HMS *Vidal*, provided these recollections for us: 'Gales were sweeping in from the west one after another as we steamed out to Rockall Bank, but our Meteorological Officer forecast a three-hour lull which the Captain decided would be sufficient — "After all, so long as you can land you can perform the annexation ceremony, and if by then it's too rough for the helicopter to operate, you can always jump off the rock and get picked up by a boat!" As soon as the wind eased the airlift began: four men, a flagstaff constructed on board from old motorboat shafts, together with a Union Flag, numerous buckets of quick-drying cement and an engraved brass plaque. The plaque and flagstaff were soon cemented in place, but there was no time for the cement to dry properly before we performed the ceremony taking possession of the islet in the name of Her Majesty Queen Elizabeth the Second. While the 'disembarkation' was underway, rock samples were chipped off and seaweed samples were collected and bagged; when examined on board these provided a few zoological specimens, including one crustacean and some periwinkles. Finally, after what seemed a lengthy waiting period when I was alone on the rock, the helicopter returned to collect me, and the operation concluded with a tricky landing on *Vidal* in a heavy swell and rising wind.'

Rockall is the tip of the Rockall Plateau, a microcontinent which broke off from the European continental mass during the opening of the Atlantic. Denmark, which exercises sovereignty over the Faroe Islands, claims that the Rockall Plateau is the south-westerly extension of the same plateau as the Faroes, and that the two are part of the same microcontinent (Figure 3.5). The Danes thus consider that the Rockall Plateau qualifies as Danish continental shelf, and that their claim is substantiated by the existence of the regions of deep water to the north-west (between the 'Faroe–Rockall microcontinent' and the Icelandic continental shelf) and to the south-east (the Rockall Trough). Iceland's ambitious claim is based on liberal interpretation of UNCLOS definitions, particularly that regarding what constitutes the outer edge of

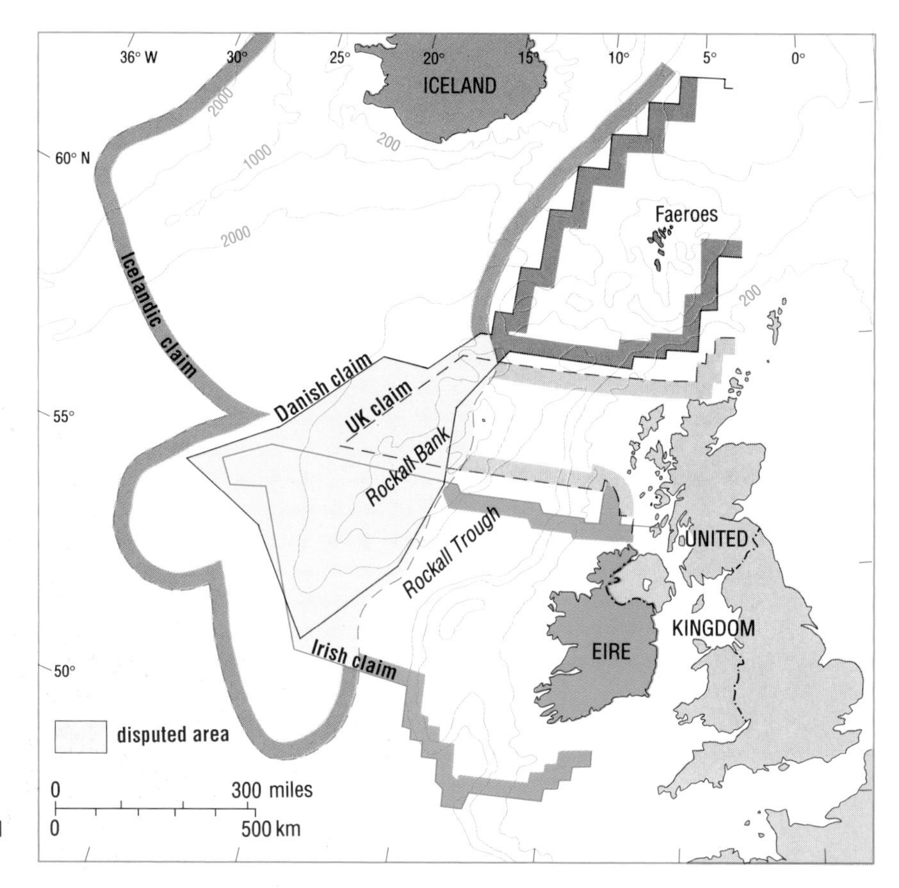

Figure 3.5 Map of part of the North Atlantic showing bathymetry and the competing claims over Rockall. Contour values are in metres. Rockall Bank is another name for Rockall Plateau.

the continental margin. The Irish Republic claimed the Rockall Plateau as an extension of the Irish continental shelf, despite the existence of the Rockall Trough, but this particular bone of contention between Britain and Ireland was settled in 1989. Britain, in any case, annexed the island in 1955 on the basis of proximity rather than geology or bathymetry.

QUESTION 3.3 If the United Kingdom cannot support its claim that Rockall is an island in the full legal sense, which maritime zone(s) may it nevertheless declare around the rock under the Law of the Sea Convention (Appendix 2)?

These examples clearly illustrate the importance that states are attaching to their exclusive economic zones. States with potentially valuable resources and/or large areas of continental shelf see the legal status of exclusive economic zones as a great opportunity to be seized; land-locked states with no continental shelf or exclusive economic zone are looking forward to the time when, under the terms of the Convention on the Law of the Sea, they will be given access to both fisheries (item 5 in Appendix 2) and sea-bed resources (through the International Sea-Bed Authority, see item 11 in Appendix 2).

3.4 A LOOK AHEAD

The evolution of the Law of the Sea since World War II reflects both a progressive erosion of the freedom of the seas under increasing pressure from national resource interests, and international attempts to regulate this trend. Its key concepts have developed over centuries, and only now, as the year 2000 approaches, is the outline of a global regime crystallizing. The distinction between the historically traditional maritime powers and the rest has become blurred. The traditional maritime powers are coming to recognize that their coastal and shelf resources have overtaken their shipping industries in commercial and political importance. This is largely due to the growth of the offshore oil industry, but at the same time the share of the world maritime transport taken by technologically advanced nations has declined in the face of competition from some newly industrializing countries. The latter are moving in the opposite direction; they are no longer archetypal 'coastal states' whose interest in the sea is confined to fishing and local trade.

UNCLOS is a remarkable document. Its 320 articles and nine annexes deal with almost every human use of the oceans and constitute a global framework for a maritime conduct of nations. The principle that 'the [International Sea-Bed] Area and its resources are the common heritage of mankind' has found expression in its text. Thus, against the odds, UNCLOS has achieved a wide measure of agreement, although some familiar problems remain (*cf.* Section 3.3). Political factors still cloud the legal issues at all levels of negotiation, but it was the primarily economic matter of sharing the resources of the deep sea-bed that blocked consensus. Hard bargaining failed to close the gap between Western industrialized nations and an informal coalition of developing countries, whose numerical superiority prevailed in the vote requested by the USA, when all efforts to mediate had been exhausted. The dispute boils down to the incompatibility of the common heritage principle with a freedom of the seas approach, in which sea-bed resources are considered available to those with facilities to exploit them. Bearing in mind that commercial deep-sea mining is very unlikely to start before the end of this century, will the advantages of a universally accepted Law of the Sea be postponed indefinitely for the uncertain prospect of wealth from the sea-bed? This is the key question that still awaits an answer.

All this is taking place within the context of a rapidly evolving human perception of the seas. The traditional idea of an ocean wilderness is being replaced by the concept of a managed ocean. The whole theoretical basis of the Law of the Sea has changed, but the practical implications have yet to be tested. Whatever the final outcome, one thing will remain true:

> 'The sea has many voices,
> Many gods and many voices.'
>
> T.S. Eliot, *Four Quartets*.

The following two Chapters provide examples of that quotation.

3.5 SUMMARY OF CHAPTER 3

1 Three geographical categories of states may be identified (island, coastal and land-locked), but within these categories the importance of the sea to a particular nation depends also on socio-economic factors such as population and degree of development. Differences between states in the different categories also determine the relative importance of the three principal maritime roles: mercantile (trading, shipping), resource-oriented (fishing, hydrocarbons), or strategic (navies).

2 UNCLOS has not yet come into force. Many of the provisions of the previous (1958) Conventions have become part of customary law, which in turn fed back into UNCLOS. Similarly, at least some parts of UNCLOS have some value as customary law, even though the 1982 Conference failed to achieve consensus and thereafter a number of major developed countries refused to sign it. The number of states that have formally ratified the Convention was still well below the required minimum of 60 in 1991.

3 UNCLOS itself is a comprehensive update in the light of developments since the first UN Conference in 1958. It attempts to establish a unified international legal regime for all maritime purposes, sets out general principles and provides a framework for development of other more specialized agreements (the umbrella function); and it is thus more than a piecemeal revision exercise.

Principal points are:

(a) Coastal states can claim a twelve-mile territorial sea and a twelve-mile contiguous zone. The coastal state has virtually complete jurisdiction over all activities in these zones, but must preserve the right of innocent passage. The concept of transit passage is introduced to cope with international straits.

(b) The introduction of the exclusive economic zone (EEZ), extending 200 miles from shore (188 miles from the territorial sea boundary), is one of the two most important developments (see also (d) below).

Control over research, 'exploitation and conservation of both sea-bed and waters is reserved' to the coastal state. Other states have freedom of navigation and overflight and freedom to lay pipelines and cables. Proposals for marine scientific research in the EEZs of other nations require the consent of the respective coastal state(s).

(c) The definition of the legal continental shelf now includes a distance limit (up to 350 miles from shore or 100 miles beyond the 2 500 metre isobath). Coastal (and island) states are in any case assured of a minimum 200-mile (EEZ) width of continental shelf and thus get approximately more equal areas of shelf per unit length of shoreline than was the case under the Convention on the Continental Shelf. The sea-bed between the outer boundary of the EEZ and that of the legal continental shelf is under the jurisdiction of the coastal state, but (i) the waters above this zone are part of the high seas and (ii) the coastal state is required to share revenue from shelf resources beyond the EEZ limit with the international community.

(d) UNCLOS made no significant changes in the law concerning the waters of the high seas, beyond what was in the Convention on the High Seas, i.e. freedom of navigation, of fishing, and of scientific research. The other major change under UNCLOS (*cf.* (b)) concerns the International Sea-Bed Area, beyond the limits of EEZs and legal continental shelves. It will come under the control of the International Sea-Bed Authority, which will regulate exploration and exploitation of deep-sea minerals and ensure equitable distribution of revenue from such enterprises among all nations, including land-locked states. This was perhaps the biggest bone of contention at the 1982 Conference and the main reason why some major developed nations refused to sign.

4 UNCLOS sets out to frame an international legal regime, but there are other agreements at both international and regional levels, outside its scope, though perhaps adoptable under its 'umbrella'. Many disputes over maritime issues have been more or less resolved in recent decades with or without recourse to the Law of the Sea; but enforcement of any settlements or maritime regulations in general may still be a problem even when the Convention does come into force. Nonetheless, it is still a remarkable document, providing a global framework for the maritime conduct of nations.

Now try the following questions to consolidate your understanding of this Chapter.

QUESTION 3.4 (a) '. . . the EEZ extends 200 miles from shore. . . .' Strictly speaking, that statement is not quite true. What is wrong with it? Does the same problem apply to the statement in Question 3.1 about the new definition of legal continental shelves: '. . . equal areas of shelf per unit length of shoreline or coastline. . .'?

(b) Find a map showing ocean bathymetry, and explain why none of the states of western South America can claim a continental shelf more than 200 miles wide; whereas around Newfoundland, Canada can claim up to 350 miles (or even more).

QUESTION 3.5 To what extent would the definition of the legal continental shelf under UNCLOS change the relative placings (and the numbers) in Table III of Appendix 1?

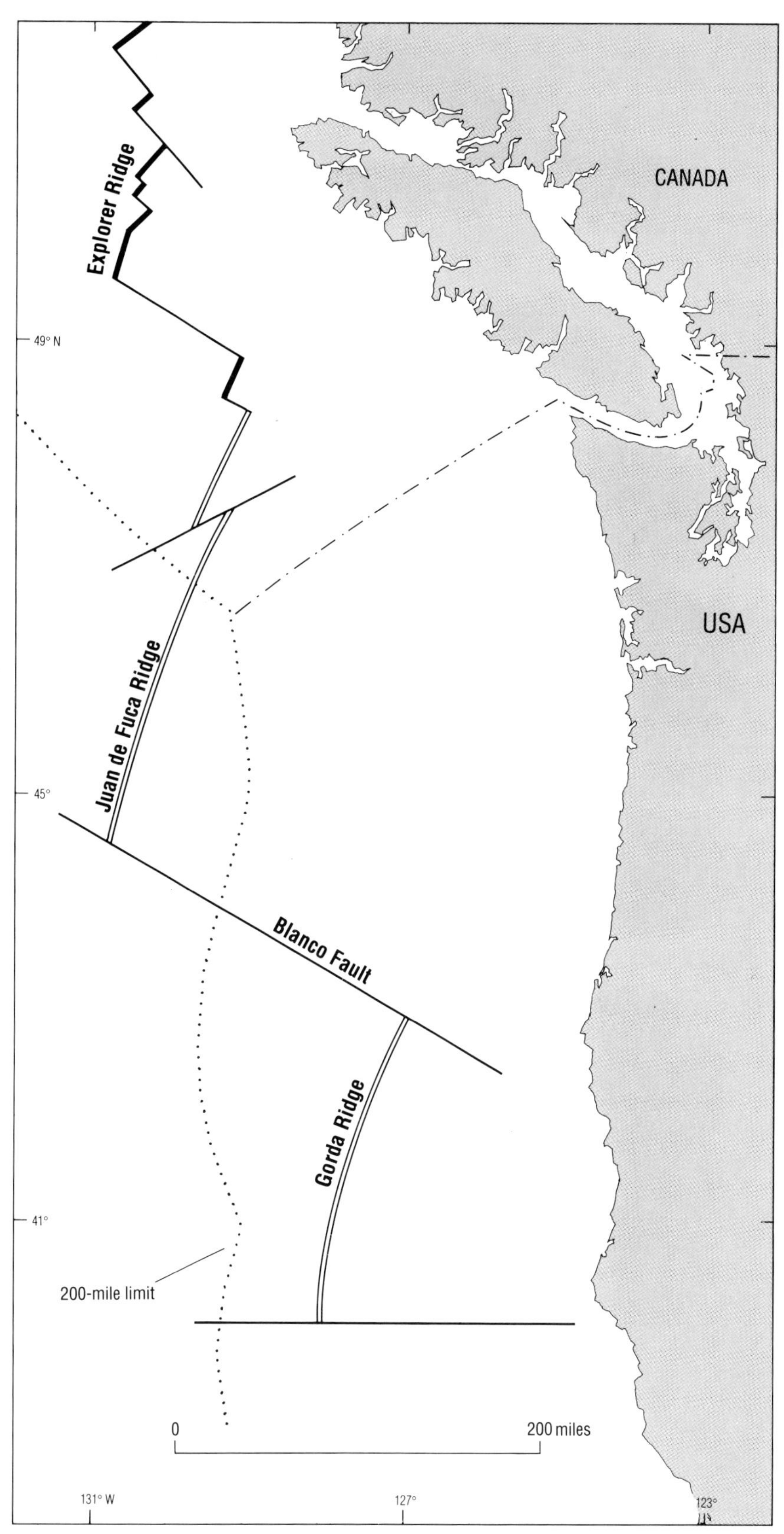

Figure 3.6 Part of the East Pacific Rise off western North America, showing the 200-mile limit (dots) and the offshore boundary between US and Canadian jurisdiction (dashes). (For use with Question 3.6 overleaf.)

QUESTION 3.6 Figure 3.6 shows the position of some segments of the East Pacific Rise off western North America. Hydrothermal sulphide deposits are known to occur at places along the Gorda and the Juan de Fuca Ridges.

(a) What are the rights of the USA and Canada in respect of the exploitation of those deposits that lie landward of the 200-mile line?

(b) The southern segment of the Juan de Fuca Ridge lies outside the 200-mile line on Figure 3.6. The average depth of the East Pacific Rise crest is about 2 500 metres, sloping away to deeper water on either side. Explain whether deposits in this segment fall within the legal continental shelf of the USA (or Canada), according to the definition *either* (i) in the Convention on the Continental Shelf *or* (ii) in the relevant part of UNCLOS.

(c) Does your answer to (b) help to explain why a revised definition of the legal continental shelf was felt to be necessary?

(d) Can the USA legitimately lay claim to these deposits under the terms of UNCLOS?

QUESTION 3.7 Look back to the definition of living resources that may be harvested under the continental shelf provisions of UNCLOS (see the relevant part of Section 3.1.2). Which of the following would qualify as exploitable under the criteria given there: lobsters, swimming crabs, spider crabs, clams, scallops, razor shells?

CHAPTER 4 THE ARCTIC OCEAN: ICE, OIL AND SOVEREIGNTY

The Arctic Ocean is sometimes called the 'polar Mediterranean'. A region of ocean more than four times the area of its temperate counterpart, it is almost completely surrounded by land (Figure 4.1), so forming a natural antithesis to the Antarctic continent, completely surrounded by the Southern Ocean.

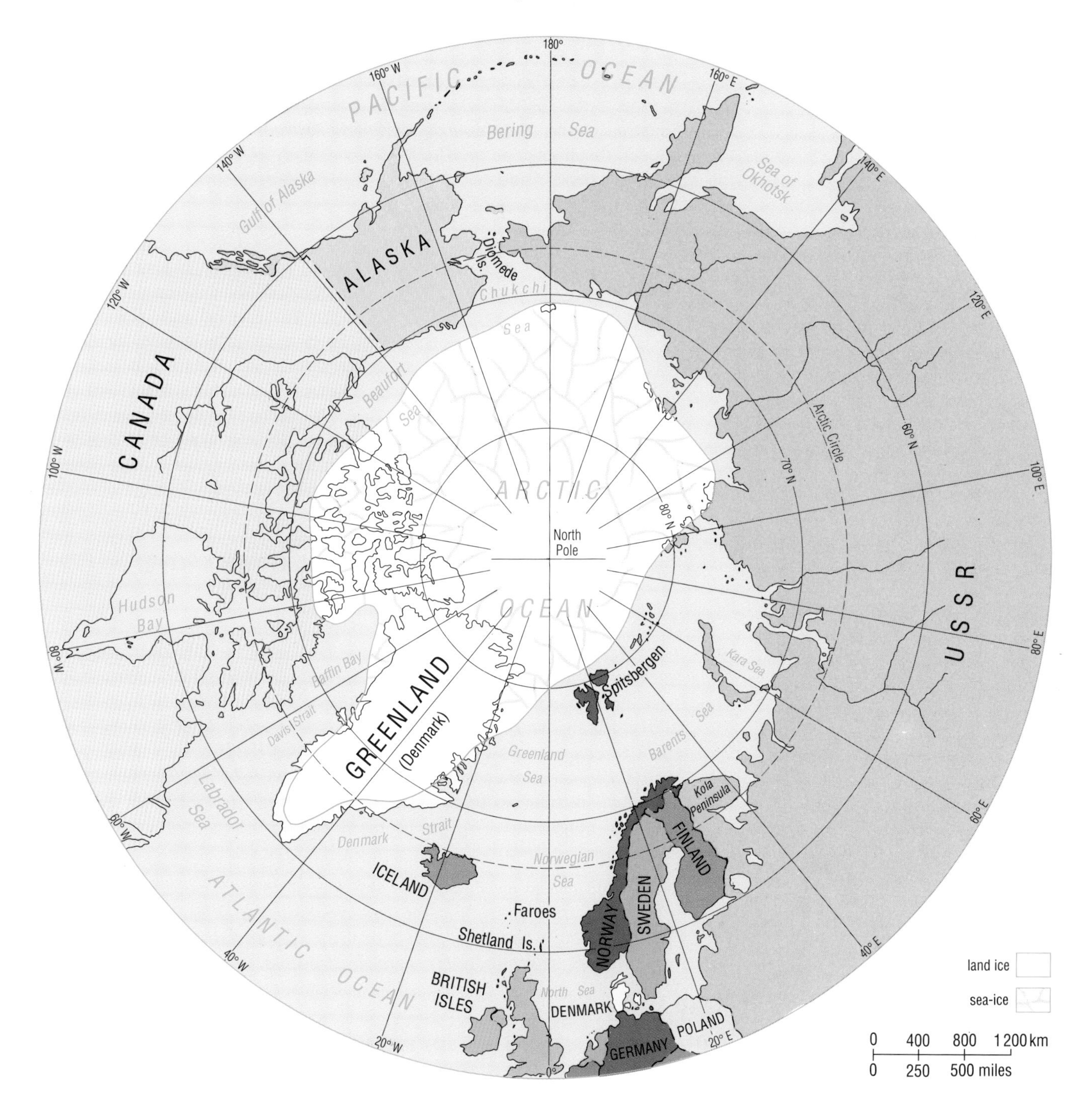

Figure 4.1 Geographical and political setting of the Arctic Ocean. The approximate extent of permanent land and sea ice cover is shown schematically in this and subsequent Figures.

For a number of decades, the Arctic Ocean and its adjacent seas* have been important in human affairs, largely for strategic reasons. The United States and the Soviet Union face one another across the basin, and the shortest air route between them is across the North Pole. At the narrowest part of the Bering Straits, the United States (Alaska) and the Soviet Union are only 57 miles apart; the Diomede Islands in the central part of the Straits are partitioned between the two powers, with Big Diomede (Ratnova Island) belonging to the Soviet Union and Little Diomede to the United States. Furthermore, the main seaway for the Soviet Northern Fleet, based on the Kola Peninsula, is through the Norwegian Sea and thence through either the Denmark Strait or the Faroes–Shetland Gap (Figure 4.1), and so this region has been a front line for NATO defences.

However, it has become apparent that the region is rich in natural resources, and the Arctic of the 21st century may well be a focus of industrial, rather than military, activity. Taking hydrocarbon reserves as our main example, we will look at the challenges facing those who wish to exploit these resources, along with the environmental pressures such exploitation will certainly bring about.

The question of what rights various states have to exploit marine resources, and to transport them by certain sea routes, is settled through international law. Similarly, measures to protect the environment can only be successful if they are based within an accepted legal framework. We will look at these legal aspects in Section 4.2. Meanwhile, we begin with a brief overview of how and when human communities became settled around the rim of the Arctic Ocean. We do this for two reasons: first, the people who will be most directly affected by development of the Arctic are the indigenous populations; secondly, the lifestyle of these people, evolved over a long period, is an important factor in the debates concerning sovereignty over ice-covered Arctic waters.

4.1 HUMAN SETTLEMENT AROUND THE ARCTIC

Mention of the Arctic conjures up images of tundra and polar desert surrounding an icy sea—but the Arctic has not always been like this. The Earth's 'normal' climatic state seems to be tropical to subtropical from the Equator to the poles. This normal, warm situation is at times interrupted by Ice Ages—periods lasting some millions of years, during which extensive parts of the Earth are covered by ice-sheets. During Ice Ages, these ice-sheets alternately grow and decline: times of maximum ice extent are referred to as glacial intervals and times of minimum extent as interglacials. We are presently still in the Pleistocene Ice Age, but enjoying the relative warmth of an interglacial.

As the continental ice-sheets wax and wane, so water is removed from or added to the oceans, and the sea-level rises and falls (Figure 4.2). Climatic changes and their effects on ice cover and sea-level have been major factors determining the pattern of human settlement around the Arctic.

*In this Case Study, we use terms like 'Arctic seas' to mean the Arctic Ocean and its adjacent seas; we therefore sometimes include areas which geographers classify as 'sub-Arctic'.

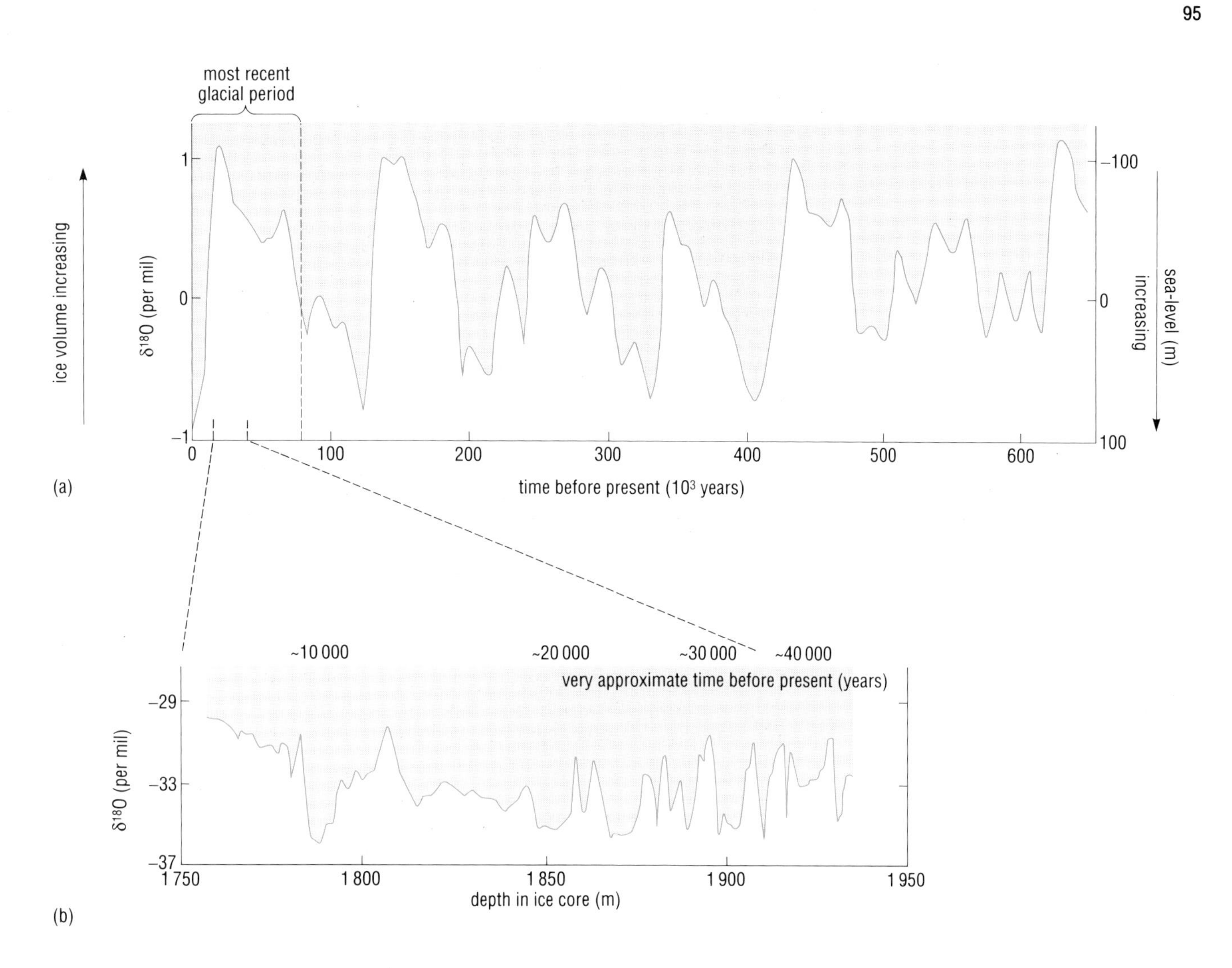

Figure 4.2 The variation in the amount of ice in Northern Hemisphere ice-caps (a) over the past 600 000 years, (b) between about 9 000 and 50 000 years ago.

Both curves actually show the variation of $\delta^{18}O$, which is a measure of the oxygen isotope ratio ^{18}O:^{16}O. A change in $\delta^{18}O$ of 1 per mil corresponds to a change in sea-level of about 100. In (a), $\delta^{18}O$ has been measured in the skeletal remains of planktonic foraminiferans and the curve is a composite for several deep-sea cores. (Note: Some deep-sea cores indicate that the total variation in sea-level, shown here as 200 m, could have been only 100–150 m.) In (b) $\delta^{18}O$ has been determined at different levels in the ice of a core from the Greenland ice-cap. Also, in (b) $\delta^{18}O$ values have been calibrated against a standard, whereas in (a) the zero is arbitrary. The main point is that the more positive the value, the more ice there was.

4.1.1 EARLY MIGRATIONS

Human beings first penetrated into the Arctic regions 40 000–35 000 years ago. These first settlers were Neanderthal: their artefacts have been found extensively in Eurasia, but not in America, even though there were a number of periods when sea-levels were low enough to permit passage over the land bridge across the Bering Straits (*cf.* Figure 4.2).

However, with the marked decline in temperatures which occurred about 25 000 years ago, not only did the sea-level fall further (to about 200 m below its present level; *cf.* Figure 4.2(a)), but the Arctic Ocean became a frozen inland sea. Human beings—no longer Neanderthal, but modern man—crossed over from Asia into Alaska. Some groups made their way south, eventually reaching South America, but a major part of the population became trapped in Alaska by the growth of the North American ice-sheets. Following the glacial maximum 20 000–18 000 years ago (*cf.* Figure 4.2(a)), the climate began to warm, and by 11 000 years ago the Bering Straits were once again a maritime passage.

Over the succeeding millennia, warming continued, although there were still marked climatic fluctuations. Between about 4 000 and 3 000 years ago, during a relatively warm period, there was a spread of hunter-gatherer peoples eastwards from Siberia/Alaska all the way to northern Greenland. This immigration into Arctic Canada, which was the first since the disappearance of the continental ice-cap, was of people who lived mainly on caribou and musk-ox, seasonally supplemented by fish.

During the relatively cold period which followed, various groups developed ways of living suited to the local environment. In the eastern Canadian Arctic, seal-hunting became important, perhaps because land animals became scarcer and whales were excluded from the region by sea-ice. The importance of seals led to the practice of building winter snow houses or igloos on the ice so that the hunters could be closer to their prey. Meanwhile, the Alaskan people became more and more dependent on whaling, and eventually developed a way of life known as the Thule culture. It was this culture which was to spread swiftly eastwards during the unusually warm period between about 1000 and 1200 AD.

The swift spread of the Thule culture from northern Alaska to Greenland occurred at a time when the waters between the Canadian Arctic islands were sufficiently ice-free in summer to allow not only the passage of boats but also migrations of bowhead whales. Subsequent deterioration of the climate meant that the waters of the Canadian Arctic once again became impassable to whales; as a result, the Canadian population returned to seal-hunting while the hunting of whales and other marine mammals from boats became confined to the waters off Alaska and Greenland (Figure 4.3).

Figure 4.3 Hunting for seals and whales in the 18th century. Note that while seals may be hunted from the sea-ice, whales were only available after the pack-ice began to break up in spring. The small boats are kayaks, and the larger open boats are known as umiaks.

Figure 4.4 The end of a battle between Inuit and Norsemen. Note that the Norseman is wearing not furs but traditional Norse woven clothing, which would have been unsuitable for the climatic conditions. Such battles occurred because when crop-growing and husbandry became difficult the Norsemen turned more to hunting and this brought them into conflict with the Inuit who at that time were moving south in pursuit of seals.

In this brief discussion of the spread of peoples around the western side of the Arctic (the region we will be considering in Section 4.3), we have tried to convey the extent to which the pattern of human settlement in the region has been determined by climatic conditions. Mention should also be made of the exploits of the Norsemen who, under Erik the Red, sailed to south-western Greenland and, from about 1000 to 1500 AD, sustained a colony there. The Norsemen arrived towards the end of the most recent 'warm period', and it is thought that they eventually perished because, when the climate deteriorated, they failed to adapt their lifestyle sufficiently (*cf.* Figure 4.4).

4.1.2 PRESENT-DAY CIRCUMPOLAR PEOPLES

Under present climatic conditions, the polar ice-cap expands and contracts with the seasons, but much of the pack-ice survives through the summer months (Figure 4.5). As a result, a large part of the Arctic Ocean is covered with thick, rugged, multi-year ice. The pack-ice shifts with winds and currents, but many of the Arctic islands are either seasonally or permanently fused together by fast (i.e. stationary) ice, which is frozen to the land and remains in place through the winter, sometimes even surviving through the summer. These unusual and harsh conditions shape the lives of the large numbers of indigenous Arctic peoples who live close to the sea and have a way of life which is dependent on marine resources.

Figure 4.5 Near-minimum (a) and near-maximum (b) ice cover in northern high latitudes, as determined using microwave measurements obtained from the *Nimbus* satellite programme. The different colours represent the percentage of sea-surface covered by ice: a pinkish-red tone indicates 100% coverage, while a light tone represents a coverage of 20% or less. Image (a) is for 1–7 September 1979 and image (b) is for 3–7 February, 1979.

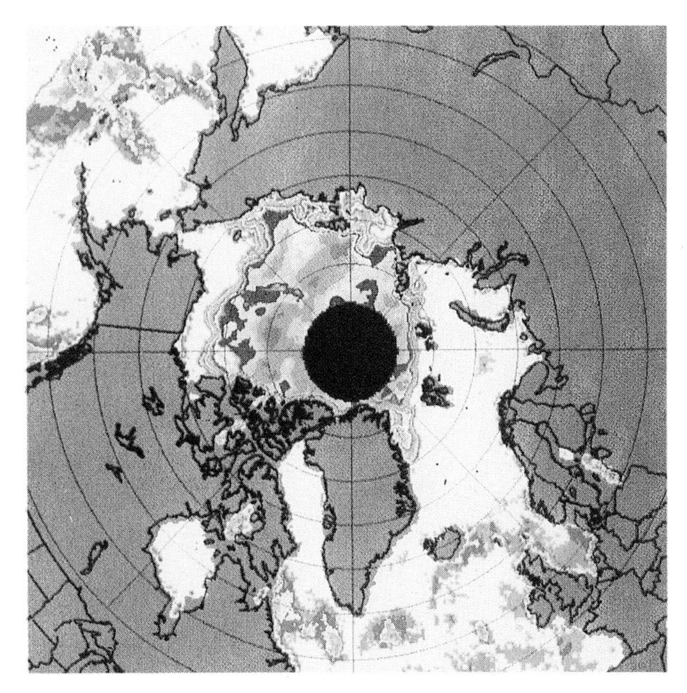

(a)

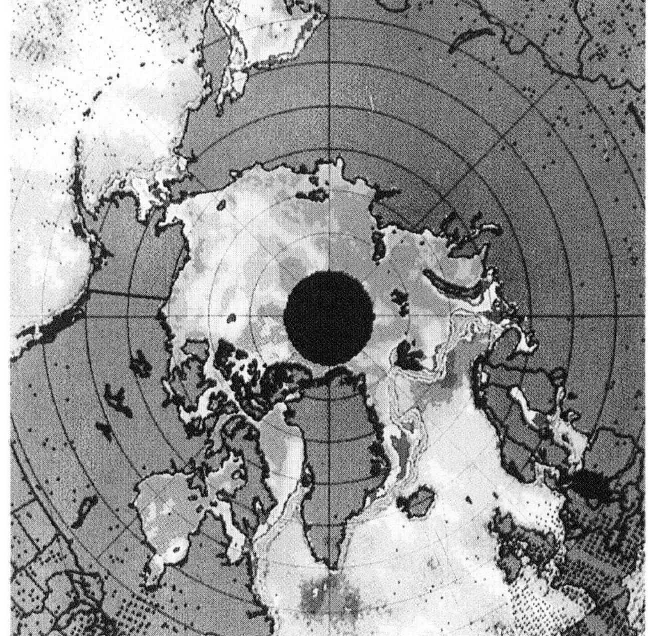

(b)

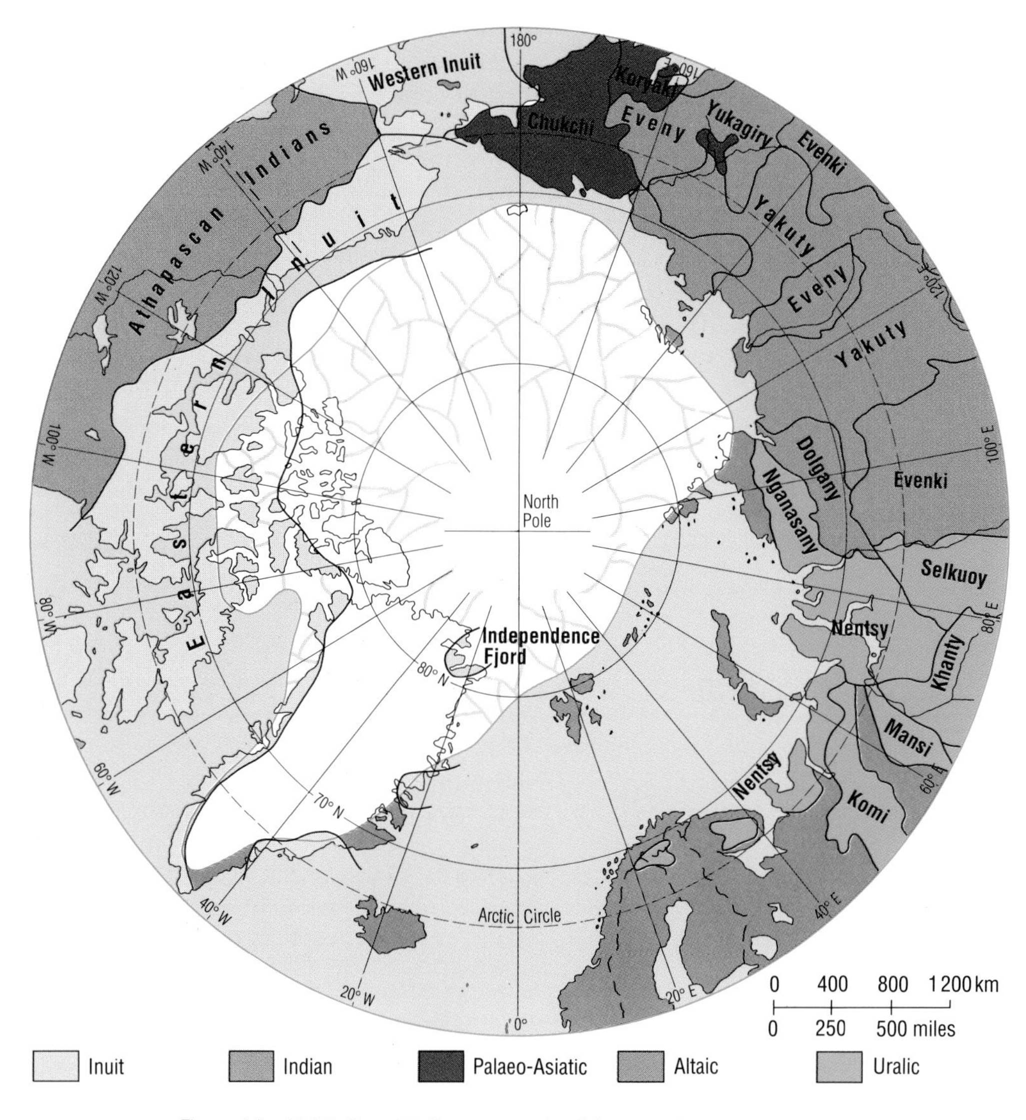

Figure 4.6 Distribution of indigenous peoples living around the Arctic at the present day.

With the exception of Iceland, all the states which border the Arctic have indigenous peoples (Figure 4.6). The Lapps of Scandinavia have a population of about 30 000, many of whom traditionally depend on the fish, seals and whales of the Barents Sea. However, the inshore artisanal (non-commercial) fishing of the Barents Sea collapsed in the late 1980s, and fishing there is now mainly by large deep-sea trawlers and drifters. Some Lapps live in what is presently the northern USSR, as do a number of other different ethnic groups. The people who live in eastern Siberia, Alaska, Canada and Greenland are the Inuit. In all, the Inuit number about 100 000. Of these, 30 000 live around the coast of Alaska, 25 000 in small scattered communities along the northern coastline of Canada, and about 43 000 in Greenland (where they make up about 85% of the total population).

Greenland is territorially part of Denmark, though it was granted Home Rule in 1979. Because of the Greenland ice-sheet, only a small part of the island is habitable, and the population is largely concentrated on a narrow strip of the south-west coast facing the Davis Strait, Baffin Bay and the Atlantic Ocean. The economy is heavily dependent on marine resources, particularly commercial catches of cod, halibut and deep-water prawns. There is also extensive harvesting of marine mammals, both for extra income and subsistence food.

The majority of Canadian Inuit live in the North-West Territories. Until several decades ago, they lived in widely scattered hunting camps and followed their traditional lifestyle, completely dependent on natural resources for food, clothing and housing. The Canadian Government then persuaded them to come to small centralized communities where health, education and other services were available. However, loss of the link with the natural environment was deeply felt, and—with the support of the Canadian Government—many Inuit have returned to their traditional homelands, although now equipped with rifles and motorized snowmobiles.

The native peoples of Canada—the Indians and the Inuit—have been engaged in negotiations over land ownership and rights to wildlife harvesting (management and exploitation) since 1973. Their claim for aboriginal rights, based on native use and occupancy since time immemorial, extends to marine areas, in particular to land-fast ice which is regarded as an extension of the land and a hunting ground for marine mammals (Figure 4.7).

Seals form the most important part of the Inuit economy and diet. For many communities, the ringed seal is the most commonly caught marine mammal (as, for example, in Lancaster Sound; see Table 4.1). Narwhals and walruses have become less important as sources of food, although they are still valued for their ivory and the prestige of the hunt. Beluga whales and polar bears are also hunted, though in relatively small numbers.

Table 4.1 Estimates* of marine mammals harvested by three communities in Lancaster Sound during 1979.

Species	Clyde River (87 hunters)	Grise Fjord (22 hunters)	Pond Inlet (122 hunters)
Ringed seal	4 733	686	2 487
Bearded seal	5	25	38
Harp seal	4	166	21
Narwhal	5	15	139
Polar bear	21	24	16
Walrus	—	9	14
Beluga whale	—	14	9

* For various reasons, these estimates may not accurately reflect actual abundances. For example, natural populations fluctuate; also, the extent to which the Inuit depend on marine mammals varies from place to place and year to year.

All of these creaures depend ultimately on the primary productivity of algae, growing either planktonically in the upper layers of the ocean or on the underside of the sea-ice (Figure 4.8). Although the annual phytoplankton bloom is very short—only a few weeks in the high Arctic—it supports a vast increase in zooplankton production. This supports large numbers of migratory mammals, such as whales and walruses, which make their way into the Arctic as the pack-ice breaks up

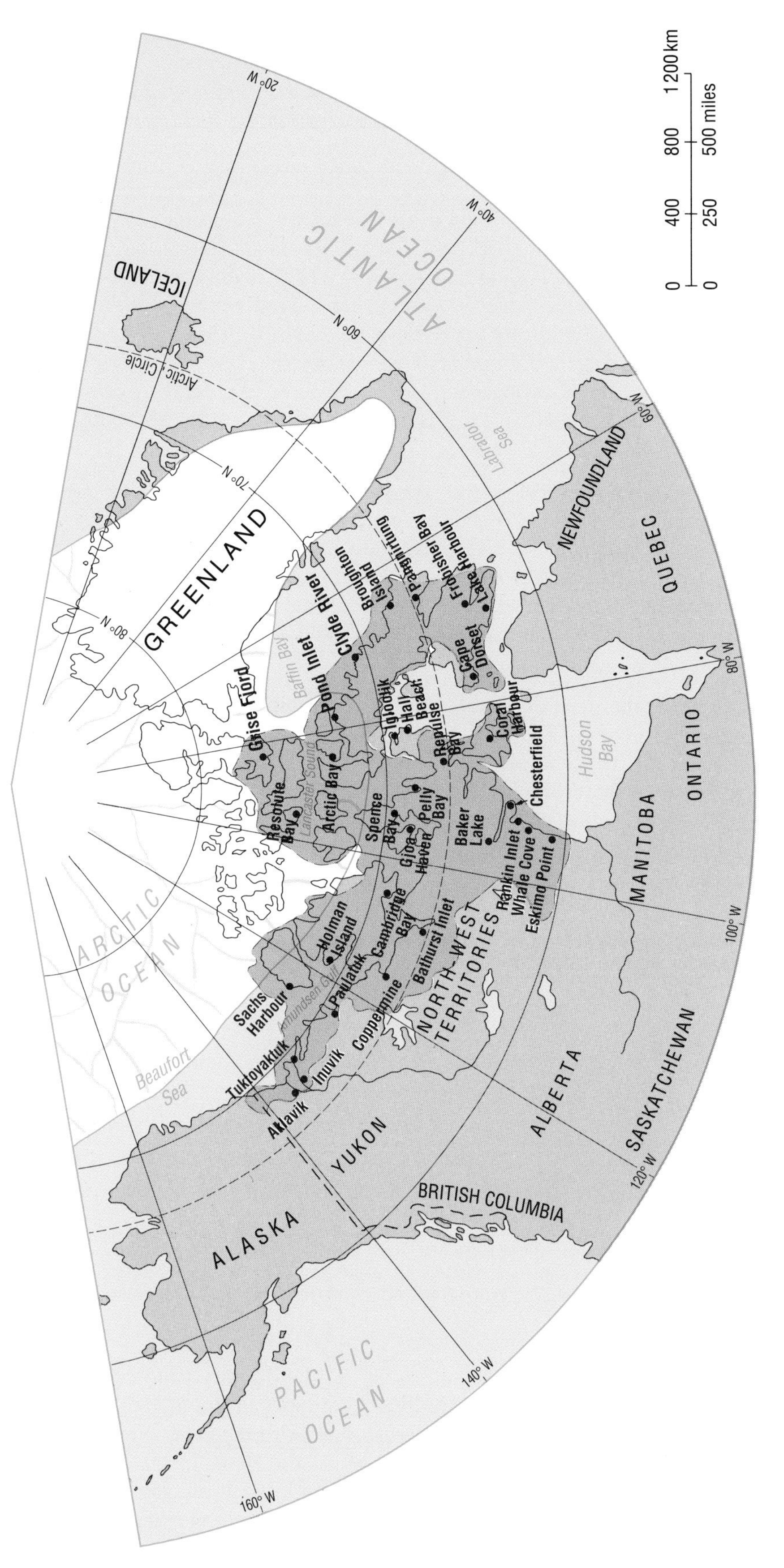

Figure 4.7 Areas of sea-ice and land used by the Inuit for hunting. Dots show the locations of the main settlements; the communities referred to in the text are shown in larger type.

Figure 4.8 (Opposite) Food web for Arctic waters (not to scale).

Note that not all species and links are shown.

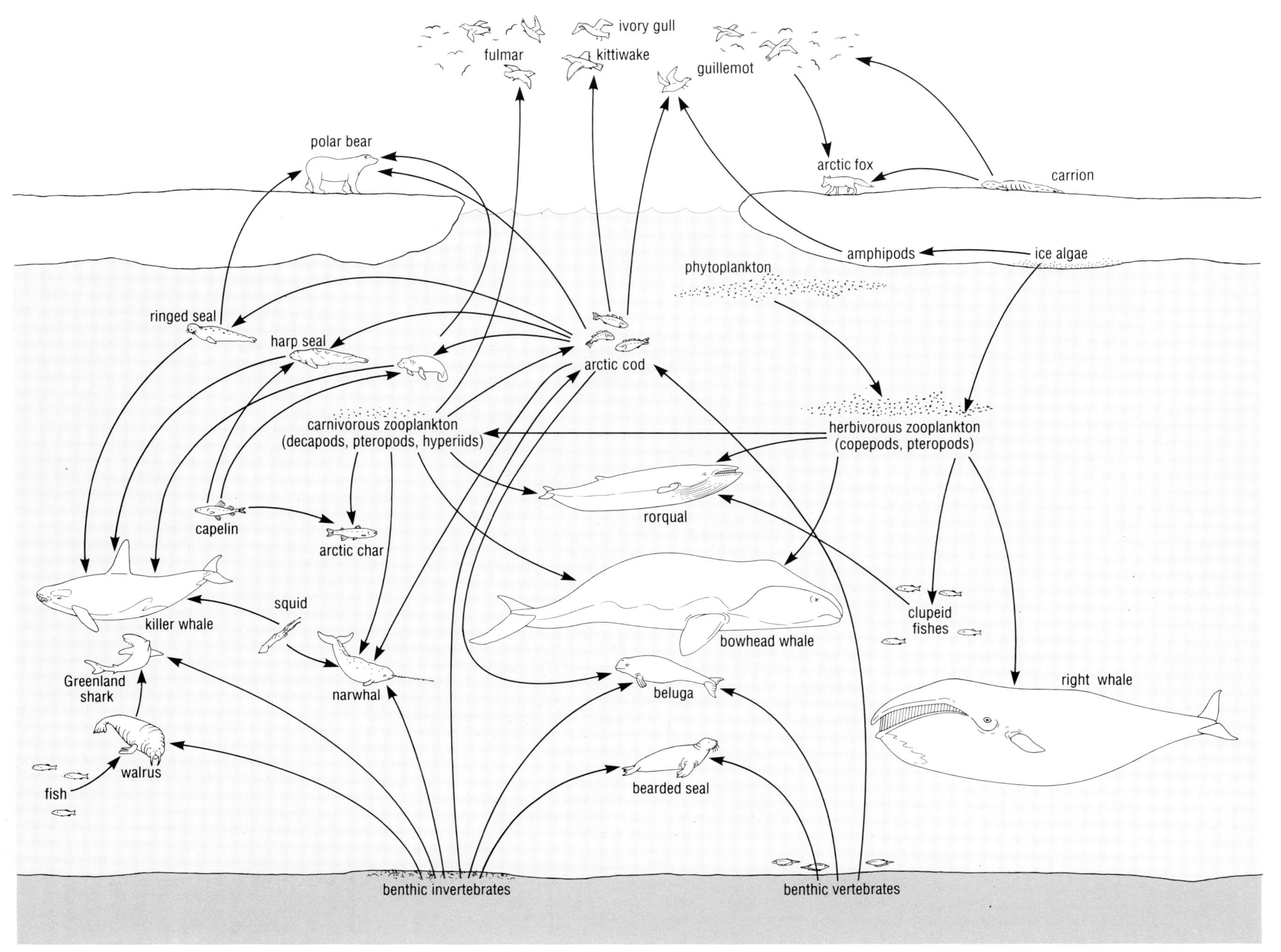
ivory gull
fulmar
kittiwake
guillemot
polar bear
arctic fox
carrion
amphipods
ice algae
phytoplankton
ringed seal
harp seal
arctic cod
carnivorous zooplankton
(decapods, pteropods, hyperiids)
herbivorous zooplankton
(copepods, pteropods)
rorqual
capelin
arctic char
squid
killer whale
clupeid
fishes
bowhead whale
Greenland
shark
narwhal
beluga
right whale
walrus
fish
bearded seal
benthic invertebrates
benthic vertebrates

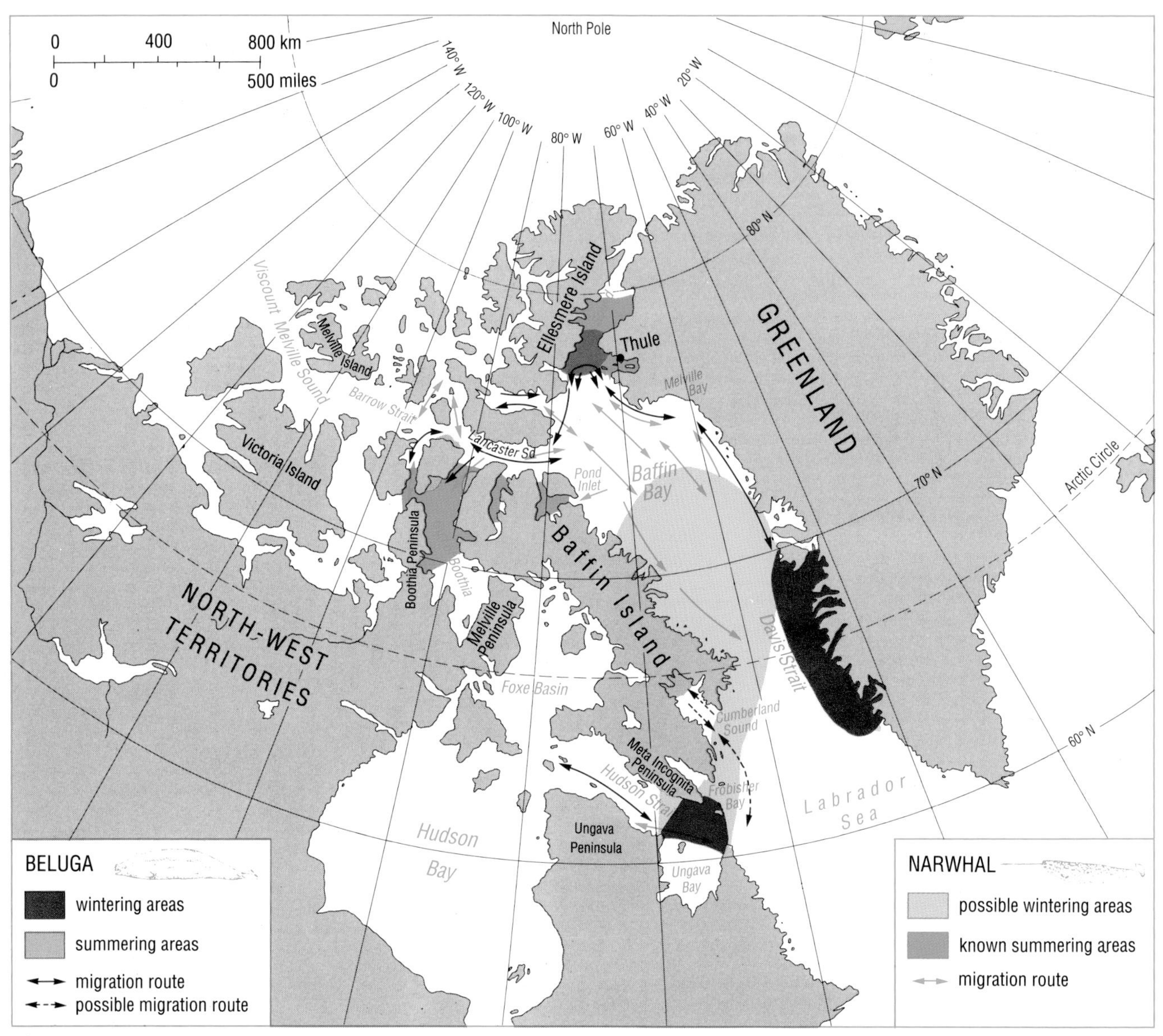

(a)

Figure 4.9 Examples of migration routes of marine mammals into the eastern Arctic.
(a) Beluga (white) whale and narwhal. Toned areas represent regions where the mammals are known or suspected to spend the winter, and the darker areas represent regions where they congregate in summer.

(Figure 4.9). Such large concentrations of animals, often brought together from long distances, have proved so convenient for commercial exploitation that some populations have become severely depleted. As far as subsistence exploitation by indigenous people is concerned, high seasonal abundances of certain food species is a bonus; for year-round survival, however, the Inuit must depend on non-migratory species like the ringed seal (*cf.* Table 4.1).

The Inuit do not live entirely on marine mammals. They also hunt terrestrial mammals such as caribou and musk-ox, as well as migratory birds and their eggs. Fish are also an important part of their diet, particularly mature Arctic char which every summer migrate from freshwater lakes and rivers to feed in the sea.

By reference to Figure 4.8, what further animal is essential to the Inuit's food supply (although in fact they rarely consume it directly)?

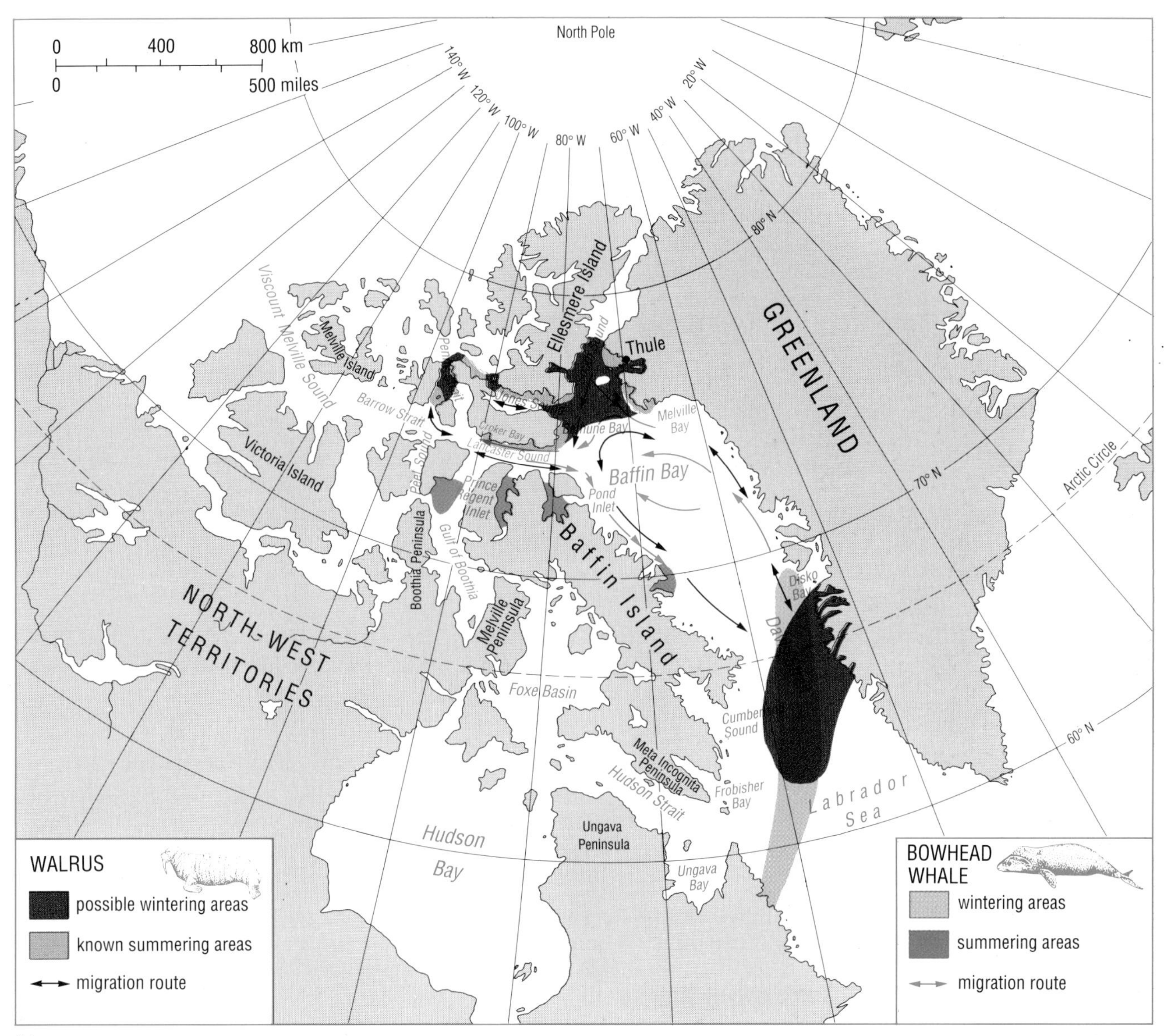

(b)

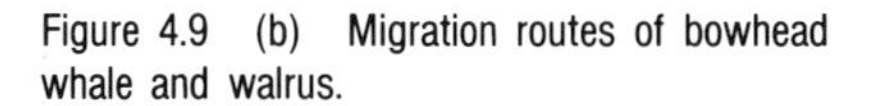
Figure 4.9 (b) Migration routes of bowhead whale and walrus.

The Arctic cod. As indicated by Figure 4.8, this is an important food source for a wide variety of birds and mammals, many of which are exploited by the Inuit. Its importance in the Arctic food web results partly from the fact that it is adapted to live and feed in a wide range of habitats: at the surface and under sea-ice; at the edge of land-fast ice; in shallow coastal waters in summer; and at depths of more than 900 m. Arctic cod are distributed throughout the Arctic, and are extremely abundant. During the long winter they are widely dispersed, but in summer they congregate into enormous schools—all the more convenient for the predatory seals, as well as the summer-visiting whales: belugas, narwhals and bowheads (*cf.* Figure 4.9).

For many Inuit, the dependence on marine resources is declining, as they come under increasing pressure to adopt a wage economy. This is particularly true for the Inuit living on the borders of the Beaufort Sea, for reasons which will become clear in Section 4.3.

4.2 THE ARCTIC SEAS—WHO OWNS WHAT?

In this Section, we look at the various arguments that have been put forward concerning how the Arctic seas might be divided up between the countries which surround them. For convenience, we will concentrate particularly on the legal status of the waters to the north of Canada.

Looking at a map of the Arctic region (e.g. Figure 4.1 or 4.10), you might assume that the islands to the north of Canada are Canadian, and that, as a result, the waters around them are also Canadian: a typical inhabitant of Canada might well make the same assumption.

The question is: to what extent is this view supported by international law?

There are a number of separate, but interrelated philosophies that can be employed to support Canadian sovereignty over the Arctic islands and a large area of the Arctic seas. The first is based around what has come to be known as 'the sector theory'.

4.2.1 THE SECTOR THEORY

According to the **sector theory**, the polar regions should be divided up by lines of longitude and the sizes of sectors allotted to individual states should be determined by the frontiers of land territories bordering the area.

The idea of dividing up the globe by means of lines of longitude dates back to at least the 15th century (*cf.* Section 2.1). Lines of longitude do form convenient geographical boundaries (on paper, anyway) and have frequently been used in international treaties and legal and administrative documents.

The first use of lines of longitude to define *sectors* seems to have been made in 1878, in a joint address to the British Parliament by the House of Commons and the Senate of Canada. The British Government wished to clarify the extent of the territory (originally granted to the Hudson's Bay Company) that it had transferred to Canada in 1870 and, at the same time, to make an additional transfer to Canada of the Arctic islands to the north. The geographical limits of the lands and territories to be transferred were described as follows:

> '. . . on the East by the Atlantic Ocean, which boundary shall extend to the North by Davis Straits, Baffin's Bay, Smith's Straits and Kennedy Channel, including all the islands in and adjacent thereto, which belong to Great Britain by right of discovery or otherwise; on the North the Boundary shall be so extended as to include the entire continent to the Arctic Ocean, and all the islands in the same *westward to the one hundred and forty-first meridian west of Greenwich*; and on the North-West by the United States territory of Alaska.'

Figure 4.10 (Opposite) The Arctic islands to the north of Canada. The boundaries referred to in the 1878 address to the British Parliament, concerning transfer of territory to Canada, are shown in red.

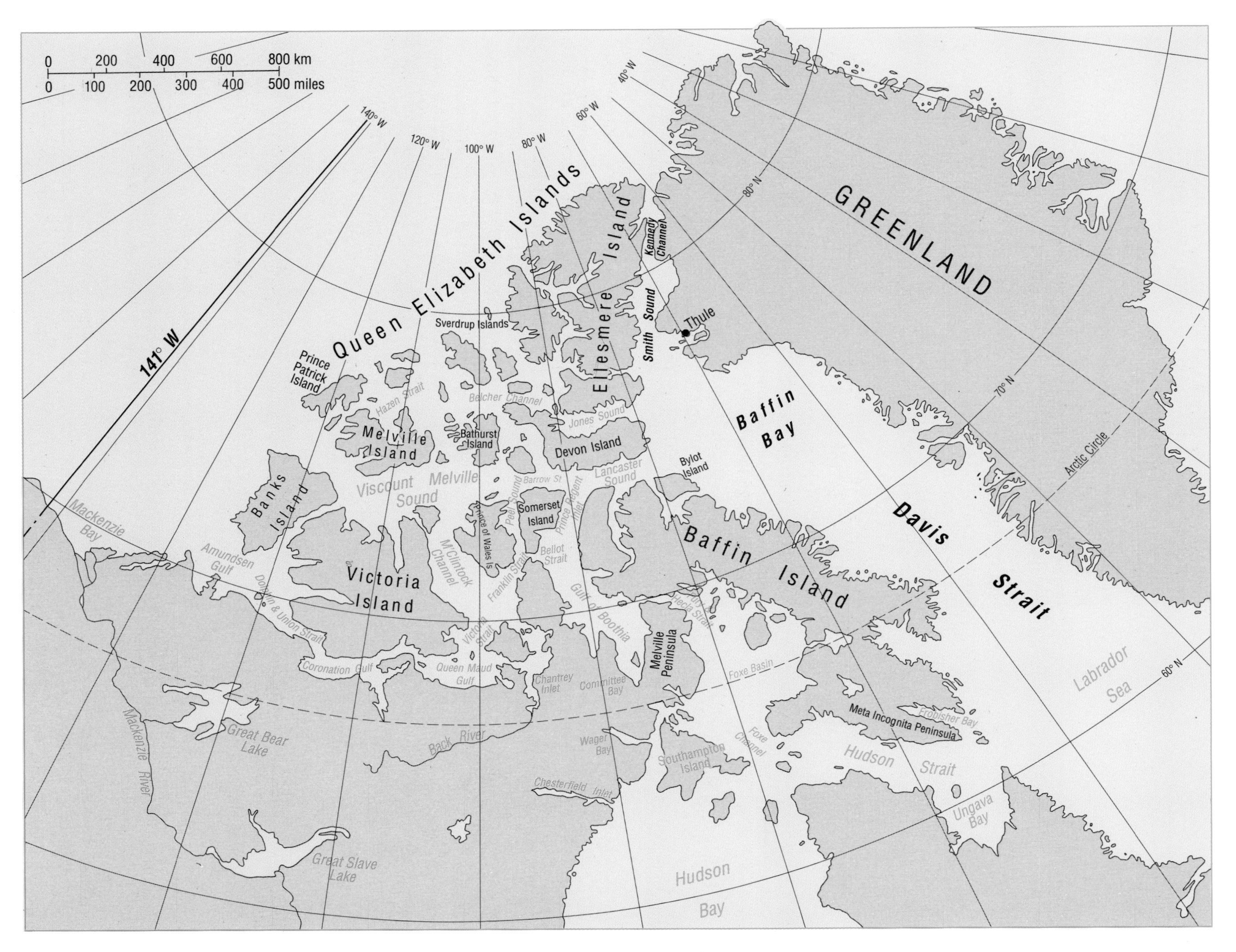
0 200 400 600 800 km
0 100 200 300 400 500 miles
141° W
140° W
120° W
100° W
80° W
60° W
40° W
80° N
70° N
60° N
Arctic Circle
GREENLAND
Queen Elizabeth Islands
Ellesmere Island
Kennedy Channel
Smith Sound
Thule
Sverdrup Islands
Prince Patrick Island
Hazen Strait
Belcher Channel
Jones Sound
Melville Island
Bathurst Island
Devon Island
Baffin Bay
Bylot Island
Lancaster Sound
Barrow St
Viscount Melville Sound
Banks Island
Peel Sound
Somerset Island
Prince Regent Inlet
Bellot Strait
Prince of Wales Is
M'Clintock Channel
Franklin Strait
Davis Strait
Baffin Island
Mackenzie Bay
Amundsen Gulf
Dolphin & Union Strait
Victoria Island
Gulf of Boothia
Fury & Hecla Strait
Melville Peninsula
Victoria Strait
Coronation Gulf
Queen Maud Gulf
Chantrey Inlet
Committee Bay
Foxe Basin
Labrador Sea
Meta Incognita Peninsula
Frobisher Bay
Great Bear Lake
Mackenzie River
Back River
Wager Bay
Foxe Channel
Southampton Island
Hudson Strait
Chesterfield Inlet
Ungava Bay
Great Slave Lake
Hudson Bay

These boundaries are shown in Figure 4.10. Note that the territory to be transferred is confined to *land*; it is not the whole sector to the east of the 141° W meridian. In the 1870s, the idea of maritime territory was effectively disregarded.

Note also that the *northern* limit of the transferred territory was left undefined. When the transfer of territories was actually made in 1880, Britain did not wish to risk purporting to transfer more lands than it had title to, and as a result the situation regarding the Arctic islands became even less clear. The Canadian Government therefore began to take various steps to consolidate Canada's claim to them, and in 1895 the unorganized North-West Territories were split into four provisional Districts of which the northernmost Franklin District was described as extending to 83.25° N. In 1897, Franklin District was redefined to include 'all lands and islands' *contained in a sector constituted by the 141st and 60th meridians*. In other words, Canadian territory was envisaged as extending to the North Pole, although the official map showing the Districts themselves did not extend quite that far north (Figure 4.11(a)).

However, in 1904 the Canadian Department of the Interior published a map which *did* show the boundaries of Canadian territory as the 141st and 60th meridians extending all the way to the North Pole. The map concerned was entitled 'Explorations in Northern Canada and Adjacent Portions of Greenland and Alaska' (Figure 4.11(b)), and demonstrated the accepted link between active exploration and national presence, and consolidation of legal title to territory. The year before the map was published, A.P. Low, of the Geological Survey of Canada, had been sent to patrol the waters of Hudson Bay and those around the Arctic islands, visiting numerous localities throughout the region and demonstrating Canada's sovereignty. At the time, this sovereignty was in fact being contested by Norway, on the basis of explorations by Sverdrup, who had completed his second Arctic expedition in 1902, and Amundsen, who had begun his three-year journey through the North-West Passage.

During 1906–7, Captain J.E. Bernier in his vessel *Arctic* made numerous landings on the Arctic islands and took formal possession for Canada at a number of places. His activities appear to have inspired Senator Pascal Poirier to propose the following resolution to the Canadian Senate in 1907: 'That it be resolved that the Senate is of the opinion that the time has come for Canada to make a formal declaration of the possession of the *lands and islands* situated in the north of the Dominion, and extending to the North Pole.' He quoted from the Charter of the Hudson's Bay Company, and referred to old maps which showed island territories discovered and taken possession of by English sailors, up to and beyond the 82nd parallel. He then went on to expound the sector

Figure 4.11 (a) 'Map shewing the Provisional Districts of Canada', published by the Canadian Department of the Interior in 1897. Franklin District, the most northerly District of the North-West Territories, is envisaged as extending off the top of the map. The establishment of the Districts of the North-West Territories was just one episode in the complicated territorial evolution of Canada.

(b) The map 'Explorations in Northern Canada and Adjacent Portions of Greenland and Alaska', published by the Canadian Department of the Interior in 1904. Virtually all maps published since then by the Department of the Interior and later by the Department of Energy, Mines and Resources have shown boundary lines extending to the North Pole. Since 1952, such boundary lines have commonly been shown on maps and charts of Canada.

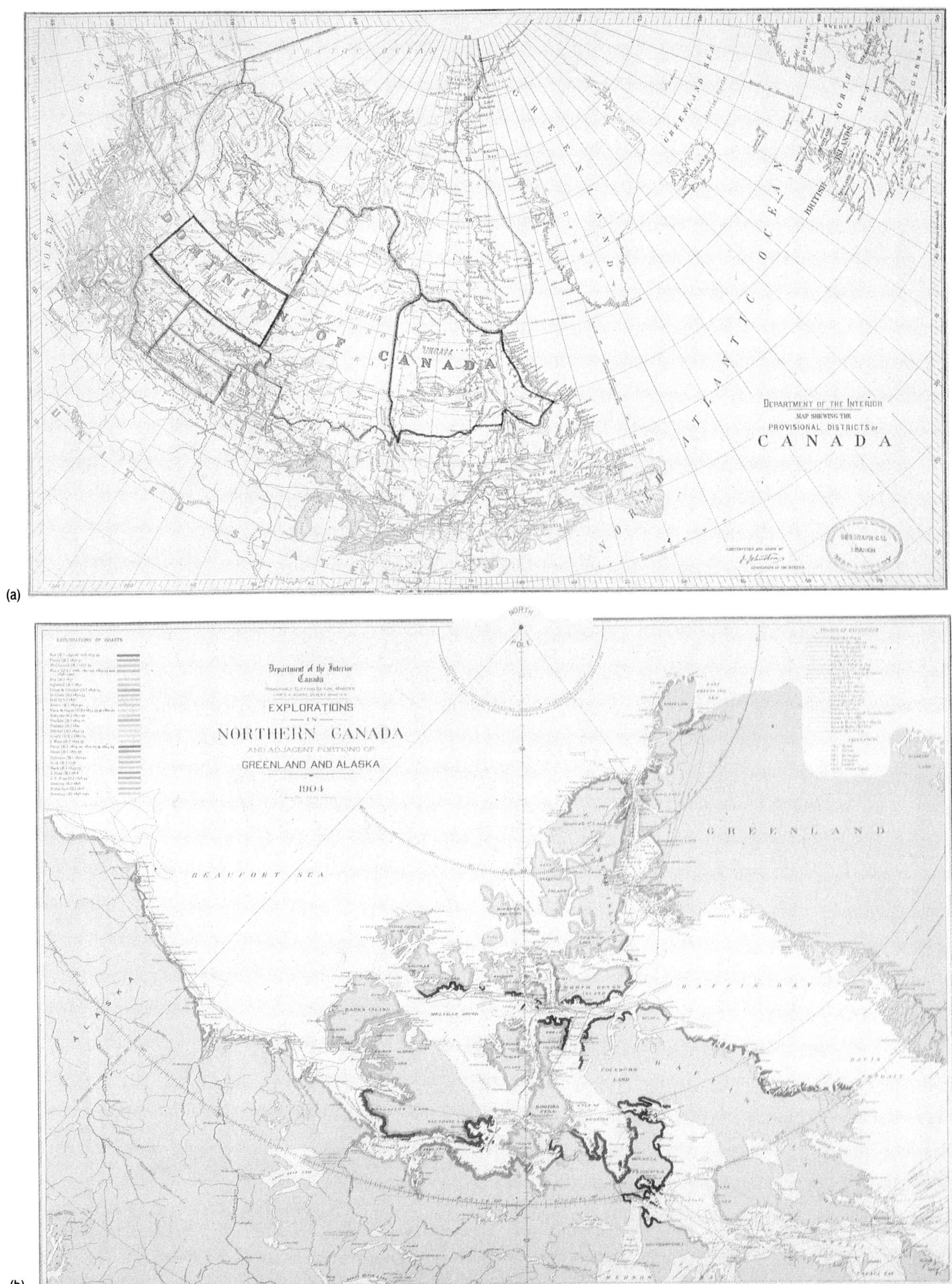
ARCTIC OCEAN
NORTH PACIFIC OCEAN
DOMINION OF CANADA
KEEWATIN
UNGAVA
UNITED STATES
GREENLAND
GREENLAND SEA
NORTH SEA
BRITISH ISLANDS
NORTH ATLANTIC OCEAN
DEPARTMENT OF THE INTERIOR
MAP SHEWING THE
PROVISIONAL DISTRICTS OF
CANADA
NORTH POLE
EXPLORATIONS OF COASTS
Department of the Interior
Canada
EXPLORATIONS
IN
NORTHERN CANADA
AND ADJACENT PORTIONS OF
GREENLAND AND ALASKA
1904
GREENLAND
BEAUFORT SEA
ALASKA
BAFFIN BAY
BANKS ISLAND
MELVILLE SOUND
DAVIS STRAIT

(a)

(b)

theory whereby: 'All the lands between the two lines up to the North Pole should belong and do belong to the country whose territory abuts up there'; he continued by applying the theory, allowing a sector each to Canada, Russia, the United States, Norway and Sweden.

However, Senator Poirier's motion was not seconded, and the Honourable Sir Richard Cartwright, the Canadian Government's representative in the Senate, felt it would be impolitic to push this line of argument. He felt that Canada had 'very reasonably good ground to regard Hudson Bay as *mare clausum* and as belonging to it, that everything there may be considered as pertaining thereto', but reserved his opinion regarding extension of Canadian territory to the North Pole, saying: 'I am not aware that there have been any original discoverers as yet who can assert a claim to the North Pole, and I do not know that it would be of any great practical advantage to us, or to any country, to assert jurisdiction quite as far north as that'.

Nevertheless, in 1909 (the year that the American Robert Peary reached the North Pole) Bernier, leading his second expedition to the Arctic, took formal possession of the whole 'Arctic Archipelago' by unveiling a bronze plaque at Parry's Rock on Melville Island. The inscription read:

> This Memorial is erected today to commemorate the taking possession for the DOMINION OF CANADA of the whole ARCTIC ARCHIPELAGO lying to the north of America from long. 60° W to 141° W up to latitude 90° N. Winter har. Melvile Island. CGS Arctic. July 1st 1909. J.E. Bernier. Commander.

Thus, although presumably restricted to land areas, Bernier's claim for Canada encompassed the whole sector referred to by Poirier in his speech, and shown on the map of 1904.

This brief discussion of the development of the sector theory as it pertains to the Arctic illustrates the three separate threads which run through the arguments made in support of the theory. The first is that lines of longitude form legitimate boundaries, particularly when the lines of longitude in question have already been used in international legal transactions. By itself, this is a fairly weak argument for the sector theory, and is generally only used along with one or both of the other two lines of argument. These depend on the concepts of *contiguity* and *customary law*.

Contiguity as a basis for the sector theory

As Senator Poirier outlined in his speech to the Canadian Senate, English sailors had discovered various Arctic islands and claimed them for the English Crown. In the 16th and 17th centuries, this 'acquisition by discovery' was the normal means whereby maritime powers obtained initial title to new territories. However, in time, the Roman Law principle for the acquisition of property was adapted for the acquisition of territory in international law. According to this principle, legal possession of territory results from a clear intent to occupy the territory *combined with* the actual occupation of a well-defined region. As actual occupation of, or display of sovereignty over, an entire territory was in many cases difficult or even impossible, the doctrine of **contiguity** or geographic proximity was developed. According to this doctrine, *the effective occupation of part of a region or territory gave title to the whole of the unoccupied region proximate enough to be considered a single geographic unit with the occupied portion.* When applied to the polar regions, this doctrine necessarily results in sector-shaped claims.

The issue of whether contiguity can act as a legal basis for territorial sovereignty was aired thoroughly in 1928 in the *Island of Palmas Case*. This concerned an isolated inhabited island, approximately 45 miles from the Philippine Archipelago, which was claimed by the United States, on the basis of contiguity to the Philippines, but which had been administered by The Netherlands. The arbitrator awarded the island to The Netherlands on the basis of its manifestations of sovereignty even though, given the remoteness of the island, these were, of necessity, limited. While admitting that contiguity could have a limited application at the time of taking possession because this 'can hardly extend to every portion of territory', the arbitrator clearly indicated that sovereignty could only be *maintained* through 'the display of sovereignty as a continuous and prolonged manifestation which must make itself felt through the whole territory'. In his ruling, the arbitrator specifically stated that contiguity *per se* cannot give title to an island outside territorial waters.

In 1933, the relationship between contiguity and sovereignty was again discussed at length during the course of the *Eastern Greenland Case*, between Denmark and Norway. Norway alleged that Denmark was relying on the doctrine of contiguity to avoid having to show effective occupation of the eastern part of Greenland. Denmark vigorously denied this, and during the argument both states denounced contiguity as a possible legal basis of territorial sovereignty. In the event, the International Court's judgement was in favour of Denmark, largely because the Danish claim was held to have been uncontested for 550 years (despite the fact that for several hundred years there had been no contact between Denmark and Greenland). No mention was made in the ruling of contiguity. Significantly for issues of sovereignty in the Arctic, the Court considered the limited show of Danish authority in the uncolonized part of Greenland to be acceptable, given the harsh and inaccessible nature of the terrain.

In international disputes, a legal ruling has only once been based around contiguity. This occurred in 1870, when the island of Bulama, adjacent to the mainland of Portuguese West Africa (now Angola), was awarded to Portugal because it had been discovered by the Portuguese and was situated within Portuguese territorial waters. This situation, of an island in the territorial sea of the claimant state, and that of an archipelago or closely related group of islands, are in fact regarded as the two most acceptable applications of the doctrine of contiguity.

QUESTION 4.1 (a) Look at Figure 4.10. To what extent can contiguity be used to justify Canadian territorial sovereignty over the islands between the north of Canada and the North Pole?

(b) To what extent can this argument be extended to include the waters between and around these islands?

(c) What implications does this have for contiguity as a justification for the application of the sector theory in the Arctic?

The sector theory and customary law

The third possible justification for the sector theory would be if it were accepted as part of *customary law* as evidenced through state practice. So, how do the states involved actually *behave* as far as the sector theory is concerned?

Figure 4.12 (a) Delimitations of territory claimed in the Antarctic, and locations of research stations. Note that Norway's territorial claim is not sector-shaped, and there are no claims by either the US or USSR.

The sector theory was devised for use in the Arctic but has also been applied in southern polar regions. Figure 4.12(a) shows the sectors of the Antarctic region claimed by various states; Figure 4.12(b), for comparison, shows the Arctic sector lines proposed by the Soviet jurist W.L. Lakhtine in 1929.

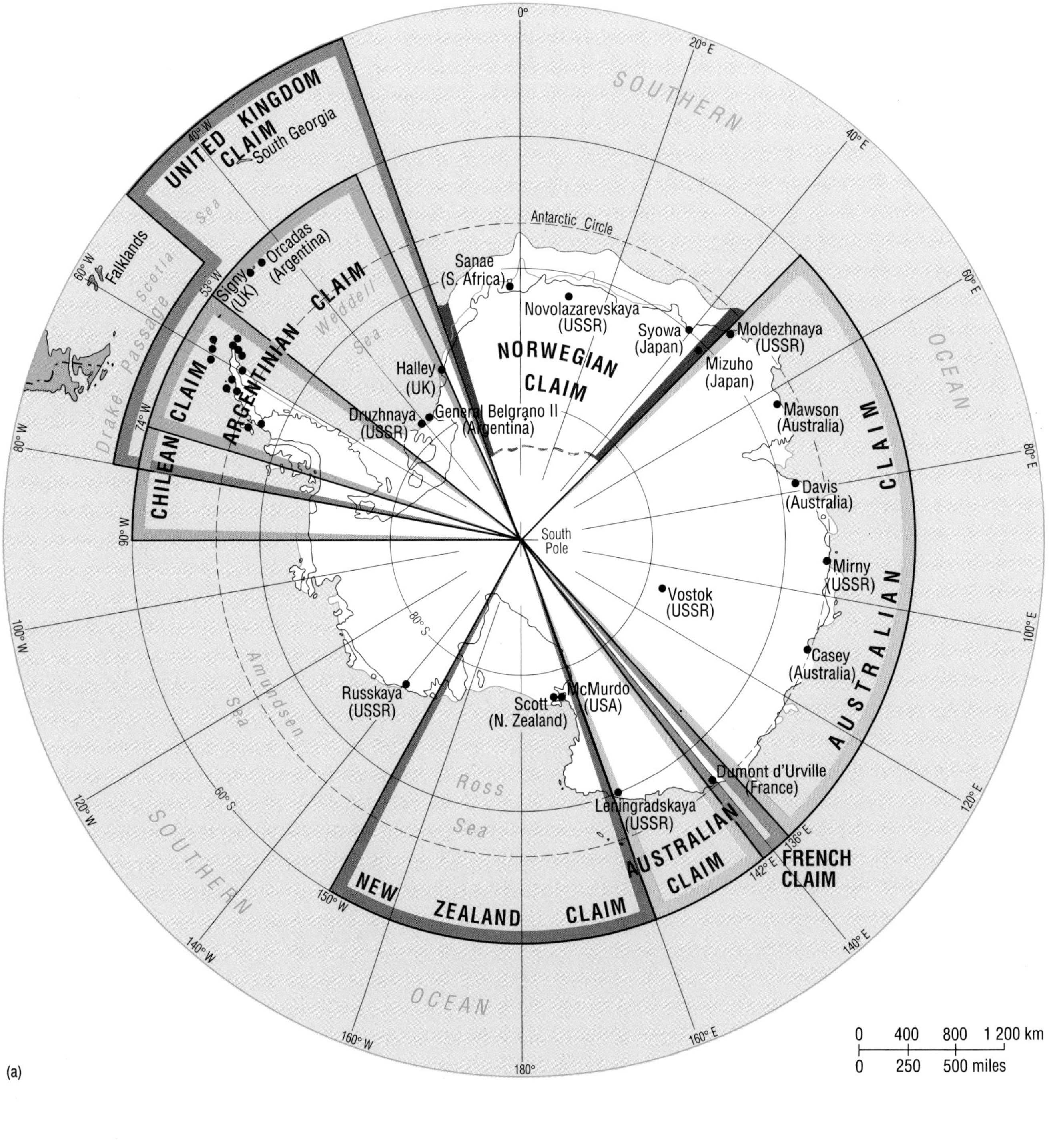

(a)

QUESTION 4.2 Study Figure 4.12(a) and (b). How do the situations in the Arctic and Antarctic compare with one another?

Despite the differences between the two situations, the Antarctic is clearly the only part of the globe comparable with the Arctic, so state practice here contributes to customary law concerning the sector theory. In summary, the situation is this. Apart from those by Argentina and Chile, Antarctic claims are based on discovery (although sectors claimed include substantial prior claims by nationals of other countries). Chile and the United Kingdom claimed their sectors on the basis of contiguity *combined with* effective occupation. Most of the other states have defined

Figure 4.12 (b) The Arctic sector lines proposed by the Soviet jurist W. L. Lakhtine in 1929.

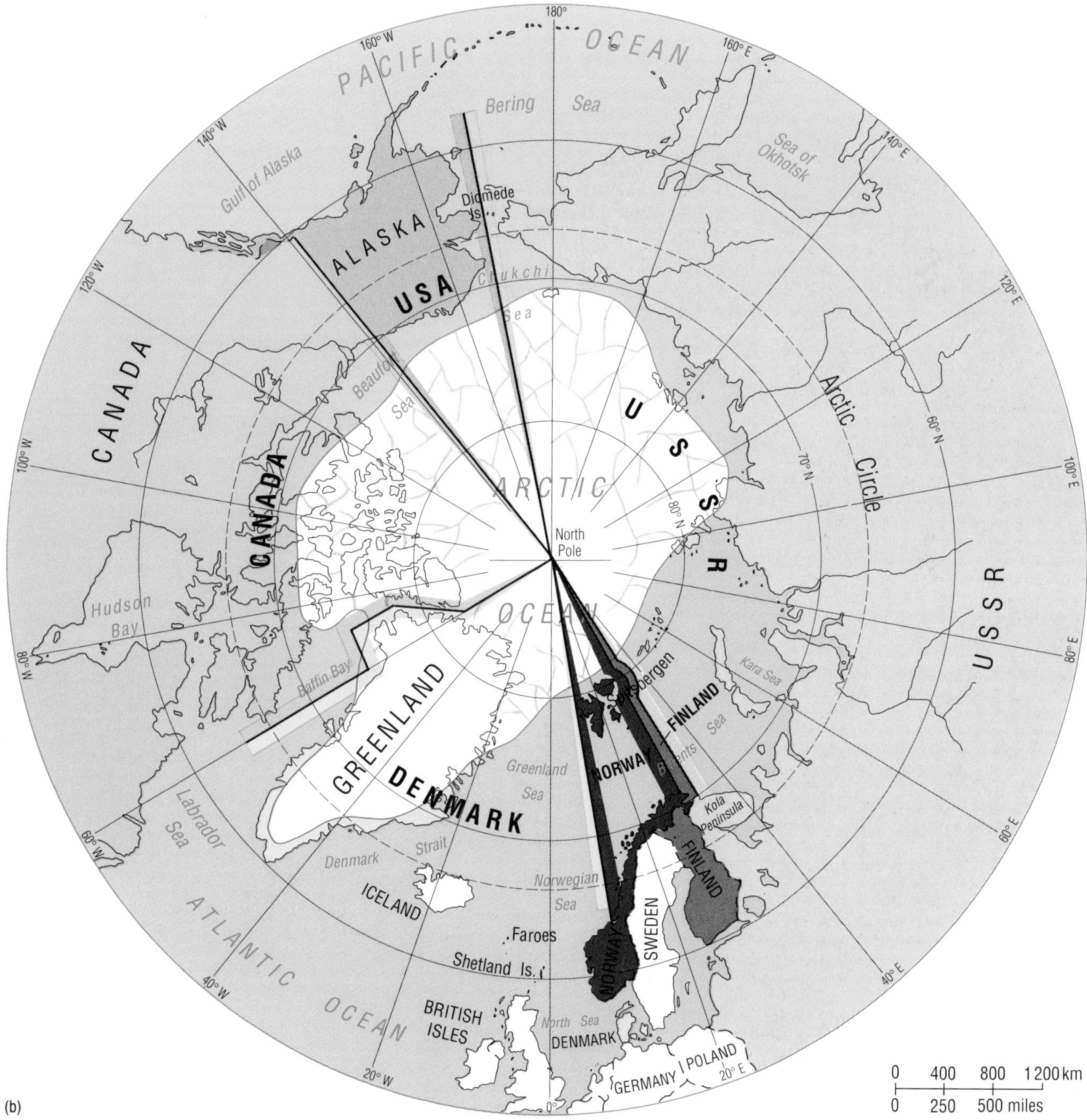

(b)

their boundaries by lines of longitude, and have not necessarily invoked the sector theory as such, although as their claims are related to their occupation of the Antarctic coast, or of territories to the north, they are effectively relying on the idea of contiguity. The occupied coastal regions are 'populated' only by scattered groups of visiting scientists, but as we have seen, in international law, in remote and hostile regions 'effective occupation' can consist of very little. However, there is a requirement for *some* indication of control of territory, and for much of the area covered by the sectors it is difficult to see how any control could be implemented. This, of itself, is a strong argument against the sector theory—an argument which can be used almost as strongly with respect to the high Arctic.

As shown in Figure 4.12(a), Norway's claim (made largely to protect its whaling industry) is the exception. Norway strongly opposes the doctrine of contiguity and its application as the sector theory and so has not extended its claim to the South Pole (or given it a northern boundary).

The United States has often expressed opposition to the sector theory and neither it nor the Soviet Union has claimed sectors of Antarctica; both reject the claims of other states. So far as the Arctic is concerned, the Soviet Union adopted a formal Decree in 1926, in which it claimed all lands and islands between the meridians of 32°4′35″ E and 168°49′30″ W. The Soviet Government has never extended this claim to include all the *waters* between the two meridians, although individual Soviet jurists have done so. In fact, such a claim would be inconsistent with the continued Soviet use of drifting ice-based scientific stations (Figure 4.13). The United States, too, has operated large numbers of drifting ice stations regardless of Arctic sector boundaries.

Denmark, like Norway, avoids *relying* on the sector theory to support its Arctic territorial claim, and both countries have opposed its use by other states. For example, Norway strongly opposed Canada's claim to sovereignty over the Sverdrup Islands (see Figure 4.10) while Denmark stated that it considered Ellesmere Island, which lies to the north of

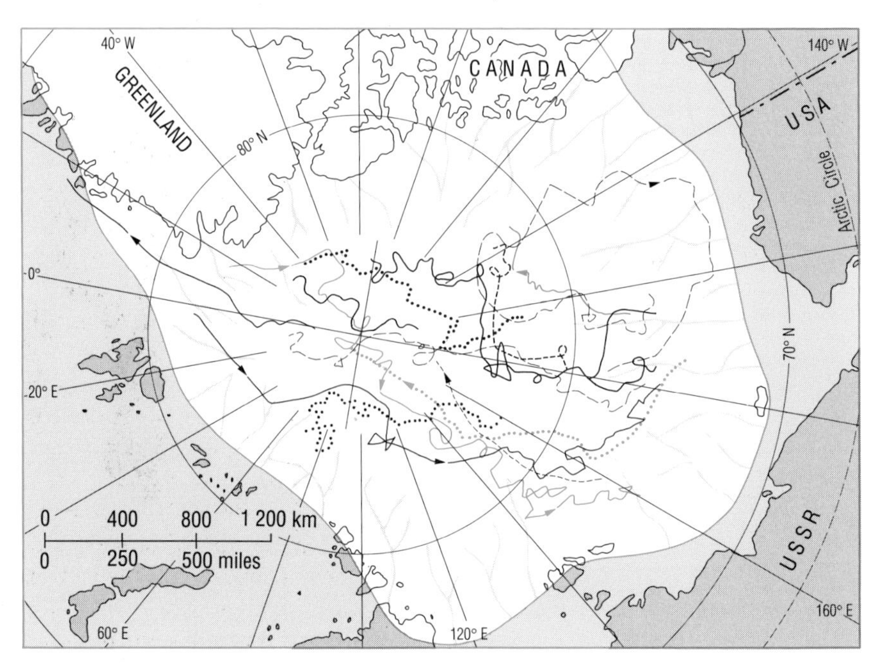

Figure 4.13 Drifting ice stations used by the USSR between 1937 and 1973. By openly operating such ice stations, both the USSR and the US effectively demonstrate that they do not recognize the sector theory.

both Canada and Greenland (which is under Danish sovereignty), as no-man's land. Although Canada itself has taken a number of official steps which indicate reliance on the sector theory, it has never adopted any law or order in Council claiming a sector; whether, and how, the sector theory has been used seems to have depended on the needs of the moment.

So, division of the Arctic into sectors cannot be justified on the basis of long-standing practice by the states involved. We have also seen that it cannot be justified by considerations of contiguity, although this does have a limited application with respect to certain remote territories and islands. Furthermore, the historic use of meridians as international boundary lines does not of itself justify the sector theory.

Most importantly in the present context, when the sector theory *has* been used to claim sovereignty over territory in the Arctic, the territory involved has always been land rather than sea.

Why is this state of affairs compatible with what would be supported by international law?

Because in claiming sea territory all the way to the North Pole, states would be claiming sovereignty over waters (albeit ice-covered) which would be sufficiently far from land to qualify as *high seas*, which cannot be appropriated by any one state (Sections 2.1 and 2.2.3).

However, under the Law of the Sea Conventions of 1958 and 1982, there are two arguments that Canada could use to claim that a large area of sea (i.e. beyond the twelve-mile territorial sea limit) around the islands to its north (*cf.* Figure 4.10)—the Canadian Arctic Archipelago—falls completely under Canadian sovereignty.
Before moving on to the next Section (and without reading ahead), try to suggest what these two arguments might be, bearing in mind what you read in Section 2.2.1.

4.2.2 THE WATERS OF THE CANADIAN ARCTIC ARCHIPELAGO

> 'Canada ... claims that the waters of the Canadian Arctic Archipelago are internal waters of Canada, on an *historical basis*, although they have not been declared as such in any treaty or by any legislation.'
>
> Statement made in 1973 by the Legal Bureau of Canada's Department of External Affairs.

The doctrine of historic waters developed out of the idea of historic bays, which had emerged during the 19th century (Section 2.2.1). Not all states accept the validity of 'historic' claims, and the definition of historic waters has never been spelled out in any convention. The Convention on the Territorial Sea (1958) recognized their existence, but only by referring to *exceptions* to the general rule: first, bays for which the maximum 24-mile closing line rule was inappropriate; and secondly, territorial waters which came under the jurisdiction of a particular state solely 'by reason of historic title', so that the boundary between these territorial waters and those of another state would need to be drawn in a way that was at variance with the normal rules. The Third Law of the Sea Conference did not discuss the question of historic waters, and so the 1982 Convention simply reproduced the 1958 provisions. Note, however, that of the two exceptions mentioned above, one equates historic waters with *internal waters*, while the other indicates that historic waters can be *territorial seas*.

QUESTION 4.3 (a) What is Canada's largest and best known 'historic bay'? Did Sir Richard Cartwright regard this bay as being Canadian internal waters?

(b) What significant difference is there, for foreign states, between the legal status of internal waters and that of territorial seas?

Despite the inconsistency highlighted above, historic waters are generally equated with internal waters. This view was endorsed by the International Court of Justice in 1951, in what is known as the *Anglo-Norwegian Fisheries Case*; see later. Because historic waters are the exceptional cases rather than those covered by international law, the burden of proof of historic title rests with the coastal state. In order for the coastal state to claim sovereignty in this manner, it must fulfil the following requirements:

1 Exercise of exclusive authority and control over the maritime area claimed, including the expulsion of foreign ships if necessary.

2 Usage over a long period of time, the actual length of which depends on the circumstances.

3 Acquiescence by foreign states, particularly those clearly affected by the claim.

To aid its case for sovereignty, the coastal state can argue that jurisdiction over the waters in question is in its 'vital interest', for its economic survival, its national security, or both. The Soviet Union used this argument in 1957 when it claimed the Bay of Peter the Great on the Sea of Japan as an historic bay.

So, how strong *is* Canada's claim that the waters of the Canadian Arctic Archipelago are Canadian historic waters?

So far as long usage (requirement 2) is concerned, it is true that before 1880 the waters of the Archipelago had been almost completely the preserve of British whalers and explorers, as well as the Inuit, the indigenous population. After 1880, most of the waters were patrolled by Canadian ships. However, we have already seen that both the British and their 'heirs'—the Canadians—claimed *sovereignty* only over *land*.

Canada has demonstrated a fair degree of 'authority and control' (requirement 1), given the area involved and the environmental conditions. In 1906, legislation was adopted requiring whalers to obtain a licence in order to hunt in Hudson Bay or the territorial waters north of the 50th parallel; this was enforced until whaling came to an end around 1915. In 1922, the Eastern Arctic Patrol was established (this relied heavily on the Canadian Mounties), and after the Second World War, the Canadian Coastguard was set up to provide ice-breaking services and supply Arctic communities with provisions. In 1977, a system was instituted whereby all ships intending to enter the waters of the Archipelago were requested to report to the Canadian Coastguard (the 'NORDREG' system).

For various reasons, these administrative activities do not convincingly indicate Canadian sovereignty over the waters of the Archipelago. For example, the whaling licences specifically referred to '*territorial* waters' north of the 50th parallel, and at the time these extended three miles from the Canadian coast. Furthermore, the NORDREG reporting system is voluntary which certainly would not be expected if they were internal waters; it would, in any case, be impossible to enforce. Most important of all, however, is the fact that the Canadian Government itself seems to have been confused as to whether the waters of the Archipelago were internal or territorial waters.

Foreign states have not always acquiesced (requirement 3) to Canadian sovereignty over the waters of the Archipelago. This is particularly true of the United States which, in 1970, made a formal protest when Canada extended its territorial sea to twelve miles and, at the same time, adopted the Arctic Waters Pollution Prevention Act. This Act enabled Canada to enforce certain pollution prevention measures both within the Archipelago and in a 100-mile strip to the north of the mainland and around the Archipelago (*cf.* Figure 4.15). Canada's reaction to the protest was, while accepting the jurisdiction of the International Court of Justice in issues of maritime law, to assert its own right to control pollution in the areas covered by the Act (referring to them as the 'marine areas adjacent to the coast of Canada'). If Canada had regarded the archipelagic waters as being clearly *internal* waters, over which Canadian sovereignty was as complete as that over Canadian land territory, it would not have seen the need to defend its right to adopt pollution legislation there, and the reservation as to the scope of the International Court of Justice would only need to have been applied to the 100-mile strip around the coast and Archipelago. In other words, Canada was itself uncertain about the precise legal status of the archipelagic waters.

For the reasons given above, Canada would not be able to make a good case for the waters of the Arctic Archipelago being *historic* internal waters. However, the role of historic waters has diminished as international law has developed. We have already seen that the 24-mile closing line rule has enabled most bay waters to be classified as internal without the coastal state having to resort to the doctrine of historic waters.

QUESTION 4.4 What other legal concept (again involving baselines) has reduced the role of historic usage, so far as coastal waters are concerned?

This baseline system (Figure 4.14) was developed by Norway at the beginning of the 19th century. During the *Anglo-Norwegian Fisheries Case* (1951), mentioned earlier, the International Court of Justice took the view that because of the sinuous and indented nature of the Norwegian coast, with its fjords and a thousand or so islands and islets, it was impossible in practice to determine the baselines by the low-water mark method (*cf.* Section 2.2.1). The Court therefore sanctioned Norway's use of straight baselines, stating that 'what really constitutes the Norwegian coastline is the outer line of the sjaergaard', the sjaergaard being the collective name for the numerous coastal islands and islets.

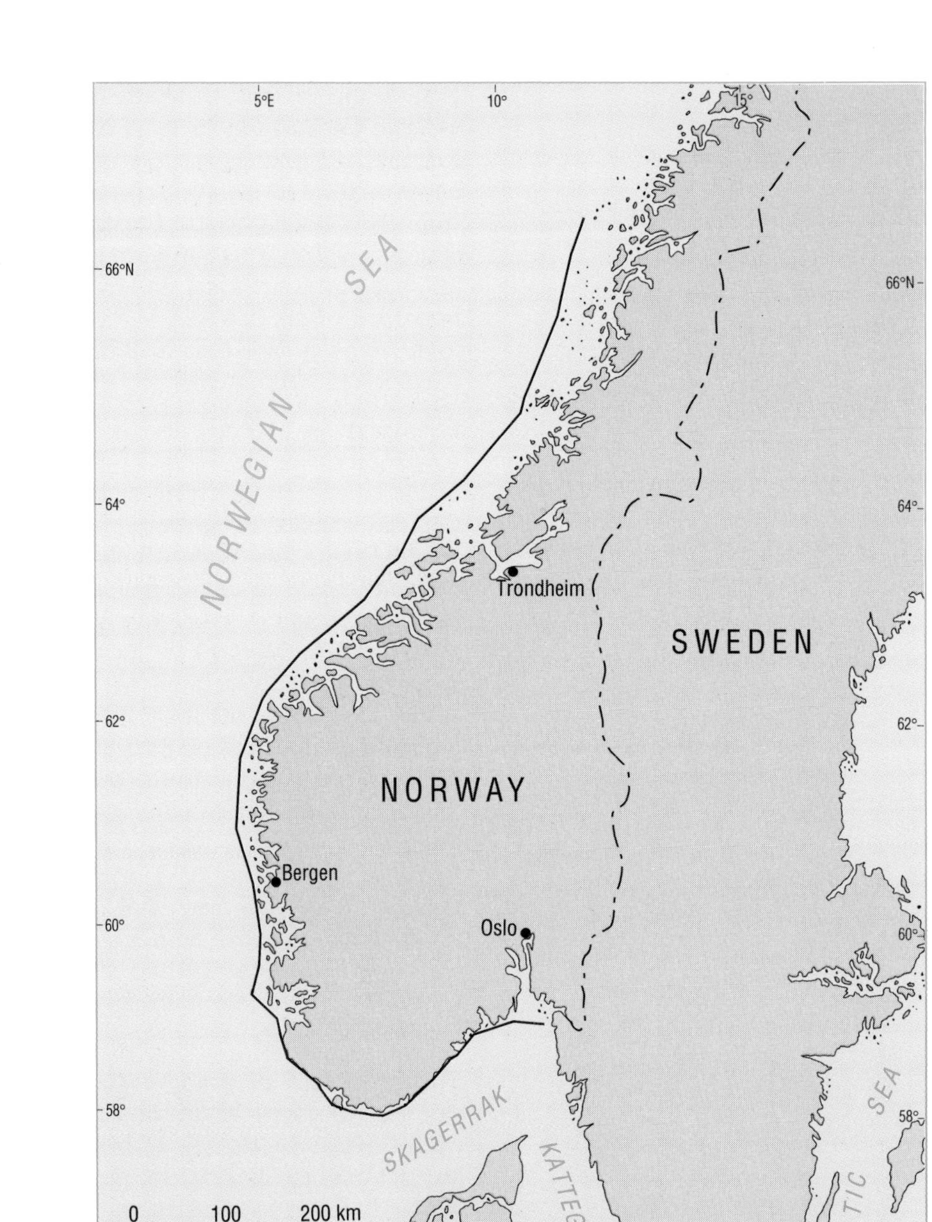

Figure 4.14 The straight baselines of south-western Norway, sanctioned by the International Court of Justice in the *Anglo-Norwegian Fisheries Case*.

Acceptance of straight baselines for deeply indented coastlines or those with a 'fringe of islands along the coast in its immediate vicinity' was later incorporated into the Convention on the Territorial Sea of 1958 and the Law of the Sea Convention of 1982. You may recall that, at the Third Law of the Sea Conference, the applicability of the straight baseline system was extended to include groups of mid-ocean islands which constitute archipelagic states (Section 3.1.2).

The Canadian Arctic islands certainly do not form a mid-ocean archipelago and do not bear much resemblance to a 'fringe of islands along the coast in its immediate vicinity' as referred to in the 1958 and 1982 Conventions. These Conventions made no provision for offshore islands which form an integral part of the territory of a continental state, but they each contain a 'let-out' clause stating that matters not covered by the Conventions continue to be governed by rules of general international law. In this case, the customary law precedent is provided by the *Anglo-Norwegian Fisheries Case*.

During the *Anglo-Norwegian Fisheries Case*, the International Court of Justice formulated three criteria or guidelines for the drawing of straight baselines, all of which were incorporated into the Conventions of 1958 and 1982 (*cf.* Sections 2.1.1 and 3.1). These criteria—of which the first two are compulsory and the third optional—may be summarized as follows:

1 Straight baselines must not depart to any appreciable extent from the general direction of the coast.

2 The enclosed sea areas must be sufficiently closely linked to the land domain to be subject to the regime of internal waters.

3 In the establishment of particular baselines, account may be taken of certain economic interests of the region concerned, the reality and importance of which are clearly evidenced by long usage.

In September 1985, Canada established a system of straight baselines for the Arctic Archipelago (Figure 4.15); these baselines came into force in January 1986.

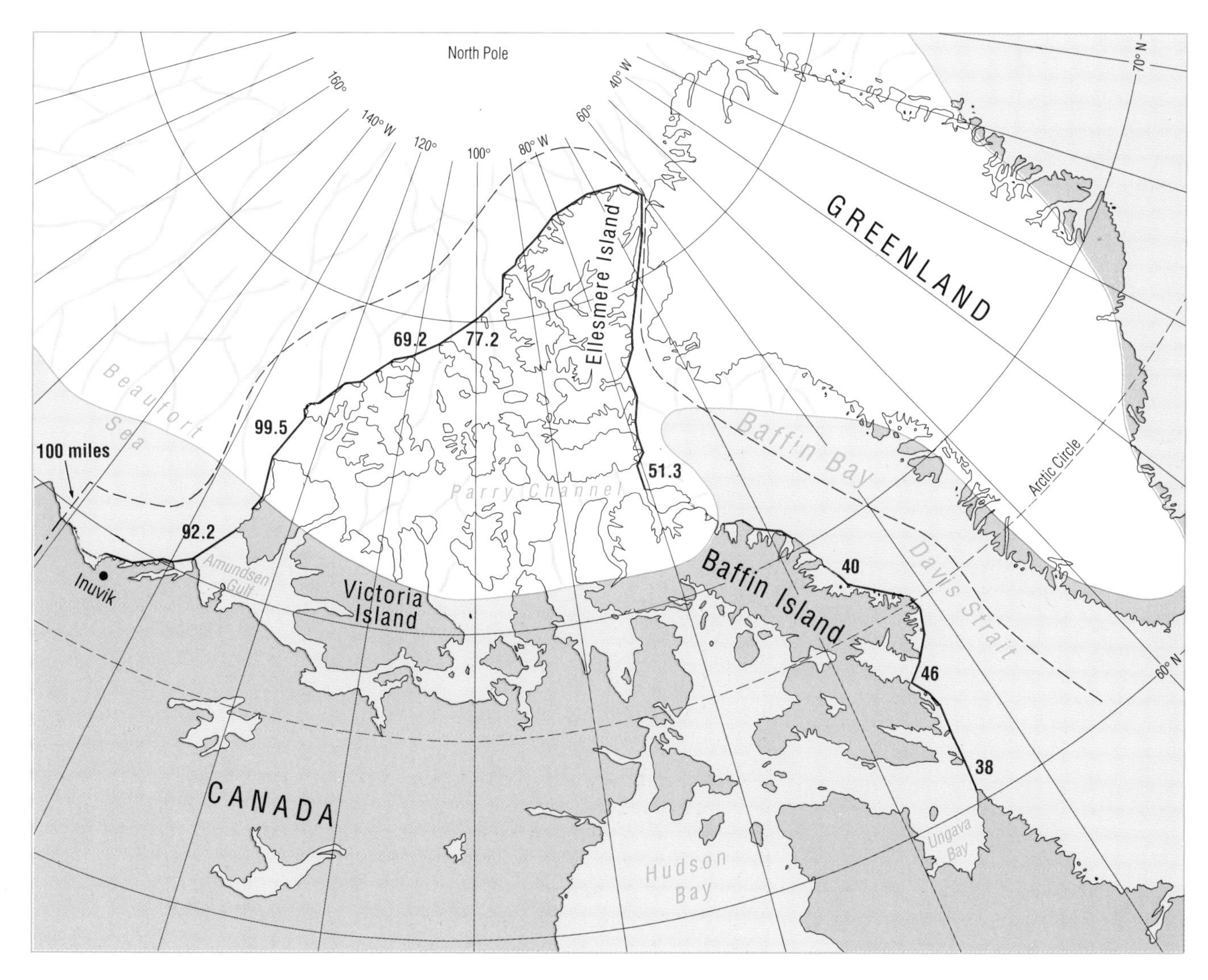

Figure 4.15 The staight baselines which Canada established around the Arctic Archipelago in 1985. The red numbers give widths across headlands, etc., in (nautical) miles. The dashed line shows the limit of the area covered by the 1970 Arctic Waters Pollution Prevention Act.

QUESTION 4.5 (a) Following the line of argument used by the International Court in the *Anglo-Norwegian Fisheries Case*, would you say that the geographical characteristics of the Canadian Arctic Archipelago justify the use of the straight baseline system?

(b) Do the baselines actually chosen: (i) comply with the compulsory criteria 1 and 2 given above; and (ii) reflect the *spirit* of the law for straight baselines?

So far as the Canadian Arctic Archipelago is concerned, there are two unusual, related, circumstances which reinforce the arguments given in the answer to Question 4.5. The first is that, for much of the year, the entire Archipelago is connected together (and hence connected to the mainland) by fast ice; even the ice in the Parry Channel is fast (*cf.* the answer to Question 4.1). This is an unusual but clear example of a strong link between land and sea (criterion 2 above). By contrast, in Baffin Bay, and even in the narrow Nares Strait, the ice is usually in motion; hence Greenland is clearly separated from Canada, as far as ice cover is concerned.

As mentioned in Section 4.1.2, the Inuit regard the land-fast ice simply as an extension of the land, to the extent that some communities live *on* the sea-ice during the winter, to be nearer to seal breathing-holes. This practice has decreased, but by means of their motorized toboggans and snowmobiles, modern Inuit hunters travel over the sea-ice even further from shore than their ancestors did. The continued protected access of the Inuit to the waters and ice between the islands of the Archipelago is essential to their economic and physical welfare, as well as to their psychological well-being. The dependence of the local Inuit on the maritime resources of the Archipelago dates back for millennia (Section 4.1), and certainly qualifies as an economic interest of long usage (criterion 3).

This long-established use of Canadian Arctic waters by the Inuit has been used as an argument for their classification as historic waters, but we have seen that, taken as a whole, the case for historic waters status is not very strong. However, the case for using straight baselines is *very* strong (*cf.* Question 4.5).

So, does it matter *how* the waters of the Canadian Arctic Archipelago come to be awarded the status of internal waters?

The answer is: yes, it does. Waters closed off by baselines primarily on historic grounds have the status of internal waters. If, however, waters gain the status of being internal waters *as a result of* the drawing of straight baselines, the situation is more complicated. Under the Conventions of 1958 and 1982, waters newly enclosed by straight baselines are internal but *subject to the right of innocent passage* if they had previously been considered part of the territorial sea or high seas (Section 2.2.1). Canada is not party to the 1958 Convention, and the 1982 Convention (which it has signed) has not yet come into force; it may therefore rely on customary law (based on the *Anglo-Norwegian Fisheries Case*) according to which such waters are internal with *no* right of innocent passage. This confusing situation will be reviewed again in Section 4.4 when we discuss the specific problems associated with the North-West Passage.

Figure 4.16 shows the positions of various maritime zones and boundaries (both actual and proposed) according to the 1978 *Polar Atlas*. It is not possible in this Case Study to discuss issues of territorial sovereignty throughout the Arctic. Suffice it to say that there are several maritime boundaries which remain unclear, including that between Norway and the Soviet Union to the east of the Svalbard Archipelago (see the striped region on Figure 4.16). Another area of uncertainty is the legal status of the sea-floor around the Archipelago itself. Under the Svalbard Treaty of 1920 (which became effective in 1924), the whole of the Archipelago, including the main island of Spitsbergen, was placed under the full and absolute sovereignty of Norway; however, the states party to the Treaty were all given equal rights to exploit the

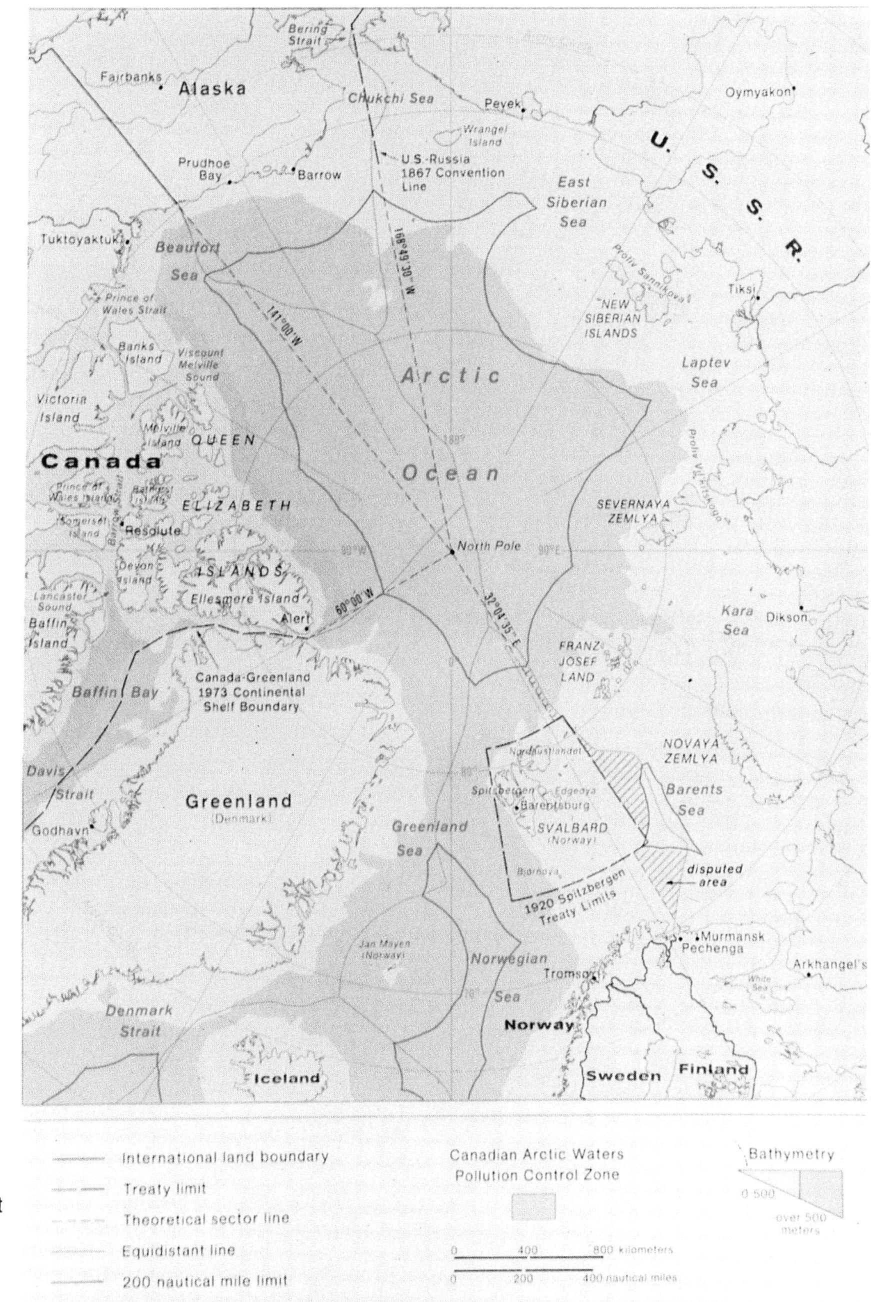

Figure 4.16 Jurisdictional lines and theoretical sector lines in the Arctic. The pink area in the region of Queen Elizabeth Islands is that covered by Canada's Arctic Waters Pollution Prevention Act (*cf.* Figure 4.15). The dashed box indicates the area covered by the Svalbard Treaty, and the 'disputed area' to the east of Svalbard is claimed by both Norway and the Soviet Union.

resources of the islands and their territorial waters. With the subsequent evolution of international law, it is now unclear whether the terms of the Treaty would extend to Svalbard's exclusive economic zone and continental shelf, or whether new arrangements would need to be made to cover the waters and sea-bed beyond the territorial sea.

4.3 HYDROCARBONS IN THE ARTIC

One of the most striking features of the Arctic Ocean, evident in Figure 4.17, is the huge proportion of its area that is continental shelf—about 70%. These shelves consist of huge thicknesses of sediment, particularly in regions which have been subsiding, and, like all continental shelves, are relatively flat. But it is important to remember that this flat topography has been imposed relatively recently.

QUESTION 4.6 When and how was this flat topography formed?

Underneath the flat topography, the sedimentary layers of the continental shelves are far from horizontal. The present Arctic region has had a long and complicated tectonic history: as the continents of North America have shifted and rotated in relation to one another, the sedimentary layers have been buckled and folded. These sediments had accumulated rapidly, burying the organic material they contained in anoxic conditions and, over time, this material matured to form oil and gas. Enormous reservoirs of hydrocarbons are now trapped amongst the folded sediments of the Arctic, both within the continental shelves and in the islands of the Canadian Arctic Archipelago (Figure 4.17).

4.3.1 PROSPECTING FOR OIL IN ICE-INFESTED WATERS

Natural oil seeps around the Arctic have been known throughout recorded history. Visiting whalers reported their existence, and the first wild-cat wells were drilled in 1902 in the Gulf of Alaska and at the western end of the Alaskan Peninsula. The most spectacular find has been onshore, at Prudhoe Bay, Alaska. This oilfield is the largest so far discovered in North America; since the late 1970s, about 1.5 million barrels of crude oil per day have been pumped along the Trans-Alaska Pipeline system to the ice-free tanker port of Valdez, in the Gulf of Alaska.

However, as shown in Figure 4.17, the greater proportion of Arctic hydrocarbon deposits occur offshore, many of them in regions which are ice-covered for all or much of the year (*cf.* Figure 4.5).

By comparison of Figures 4.5 and 4.17, which offshore hydrocarbon fields are the most accessible?

Those off the north-eastern coast of Greenland, and those in the Barents Sea, the Kara Sea, the Beaufort Sea, the Chukchi Sea and the Bering Sea. These areas are ice-free in the summer.

Exploration for oil in the Bering Sea began in 1982. Ice rarely reaches the southern parts, and although there are problems because of seismic activity and an unstable sea-floor, the weather and sea conditions are

Figure 4.17 Map showing the bathymetry of the Arctic Ocean and its adjacent seas, and the regions of the Arctic where oil and/or gas may be found; both potential and actual regions of exploitation are shown. Geologically, the Labrador Basin resembles the highly productive North Sea, and the northern part of Greenland is related to the gas-rich Arctic islands.

much like those in the North Sea and the technology employed is similar to that used in temperate waters. The same is true in the southern part of the Barents Sea.

Over much of the Arctic region, however, hydrocarbon exploitation is technically impossible at present. The permanent or seasonal ice cover prevents or hinders the usual initial stages of exploration (seismic reflection surveys from ships), and makes drilling extremely difficult and hazardous. One of the greatest problems is that the pack-ice is *permanently moving*; on short time-scales, it moves in response to shifting wind patterns, but on longer time-scales its motion is determined

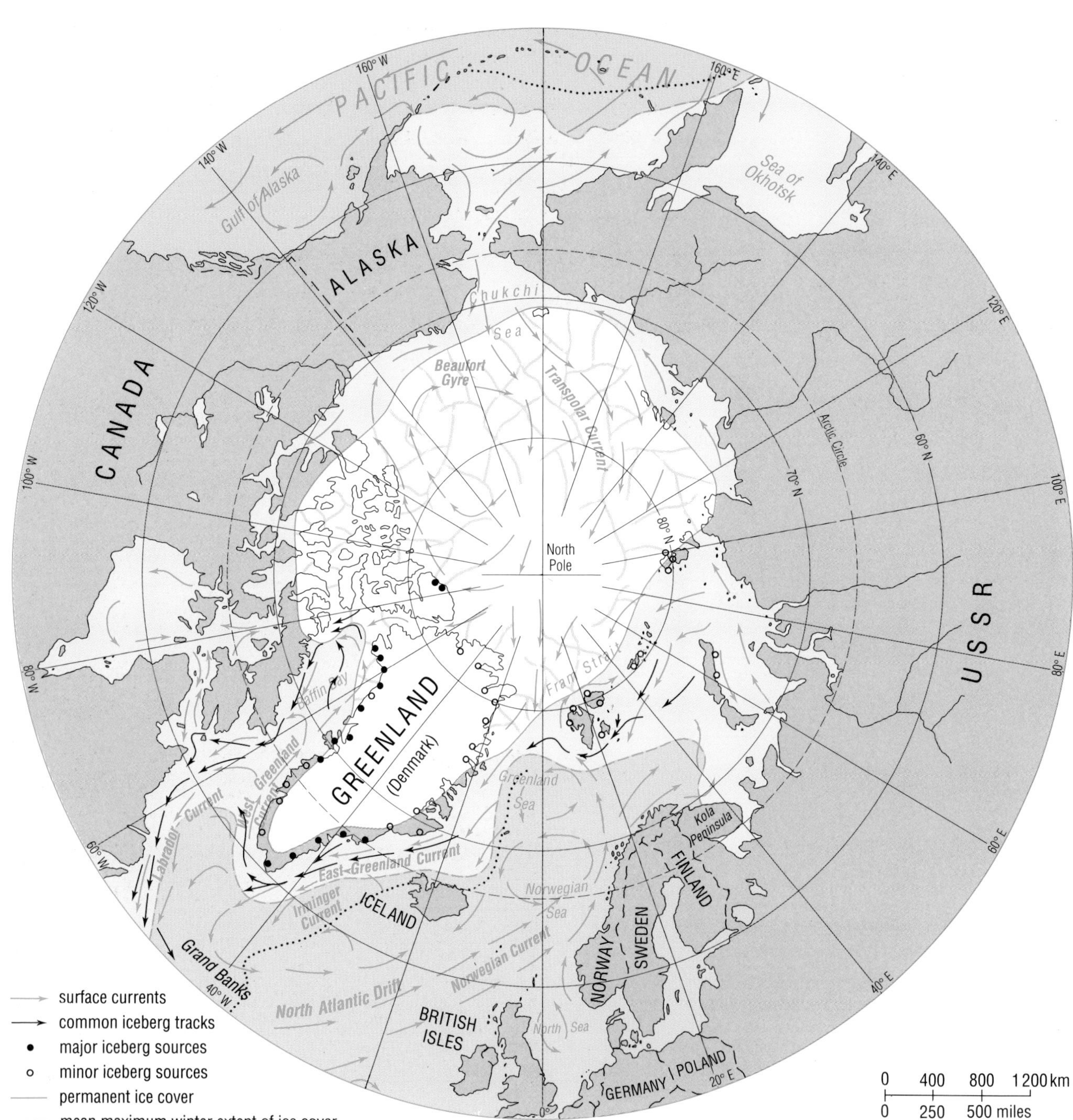

Figure 4.18 The surface currents of the Arctic Ocean and its adjacent seas, the extents of permanent ice cover and winter sea-ice (*cf.* Figure 4.5), and the sources and common tracks of icebergs.

by surface currents, which are themselves partly wind-driven*. The pattern of surface currents in high northern latitudes may be seen in Figure 4.18 which also shows the extents of permanent and winter sea-ice cover (effectively maximum and minimum extents; *cf.* Figure 4.5), and the sources and common tracks of icebergs (land ice).

* The long-term average ice-drift velocity in the Arctic is the same as the long-term average velocity of the surface water. However, the surface circulation is itself a combination of density-driven circulation and circulation driven by the wind, whose stress is transmitted downwards via the ice.

Ice motion is also affected by interactions between the floes as they jostle and grind against one another, or come together to form **pressure ridges** perhaps 10–30 m thick (Figure 4.19). The ice tends to circulate within the Arctic for a considerable time, with only a proportion being carried south each year. As a result, ice is added to the floes each winter and this, combined with successive ridging and rafting, leads to great thicknesses of **multi-year ice**. Thus, even in those areas which are only ice-covered for part of the year, the thickness of ice-floes can be considerable.

(a)

(b)

Figure 4.19 (a) A pressure ridge in the multi-year Arctic pack-ice; its full depth is several times greater than that visible above the surface.

(b) A system of **leads**. These linear regions of water are formed when the ice is torn apart by the wind; when the leads refreeze and the ice-floes are again pushed together, pressure ridges form (cf. (a)).

QUESTION 4.7 By reference to Figure 4.18, answer the following questions.

(a) Why are the waters off the eastern coast of Greenland affected by pack-ice, even in the summer months?

(b) Why is some of the thickest pack-ice to be found in the Beaufort Sea?

Since the energy crisis in the mid-1970s, fossil fuels have become tremendously important in the economic (and hence political) arena. As a result, enormous investments have been made in improving existing technology to cope with Arctic conditions, even those as bad as occur in the Beaufort Sea. Exploration here was stimulated by the discovery of oil at Prudhoe Bay and large offshore reservoirs of oil have indeed been found nearby. The water over the Beaufort Sea oilfields is generally fairly shallow (20–70 m deep, as opposed to 100–200 m over the Bering Sea oilfields), and this has enabled drilling rigs to be mounted on the sea-bed. The bases of the rigs are designed to resist or deflect the force of the shifting ice and may be either concrete constructions (Figure 4.20(a)) or artificial islands of sand and gravel (Figures 4.20(b) and 4.21).

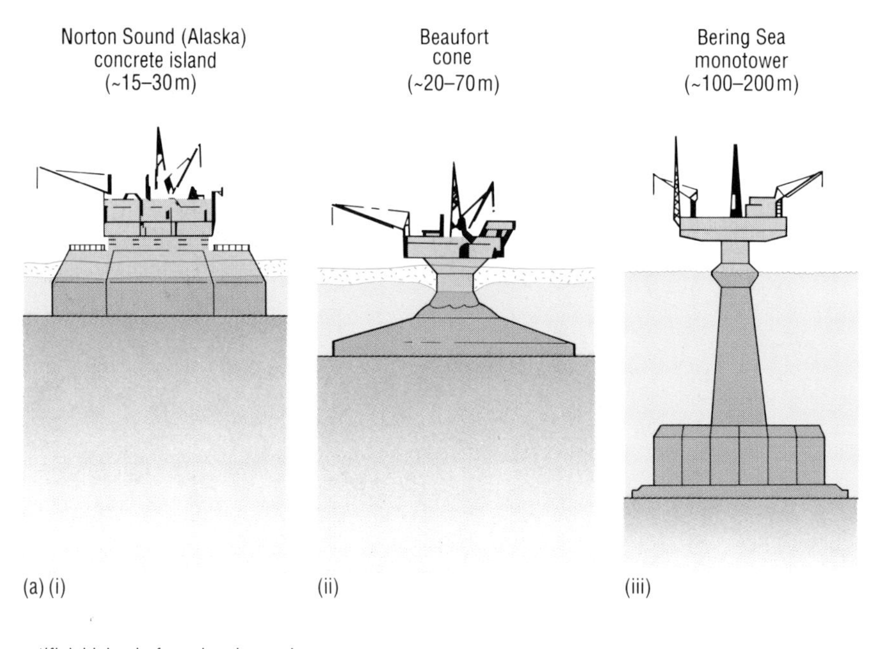

Figure 4.20 (a) Examples of steel and concrete oil production platforms used in the Arctic. The shape of the base depends on water depth and ice conditions; it may be polygonal (as in (i)) so that ice is deflected around it, or have an inverted conical top, so that ice is forced downwards (as in (ii)). Typical water depths are given in brackets.

(b) The type of platform commonly used in shallow parts of the Beaufort, Chukchi and Bering Seas. It consists of a wide, gently sloping berm made of sand and gravel dredged from the sea-floor, with a concrete caisson mounted on top. Ice impacting upon the caisson buckles or crushes, and builds up a pile of rubble on the berm. The force of ice subsequently moving in on the platform is transmitted to the berm rather than the caisson.

Figure 4.21 Artificial drilling island in the Endicott oilfield in the Beaufort Sea. This island is connected by a causeway to a production island which can be seen in the distant background.

Not all drilling rigs in the Beaufort Sea are mounted on the sea-bed: there are also ice-reinforced drilling ships and—a more recent introduction—floating platforms, moored by lines to the sea-bed (Figure 4.22). At the end of the summer, ice-breakers are used to enable the drilling ships and floating platforms to get back to harbour, thus extending the drilling season for several weeks beyond the three-and-a-half months of open water.

Although fairly large oilfields have been found off the Canadian mainland, particularly around the mouth of the Mackenzie River (*cf.* Figure 4.17), the most important discovery has been the enormous reserves of gas in and between the northern islands of the Canadian Arctic Archipelago. Here, the prime drilling season is in the winter. This is because land-drilling rigs are used, even *between* the islands, and the fast ice on which the rigs rest is at its most stable in winter. For offshore sites, the rigs rest on ice which is fast to the islands, but nevertheless moves to some extent. The ice 'platform' is artificially thickened by being sprayed with water, building up successive layers of

Figure 4.22 The floating platform *Kulluk*. As is the case with some sea-bed mounted platforms, it has a polygonal shape so that, in a sense, it has a number of bows.

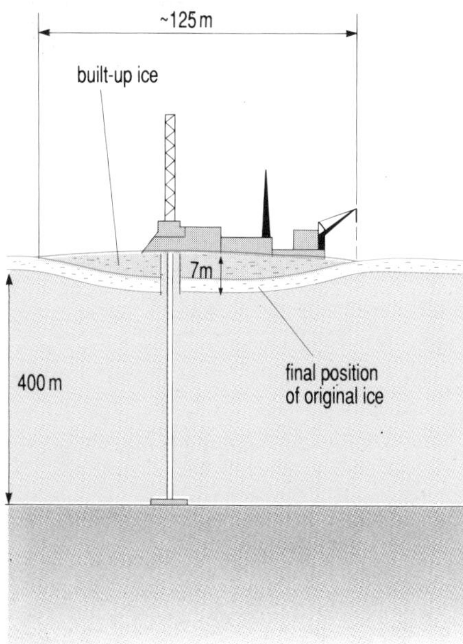

Figure 4.23 Ice platform, used particularly for drilling around the Arctic islands. The load-bearing capacity of the ice is increased, and construction time decreased, by the incorporation into the ice of urethane foam blocks; these are burnt when drilling finishes in April.

solid ice (Figure 4.23). At the latitudes of the northerly Arctic islands, there is very little precipitation, although operations are hampered by blowing snow. The greatest problem, however, is the darkness or limited daylight of the winter months.

In some parts of the Arctic, only first-year ice is generally encountered, but multi-year ice may be a problem under exceptional conditions. Such regions include the central Bering Sea, which is occasionally affected by ice carried south through the Bering Straits; Baffin Bay, which is affected by ice carried through the Nares Strait; and the central Barents Sea (see Figure 4.18). In these regions, it makes economic sense to use rigs that are strengthened to resist first-year ice, and to disconnect and move them to safety if danger threatens. However, such a strategy only works if there is an effective 'early warning system' or ice-forecasting service. Under the influence of variable winds, the edge of the loose pack-ice can move quite fast and unpredictably, and the impact of a large ice-floe could not only damage a drilling rig and cause loss of life, but also lead to oil spillage and possibly serious pollution.

Increasing use is being made of remote-sensing techniques to monitor ice movement; weather-satellite AVHRR (Advanced Very High Resolution Radiometer) images (Figure 4.24(a)) have sufficient resolution ($\sim$1 km), but their usefulness is often limited by the amount of cloud cover.

(a)

Figure 4.24 (a) AVHRR infra-red image of the Barents Sea taken during October 1987 showing the edge of the pack-ice in the vicinity of Svalbard (north is towards top right). Note the characteristic 'comma'-shaped cloud mass at the centre of an Arctic cyclonic wind system; winds to the right of the comma are southerly, and those to the left are northerly, as can be seen from the long streamers of cloud extending southwards from the edge of the ice.

(b) SAR image showing a region ~70 km across near Banks Island in the Beaufort Sea, in October 1978. The image shows ice-floes separated by either leads or very new ice. The floes seem to be made up of smaller floes which have come together to form pressure ridges (white lines).

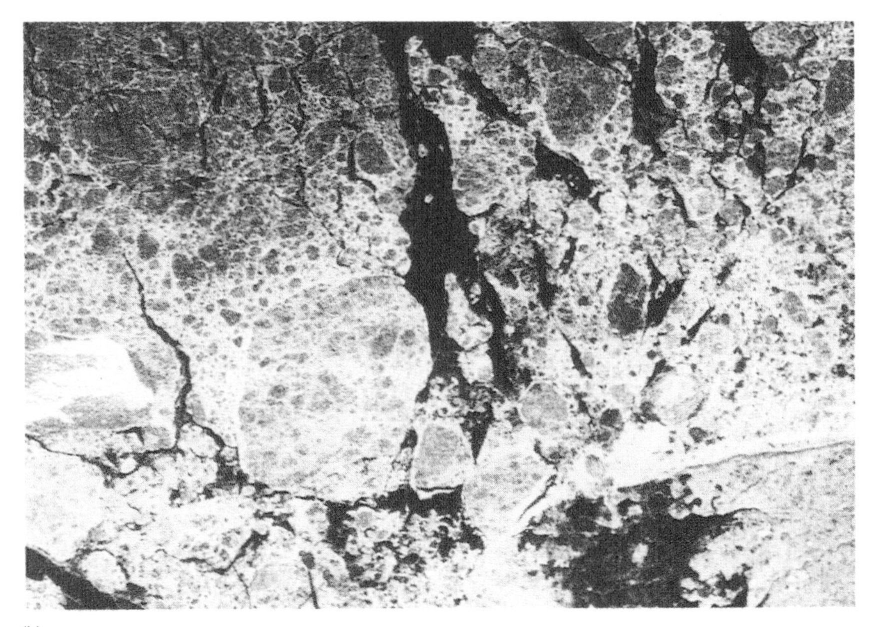

(b)

Passive microwave imagery (*cf.* Figure 4.5) does not suffer from this problem, but resolution is only ~25 km, which is not small enough to detect a band of ice moving out from the ice edge. The next generation of oceanographic satellites will carry synthetic aperture radar (SAR) packages (already used from aircraft, and on *Seasat* in the late 1970s (Figure 4.24(b))), enabling ice to be monitored in all weather conditions *and* with a high resolution. However, none of these alternatives provide a completely reliable indication of the *type* of ice being detected.

Regularly produced images of Arctic ice cover are not only of use to oil companies; they are an invaluable tool in the study of changes in pack-ice extent. The southerly limit of winter and summer pack-ice varies considerably from year to year, and from decade to decade (Figure 4.25). The 1970s and 1980s saw fairly average pack-ice conditions (although there have been some indications that ice thicknesses are decreasing in some regions; see Section 4.5).

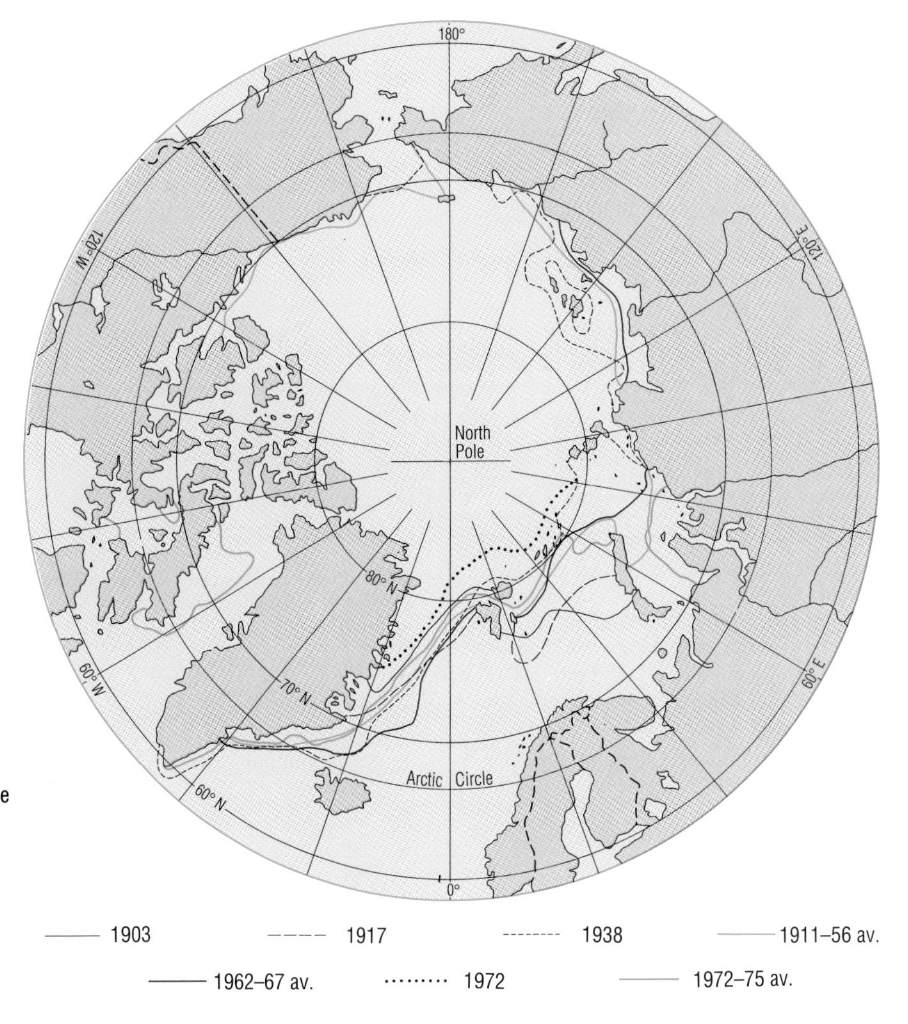

Figure 4.25 Interannual variations in the extent of summer Arctic pack-ice. A greater extent of ice does not necessarily mean that conditions were unusually severe. For example, while conditions *were* severe in the 1960s, exceptionally mild conditions during 1938 resulted in much loose pack-ice being carried around the coast of Greenland in the East Greenland Current (*cf.* Figure 4.18).

The southerly limit of winter pack-ice is constrained, even in very severe winters, by the position of the Arctic Front, i.e. by the boundary between Arctic waters and water flowing in from lower latitudes. Thus, in the Barents Sea, the southerly limit of pack-ice is determined by the path of the Norwegian Current, the downstream continuation of the Gulf Stream and the North Atlantic Drift (Figure 4.18).

What type of ice hazard, which could damage drilling rigs, have we not discussed so far? Where would be the greatest danger?

Icebergs. After calving from the Greenland glaciers, these move southwards along the eastern coast, and then northwards along the western coast: they are also a considerable hazard in Baffin Bay and off the coasts of Labrador and Newfoundland (Figure 4.18). In fact, development of the Hibernia oilfield, on the northern edge of the Grand Banks (*cf.* Figure 4.17), is awaiting the design of a production platform that can withstand iceberg impacts. One strategy for dealing with an iceberg which is threatening a production platform is to tow the *iceberg* away from the platform!

4.3.2 ECONOMIC CONSIDERATIONS

Although there are, potentially, vast offshore reserves of hydrocarbons around the Arctic Ocean (*cf.* Figure 4.17), by the end of the 1980s only the oilfields off Prudhoe Bay had proved large enough to be commercially viable. For all of the Arctic's hydrocarbon fields, the factor which will determine whether they are worth exploiting will not in the end be the cost of developing extraction technology: it will be the cost of overcoming the difficulties inherent in transporting the fossil fuels out of the Arctic.

Various methods of hydrocarbon transportation have been considered, including an enormous airship, bigger than the *Hindenburg*! One possibility is to use underwater pipelines, but laying such pipelines over a rough sea-bed and under thick ice is no easy task; also, in certain shallow regions, the sea-bed is scoured by the keels of icebergs and thick pressure ridges, and the pipelines could easily be damaged. Transportation by submarine has the obvious advantage that it would not be prevented by thick ice cover. Perhaps surprisingly, undersea routes are fairly well known, thanks to several decades of military manoeuvres and a number of exploratory voyages.

The most favoured mode of transportation is by tanker, and Panarctic Oils has been transporting crude oil from its oilfield on Cameron Island on a relatively small scale since 1985. The Arctic Pilot Project, which was set up to demonstrate the feasibility of shipping liquefied natural gas from the Arctic islands in a fleet of tankers, has been shelved until the markets for the products have been more fully investigated. In each of these cases, the route out of the Arctic is through the Canadian Arctic Archipelago—i.e. via the North-West Passage—and the companies involved are Canadian. In Section 4.4, we will look at the issues which arise when the oil company is based in, say, the United States.

Many of the Beaufort Sea oilfields are on Canada's legal continental shelf, according to the 1958 Convention on the Continental Shelf; and would be within the Canadian exclusive economic zone under the 1982 Law of the Sea Convention. However, exploration and exploitation rights have largely been sold to (non-Canadian) multinational corporations, mostly US-based but also involving Japan and countries in Western Europe (notably the United Kingdom and Germany). The Canadian Government receives royalties on oil extracted, but there are many Canadians who feel that neither the royalties nor the original fee for exploration were set sufficiently high, and that, as Canada's natural resources are its main source of economic strength, it should not part with them so easily.

Canada's situation is perhaps a warning to other, less-developed, countries with mineral resources—oil, placer deposits, manganese nodules—on their continental shelves or within their exclusive economic zones.

Why is this so?

Countries without the necessary exploration/extraction technology are likely to be tempted to lease areas of sea-bed to foreign (probably multinational) companies. While leasing secures immediate, risk-free financial gain, it could also prove shortsighted, particularly as the coastal state will not necessarily benefit from increased employment of local inhabitants—the oil, ore, etc., may well be processed elsewhere. However, it should also be said that if the state concerned had little prospect of developing the resource itself, leasing could be the best option. In this case, having taken little financial risk, it could not necessarily expect a large share in the rewards of successful exploration.

4.3.3 THE DANGER OF POLLUTION

All operations which involve the extraction and handling of oil are potential causes of oil spillage (Figure 4.26). During exploratory drilling, penetration into an oil-bearing stratum may result in a blow-out if the naturally high fluid pressure cannot be contained. Offshore drilling rigs may be damaged by massive ice-floes, driven by winds and currents. Through human error and equipment failure, oil can escape into the environment while being transported by tanker, or while being transferred into storage. Storage tanks can leak, particularly if their foundations have shifted, perhaps because the ice in the underlying ground has melted.

(a)

(b)

(c)

Figure 4.26 (a) The oil spill in the Gulf of St Lawrence, produced by the sinking of the *Kurdistan* in March 1979. The oil is amongst the ice of a broken ice-field, and is contaminating (b) the tops and (c) the rims of the ice-floes.

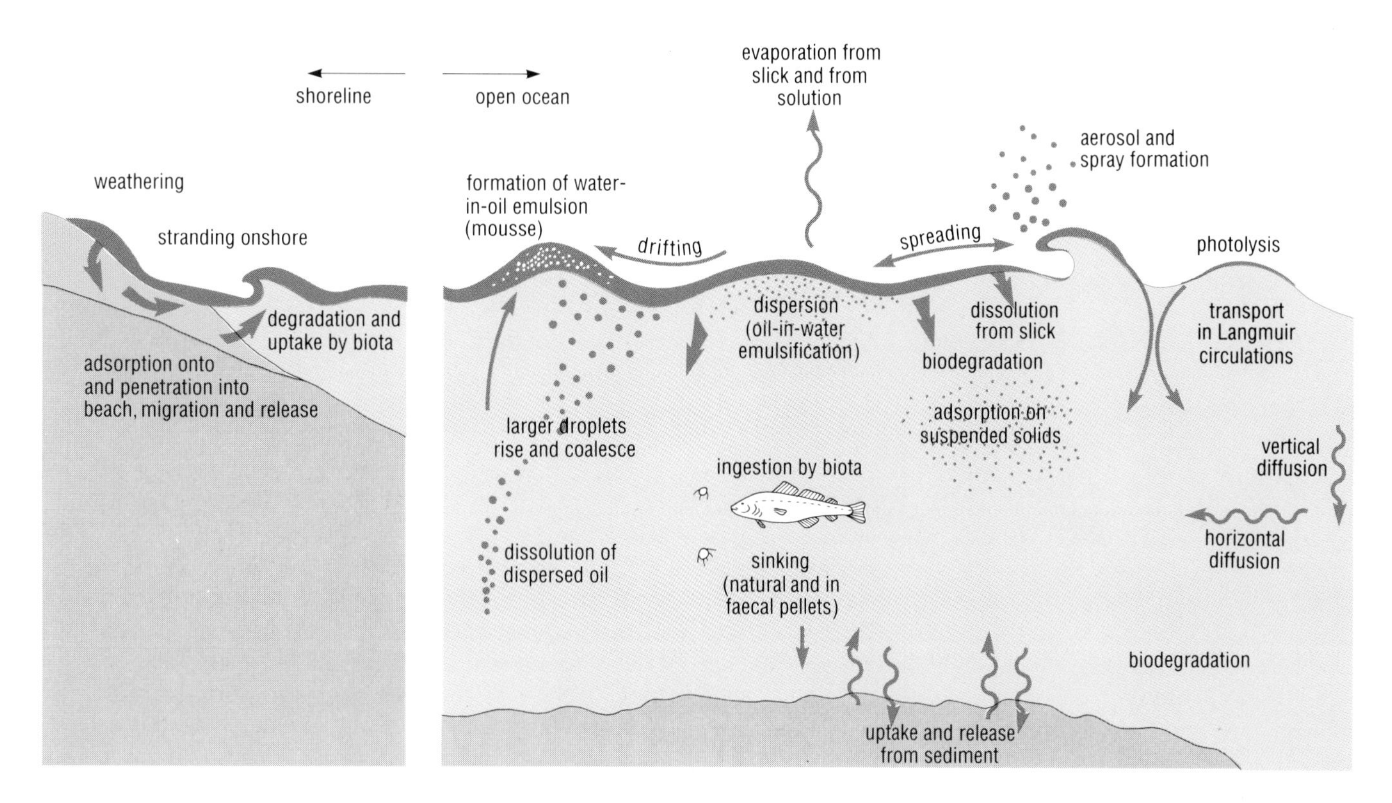

Figure 4.27 The behaviour of oil released near the surface of the ocean under ice-free conditions. Away from the shoreline, it tends to spread out over the surface, forming a layer mostly only a few microns thick.

The majority of oil spills are relatively small and, like the natural oil seeps, they seem to cause no long-term damage to the environment. However, most of the total oil spilt is lost in a small number of large and potentially devastating incidents. As you will see, exactly where and when these incidents occur determines how much harm they cause.

First, look at Figure 4.27 which shows schematically what happens to oil spilt near the surface of the ocean, in the absence of ice. On release, the oil tends to spread out rapidly over the surface of the water, and is further spread out by the action of wind. The initial fate of the oil is greatly affected by the sea-state: breaking waves propel oil directly into the water, so under certain conditions an oil-in-water emulsion or 'chocolate mousse' forms, which is very difficult to deal with.

As the oil spreads out, the more volatile hydrocarbons evaporate so that the remaining material increases in viscosity and density. Although the density of the residue is unlikely to increase above the density of water, it may increase sufficiently for the oil to be carried downwards and so contaminate plankton, fish and other organisms living in deep water or even on the sea-bed.

By reference to Figure 4.27, can you suggest two completely different ways in which oil can be carried downwards?

First, oil that has become *neutrally buoyant* can be carried downwards in regions where water is sinking. This could be, for example, in the linear convergences of wind-driven **Langmuir circulation** systems in polar regions; sinking also occurs where surface water is cooled and/or increased in salinity through ice formation, as well as in the vicinity of icebergs.

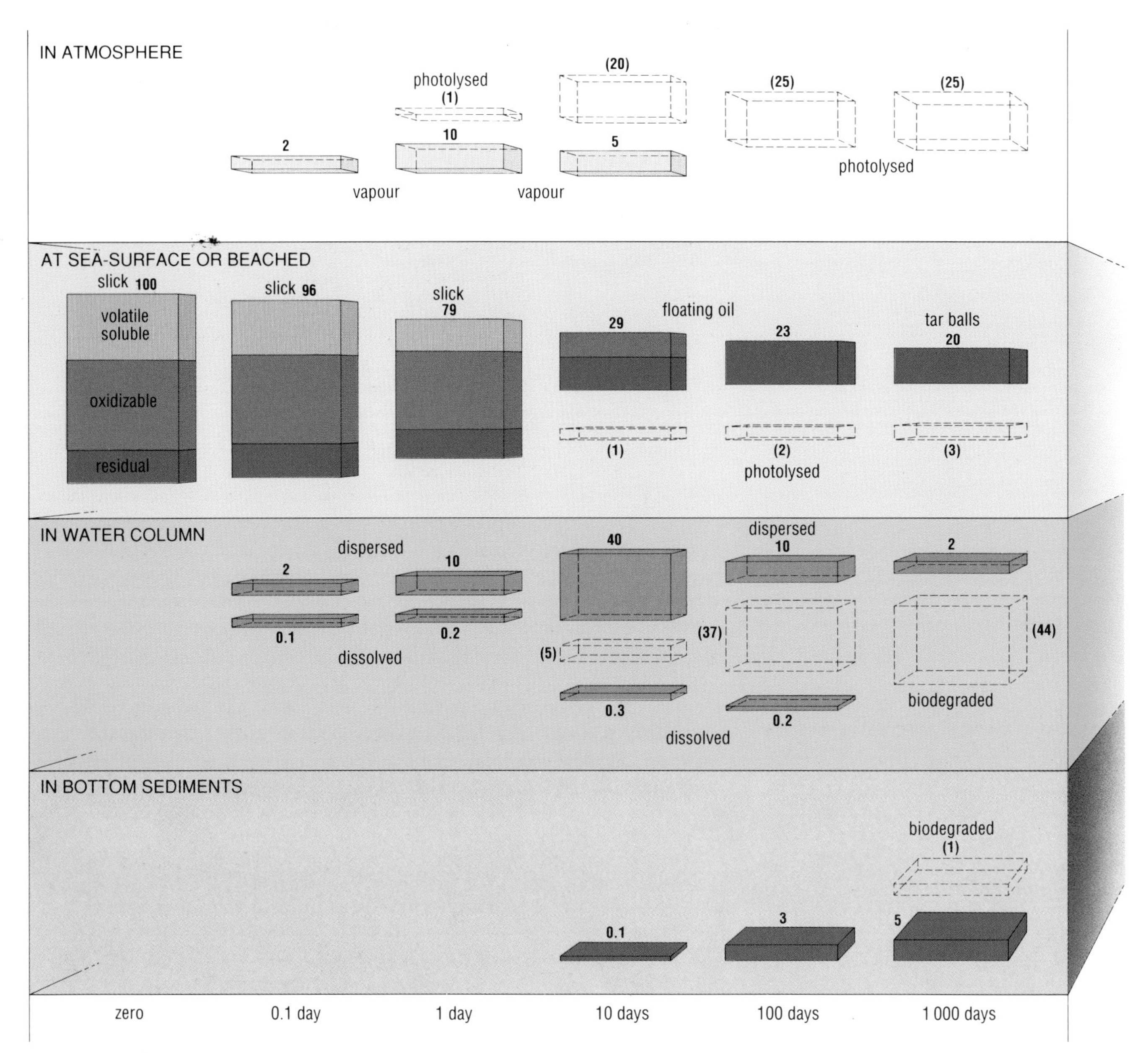

Figure 4.28 Diagrammatic representation of how a 'typical' oil spill might be expected to break down over the course of 1 000 days (about three years); in the Arctic, the period of time involved might be much longer. The numbers give the percentage of the total spill in a given state at any particular time; values in brackets indicate percentages of oil decomposed under the action of micro-organisms or light.

The second way in which hydrocarbons can sink is in association with sinking particles. After the spring bloom, these may be aggregations of dead and dying organisms and/or the faecal pellets of zooplankton. In shallow turbulent waters where bottom sediments are frequently resuspended, or in the vicinity of large sediment-laden rivers (such as the Mackenzie in Canada), large amounts of oil can be carried to the sea-bed attached to clay, silt or sand. Sometimes, an oil slick can pick up appreciable quantities of sand, and the oil–sand droplets which sink to the bottom then roll about picking up more sediment until they become immobilized. Such accumulations of oil on the sea-bed could continue to contaminate benthic organisms for a considerable time.

Figure 4.28 shows a 'guesstimate' of what happens to the oil in a 'typical' oil spill, over the course of 1 000 days. Of course, there is no such thing as a typical oil spill: the proportions of constituent hydrocarbons in crude oil vary between one reservoir and another; and the rate at which the oil is dispersed will depend upon environmental conditions—the ambient temperature and the amount of turbulence caused by waves and currents, for example. The main points to note from Figure 4.28 are that although a fraction of the oil is degraded, and yet more is dispersed, even after 1 000 days tar balls will probably remain. Under cold Arctic conditions, most of the processes leading to dispersal and degradation are slowed down, and so Arctic oil slicks will probably last significantly longer than Figure 4.28 suggests.

The fate of the oil in a spill depends on where exactly the spill occurs. Of particular concern from an environmental point of view is the release of oil and gas from the sea-bed as the result of a blow-out (Figure 4.29). Normally, the impact of an oil spill on organisms in the water column is mitigated by the tendency for oil to stay on the surface, with only a small amount dissolving or dispersing (as an emulsion); under these circumstances, the concentrations of oil in the water are fairly low and the toxic effects on marine organisms are probably not serious. In the case of a blow-out, there is considerable turbulent mixing between the rising oil and gas and the surrounding seawater, and large amounts of oil can dissolve or emulsify. As the buoyant plume of oil and gas rises to the surface, it can set up a circulatory system which may carry surface oil back down into the water column. If there is ice at the surface, the oil and gas will accumulate under it in pools or droplets (Figure 4.29); if the ice is moving, its underside can become painted with oil. As it cannot evaporate, or be burnt off, it remains in contact with the water for a very long time. If the ice cover is thickening, the oil and gas can be frozen into it, remaining trapped until the following spring when pools of oily meltwater appear on the ice surface.

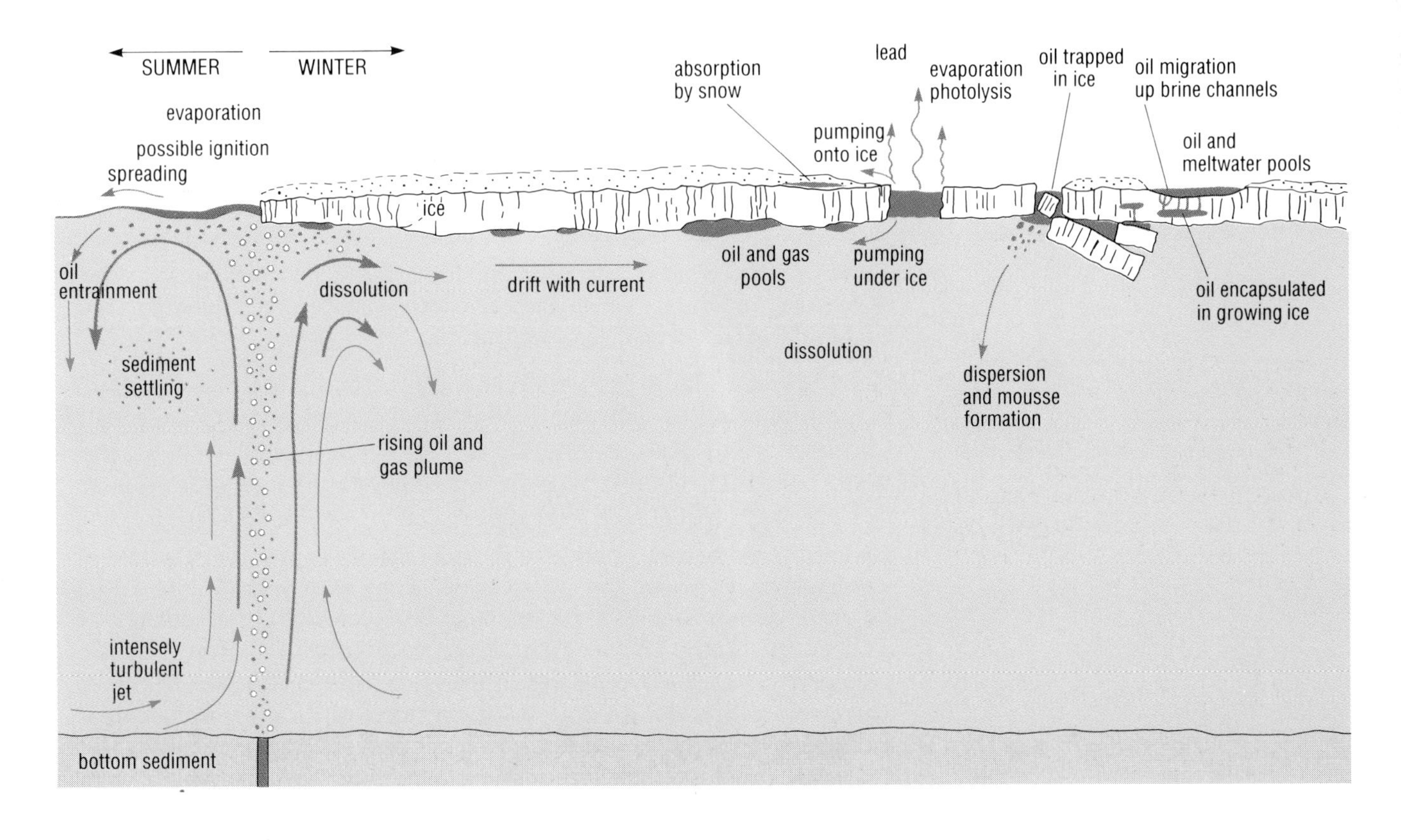

Figure 4.29 The fate of oil and gas released by a bottom blow-out in Arctic waters in summer (left-hand side) and in winter (right-hand side).

In some circumstances, the presence of ice aids the clean-up of an oil spill because the ice can provide a stable platform from which to deploy equipment. On the other hand, mixtures of oil and broken ice—which can become mixtures of oil, ice, snow, water and oil-in-water emulsion—are particularly hard to deal with (*cf.* Figure 4.26(b)).

4.3.4 THE 'FRAGILE ARCTIC' AND OIL POLLUTION

The term 'fragile Arctic' has been coined by environmentalists and the media to imply that the Arctic is more vulnerable to human interference than other parts of the globe; in particular, that its ecosystems are more likely to become irreparably damaged than ecosystems elsewhere. But is this true?

One notable aspect of Arctic ecosystems, both terrestrial and marine, is their unusual simplicity: there is a low diversity of species, and the food webs are relatively simple with few 'trophic levels' (e.g. see Figure 4.8).

Why could simple food webs render the ecosystem more vulnerable?

Because the reduction in numbers, or extinction, of a given link in the food web would be much more serious for the overall ecosystem than would be the case in systems where the animals have a wide choice of food species. Recently, however, application of mathematical techniques to ecosystem models has indicated that the reverse could be true, and that simple ecosystems are more robust than complex ones.

Any species that survives Arctic conditions is likely to be—in a sense—very tough. However, the corollary to this is that Arctic species are in fact quite vulnerable to changes in the environment to which they have become adapted. Populations that suffer a large drop in numbers as a result of (say) disease, pollution or overfishing may take years to recover, and as a result, populations of Arctic species can fluctuate markedly. The vulnerability of species tends to increase with evolutionary complexity, so that, for example, planktonic copepods are less vulnerable than fish. Most vulnerable of all are the marine mammals; and we do not know what effects removal of top predators like whales and seals might have on the ecosystem as a whole. We do know that because growth and reproduction rates are very low in the Arctic, a series of relatively minor perturbations could prevent a species recovering, and events which elsewhere would be relatively unimportant might add up to an ecological catastrophe.

Nevertheless, in the Arctic the likelihood of a 'crash' is reduced by the fact that numbers of individuals within given species are very large (as in the case of Arctic cod, for example; *cf.* Section 4.1.2). Furthermore, as Arctic ecosystems extend over large geographical areas, damage to one region may be repaired by immigration from adjacent regions.

Supporting the various marine ecosystems are, of course, the algae, which can only flourish for a very short period each year (Figure 4.30). As mentioned in Section 4.1.2, the algal flora consists of two different groups—the phytoplankton living in the water column and the ice algae, which are generally concentrated in the lowest few centimetres of the ice. There are two distinct ice floras: summer algae which grow in moving ice-floes, and winter algae which grow in fast ice. These ice algae are adapted to photosynthesis at very low light levels (they are mostly

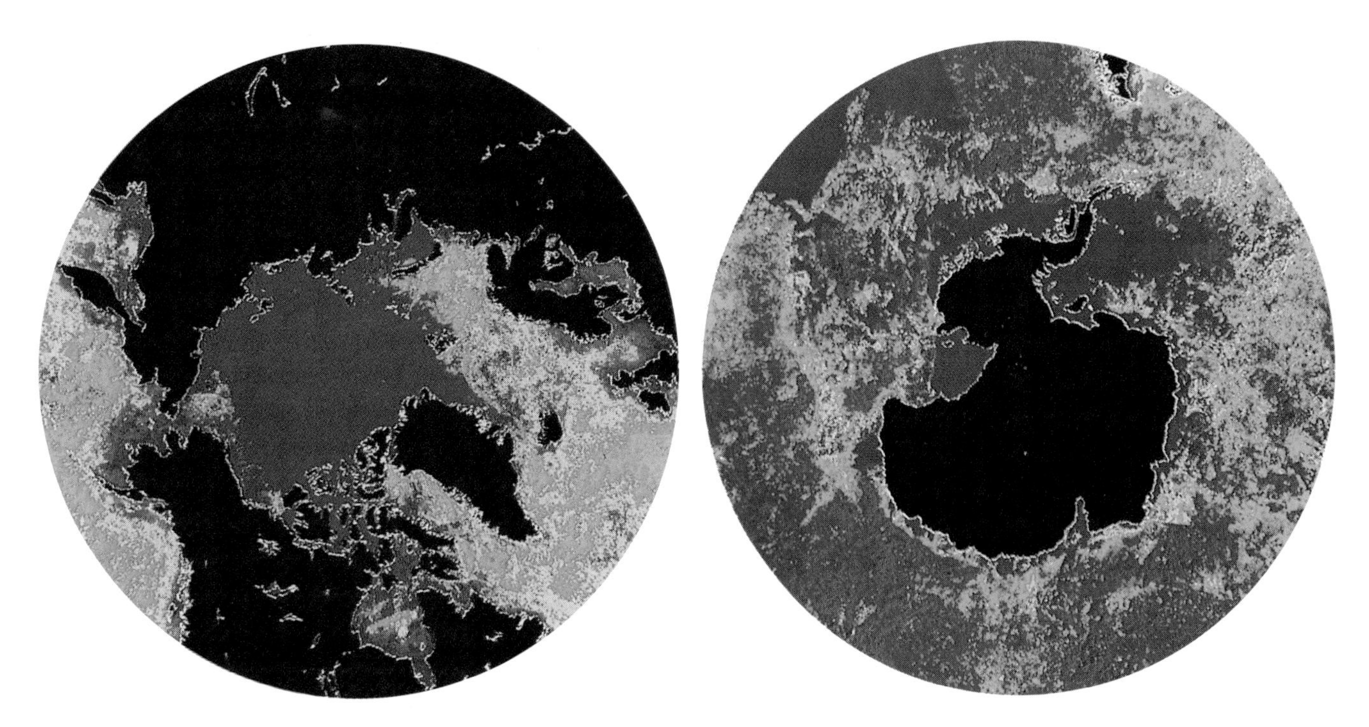

Figure 4.30 (Left) The summer phytoplankton bloom in the Arctic, as determined by the Coastal Zone Color Scanner. Grey regions are those for which there were no data, either because of cloud or ice cover, or because they were not traversed by the satellite.
(Right) The summer phytoplankton bloom around Antarctica, for comparison.

species of diatoms normally found in shallow-water benthic environments), and they use nutrients which leach down from the ice above. They have a longer growing season than the planktonic algae and may contribute as much as 25% of the total algal production of the Arctic Ocean. Algae are grazed by small invertebrates, as well as swarms of large copepods and amphipods; these in turn attract Arctic cod and other fish, as well as marine mammals and, along the ice margin, large numbers of seabirds.

Ice also has *negative* effects on primary production. First, it severely cuts down the light reaching the underlying waters, especially when it is snow-covered; thus under the Arctic ice-cap, light levels are always too low for planktonic photosynthesis to occur. Secondly, the overall result of ice formation and ice melting is the development of a surface layer of low-salinity water, about 50–75 m thick. This low-density surface layer (which is also maintained by the freshwater input from the large Russian rivers) means that the upper water column is very stable.

Is this situation favourable or unfavourable to the growth of large phytoplankton populations?

In the short term, a stable upper water column is an advantage to phytoplankton, because stability suppresses turbulent mixing and this means that phytoplankton can remain in the sunlit **photic zone** rather than be frequently carried down below it. However, a phytoplankton bloom will deplete surface waters of their nutrients; if the upper water column is very stable, vertical mixing will be inhibited, and once depleted, nutrients will not readily be replenished by the mixing up of sub-surface, nutrient-rich water. For this reason, the Arctic phytoplankton blooms which occur in spring as the leads open up are of short duration.

However, along the margins of the sea-ice, the opposite is true. For centuries, navigators have recorded how, as their ships drew near to the

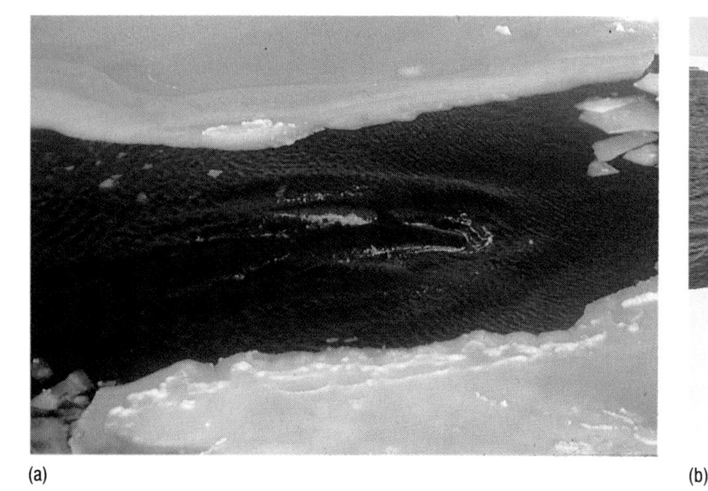

(a)

(b)

Figure 4.31 (a) Minke whale (lesser rorqual). (b) Juvenile hood seal hauled out on an ice-floe.

ice edge, they observed a marked increase in the numbers of whales, seals, birds and smaller organisms (*cf.* Figure 4.31). Only in the late 1970s was it shown that this increased abundance results from high nutrient concentrations caused by upwelling driven by winds blowing off the ice. High productivities are also found in upwellings around glaciers entering deep water, and around large icebergs, but the cause of these upwellings is not clearly understood.

A further aspect of the ice cover which has great importance for the ecology of the Arctic—and whose significance only began to be appreciated recently—is the occurrence of **polynyas**. These are extensive areas of open water which are maintained throughout winter, and which tend to recur in more-or-less the same place each year. It is thought that the surprisingly high fertility of certain parts of the Arctic—notably Lancaster Sound and, perhaps, the Beaufort Sea—is the result of upwelling in large polynyas (*cf.* Figure 4.32). These upwellings are probably partly the result of wind blowing off the ice, but are probably also an intrinsic part of the air–sea–ice interactions which cause the very existence of the polynyas. In addition to their roles as 'nutrient pumps', polynyas provide winter refuges for walruses, polar bears and some species of seals.

Returning to our theme of oil pollution, what can we now say about the dangers posed to Arctic marine ecosystems by a particular oil spill?

The message seems to be that the type and amount of damage done depends crucially on where and when the spill occurs. The worst possible scenario would probably be a large oil spill occurring in the vicinity of a polynya, or near to the ice-edge during the spring bloom. A spill occurring in mid-winter in a region of thick pack-ice would initially be less damaging, although, as we have seen, oil takes longer to break down in Arctic conditions, and the full impact of a spill would not be felt until the ice melted in the spring. By that time, the contaminated ice could have travelled a considerable distance from the site of the original incident.

A sea-bed blow-out (Figure 4.29) occurring in spring could severely affect the ice flora; the 'grazing' species vary in their tolerance to oil, but it is known that the amphipods, which form a crucial link in the food

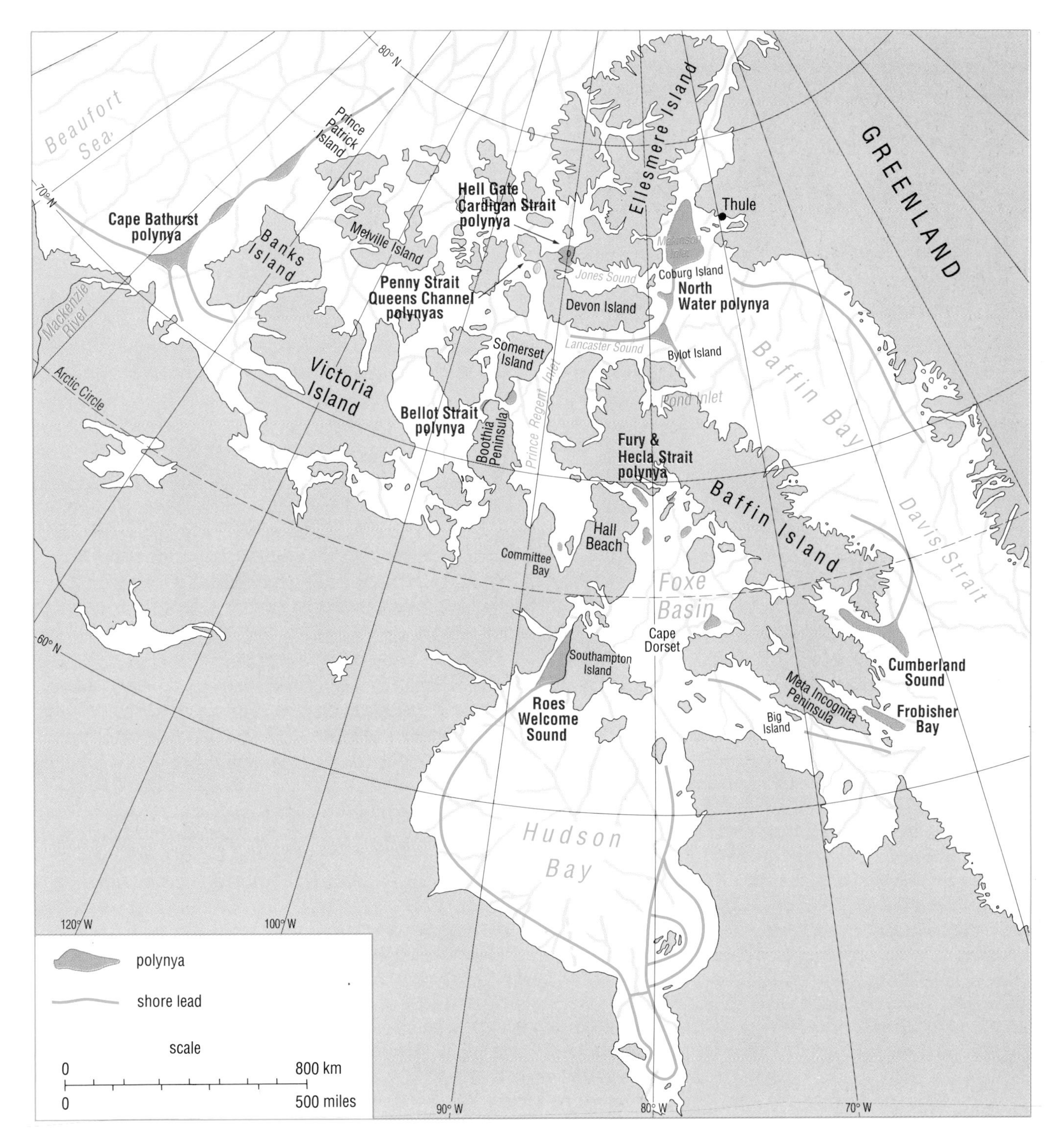

Figure 4.32 Recurrent polynyas (and shore leads) in the Canadian Arctic; these 'nutrient pumps' enrich areas of ocean much greater in area than the polynyas themselves.

web based on the ice flora, are particularly susceptible to hydrocarbons and are adversely affected by very low concentrations. Even a small oil spill could therefore severely disrupt the communities which depend on ice algae.

In spring, seabirds exhausted at the end of their migrations are particularly vulnerable; so, too, are seals and polar bears, which have low food reserves at this time. Oil spilt in the summer is more likely to reach the shore and affect birds such as terns which at that time of year would be nesting, foraging or moulting. An autumn spill in coastal waters could be disastrous for many fish fry and would also affect whales, which come inshore to breed in certain areas (e.g. the Mackenzie Delta region). In winter, seals congregate together and so even a localized spill could kill large numbers of them.

As mentioned in Section 4.3.3, oil which sinks to the sea-bed can become incorporated in the sediments. It therefore affects the infaunal communities, and, furthermore, may be periodically released to cause renewed pollution. Communities of benthic organisms have been known to remain affected at least six years after the original spill, even though, by then, no oil residues could be detected. This could be because the organisms' low metabolic rate, under the cold Arctic conditions, means that their self-cleaning mechanisms are much less effective than would be the case in lower latitudes.

As mentioned at the start of this Section, the slow growth and reproduction rates of Arctic organisms renders them particularly vulnerable to environmental stresses. Let us take one example—that of the Arctic cod, which is a staple food for many other species (Figure 4.8). Like other Arctic fish, Arctic cod grow very slowly; individuals take five years to mature and probably only spawn once in a lifetime. A severe oil spill in coastal waters could not only deplete that year's recruitment to the population, it could also destroy the entire reproductive output of the individuals which happened to spawn that year. Thus, a population of Arctic cod could take a very long time to recover from a relatively small pollution incident.

4.3.5 OTHER SOURCES OF POLLUTION

Hydrocarbons are by no means the only mineral resource to be found beneath and around the Arctic Ocean; at present, there is no commercial sea-bed mining, and relatively few of the land deposits are being exploited. However, because of the need for easy access, most extraction occurs around the coast, and there is growing concern about the effects of mining on the oceanic, as well as the terrestrial, environment.

One of the first projects to attract attention as a possible source of pollution was the 'Black Angel' lead–zinc mine at Maarmorilik on the west coast of Greenland. Before mining started, the Danish authorities had initiated chemical and biological studies of the region and set standards for acceptable changes to the environment; after mining commenced, environmental monitoring was carried out. It soon became clear that the mining operation was causing serious—and unexpected—dispersion of heavy metals (including lead, zinc, cadmium and copper) into the adjacent sea. One particular problem was that the heavy metals were leaching as dissolved ions from the mine tailings dumped in the fjord at Maarmorilik, possibly as a consequence of the low content of organic matter in marine Arctic waters. Much effort has gone into studying how the metals are dispersed and the ways in which they accumulate in seaweeds, mussels and fish, and some progress has been made in reducing the pollution.

The Arctic continental shelf has extensive deposits of sand and gravel and is rich in placer deposits of tin, gold and platinum.

How might dredging for sand, gravel and placer deposits affect the benthic communities?

The main ways in which benthic communities could be affected would probably be through physical damage, and disturbance of their habitat, rather than through toxic effects. Benthic filter-feeders in particular would suffer through their filtering mechanisms being damaged by increased sediment loads in the water.

Near habitations, the Arctic coastline is subject to the same types of pollution as occur elsewhere (i.e. municipal rubbish and sewage), but at present the pollution is mostly low level and/or very localized. Another locally generated type of pollution is that related to fishing, notably nets which entangle and eventually strangle or drown large numbers of young seals.

However, much of the pollution entering the Arctic Ocean originates well outside the Arctic Circle, and is carried in both by rivers and ocean currents. The large rivers which drain the mining and industrial centres of the Soviet Union are a particular cause for concern. They may bring in large amounts of heavy metals and other residues, but very little detailed information is yet available about the concentrations involved.

That ocean currents can transport pollution into the Arctic Ocean is clearly demonstrated by the presence there of caesium-137 which can be traced back to the Irish Sea and discharges from the Sellafield nuclear reprocessing plant on the Cumbrian coast. A more clearly visible type of pollution is the large amount of plastic debris which, having been dumped from ships far to the south, is carried into the Arctic region.

QUESTION 4.8 Over the last few hundred years, there has been a several hundred-fold increase in the concentration of lead in the ice accumulating to form the Greenland ice-sheet. Where does this lead come from, and would you expect lead concentrations in the Arctic sea-ice also to have increased?

The Arctic was once regarded as one of those remote areas of the world whose pristine, unpolluted air could be used as a 'pure standard' against which levels of atmospheric pollution elsewhere could be judged. For part of the year this is still broadly true, but from December to May there are periods when large concentrations of particulate matter accumulate in the lower atmosphere, acting as nuclei for ice crystals and scattering the light to such an extent that even in the absence of blowing snow, visibility is cut down to 10 km. This is the phenomenon known as '**Arctic haze**'.

The particles in Arctic haze contain significant amounts of anthropogenic materials, notably soot (i.e. carbon), trace metals, nitrate and, especially, sulphate, much of which probably crossed the Arctic Circle as SO_2 gas. The particles are deposited from the atmosphere, either as dust or in precipitation, and eventually enter the Arctic Ocean, perhaps after several years in the ice cover.

Given their composition, what effect will the addition of these particles have on the chemistry of surface waters? And will this effect be 'diluted' by mixing of surface waters with the bulk of the ocean waters beneath?

Deposition of particles will clearly cause the concentrations of trace metals and other pollutants in surface waters to increase. Also, because of the particles' sulphate content, surface waters will also increase in acidity (i.e. their pH will fall). It is not known what effect this will have on the marine ecosystem.

Because of the strong halocline, turbulent mixing in the surface waters of the Arctic tends to be suppressed (see Section 4.3.4), so that atmospheric pollutants—those which cause Arctic haze, and others such as PCBs (polychlorobiphenyls)—tend to remain in the top 100 m or so. This is also true of the pollutants supplied in river water, which itself helps to keep surface waters at relatively low salinities.

By studying the paths of air masses and analysing the content of airborne droplets and particles over the northern continents, it is possible to deduce the sources of the atmospheric pollution reaching the Arctic. As shown in Figure 4.33, much of the airborne pollution originates in the Soviet Union, Eastern Europe or the eastern United States; a small proportion can be traced back to South-East Asia.

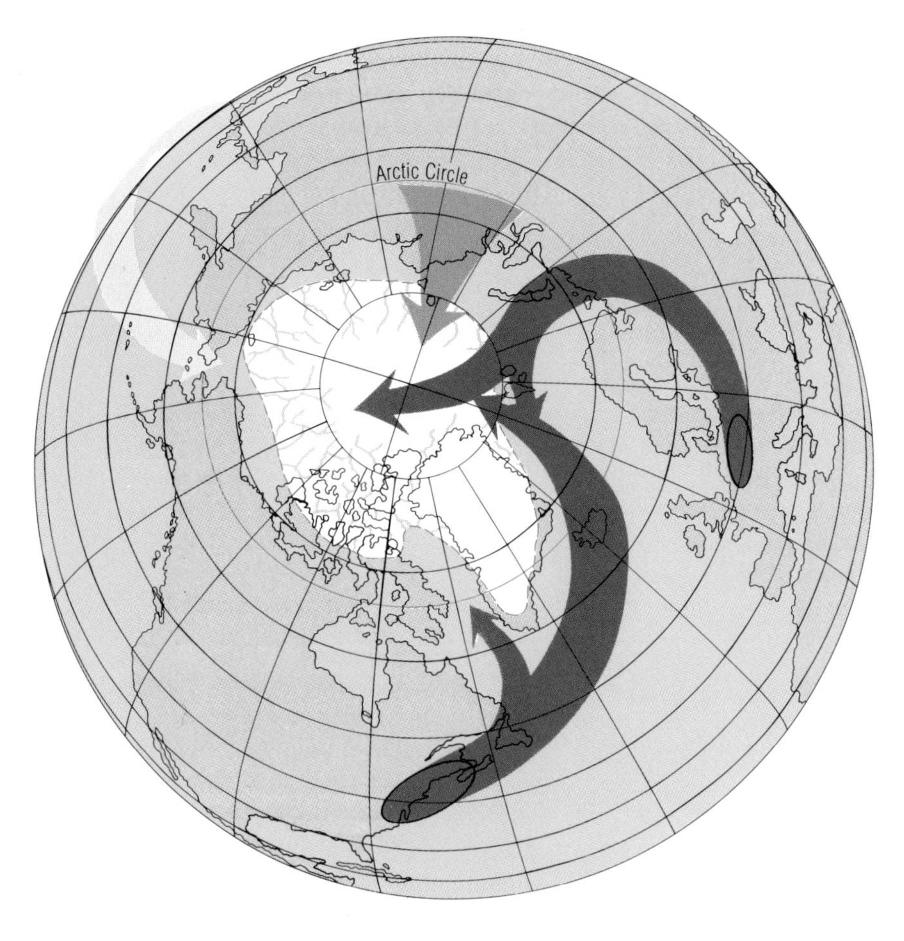

Figure 4.33 Likely paths of air masses bringing atmospheric pollution to the Arctic region. The air flow from South-East Asia carries the least pollution, partly because the region produces relatively little of it, and much has been washed out over the Pacific. The path from the USSR is speculative.

As global pollution increases, so will pollution in the Arctic, and many as yet unthought-of problems will have to be faced. It has already been observed that Arctic lichens concentrate radioactive caesium (fallout from Chernobyl) to an unexpected extent, and similar unpredicted effects could occur in marine organisms.

4.4 THE NORTH-WEST PASSAGE

When, in 1493, Pope Alexander VI declared that the world's undiscovered regions should be split between Spain and Portugal (Section 2.1), the other European powers sent their ships westwards and northwards in an attempt to locate a new route through to the Orient. Their aim was not to find and explore new lands, but to maintain commerce with Asia; in particular, they wanted to buy silks, and the spices that people depended on to make palatable the preserved meat they ate during the winter months. The land route via the Mediterranean had been cut off in 1453 when the Turks took Constantinople, and when Egypt fell in 1517 the Mediterranean sea-lanes were also closed.

The Dutch and English were the most successful explorers of the Arctic. In the mid-1500s, English voyages into the eastern Arctic resulted in lucrative trade with Russia but expeditions in search of a North-East Passage got no further than the entrance to the Kara Sea (Figure 4.1). After an unsuccessful exploratory voyage in 1580, the English abandoned the search for a North-East Passage and left the field open to the Dutch. Four years earlier, in 1576, Martin Frobisher had led the first expedition into the Arctic Archipelago, in search of a **North-West Passage**.

By now, the need for a sea passage from Europe to Asia had become less urgent because, in 1571, the naval battle of Lepanto had broken the Ottoman hold on the Mediterranean. Most European merchants were happy to use the old tried and tested routes, but the Dutch still persisted with the search for a North-East Passage and the English with the search for a North-West Passage.

There is not space here to describe the voyages of exploration and adventure amongst the Arctic islands. Figure 4.34(a) shows the routes opened up between 1576 and 1847; these were greatly augmented after 1847 during the extensive search for Franklin's party, all of whom perished like many before them (Figure 4.35). That so many of the expeditions were frustrated in their purpose owed much to the fact that they sailed into the Arctic far too early in the year and left too soon, not realizing that pack-ice conditions are least severe in August and September (*cf.* Figure 4.5(a)).

The term 'North-West Passage' is now used to mean any route between the Labrador Sea and the Beaufort Sea, via the various passages between the Arctic islands (Figure 4.34(b)). Route 1, between Prince of Wales Strait in the west and Lancaster Sound in the east is the most suitable for deep (~ 20 m) draft ships. An alternative is Route 2, through M'Clure Strait: it has adequate water depths, and large leads occasionally develop along the northern shore of the Strait and along the west coast of Banks Island. However, the Strait is normally blocked by pack-ice and has seldom been navigated by a surface vessel. Route 5A is navigable by deep-draft ships, but the presence of shoals in Fury and Hecla Strait makes it a less-attractive proposition than Lancaster Sound (through which all the other routes pass). Finally, Routes 3 and 4 involve shoals and numerous narrow straits, and so are suitable only for small vessels with drafts less than about 5 m.

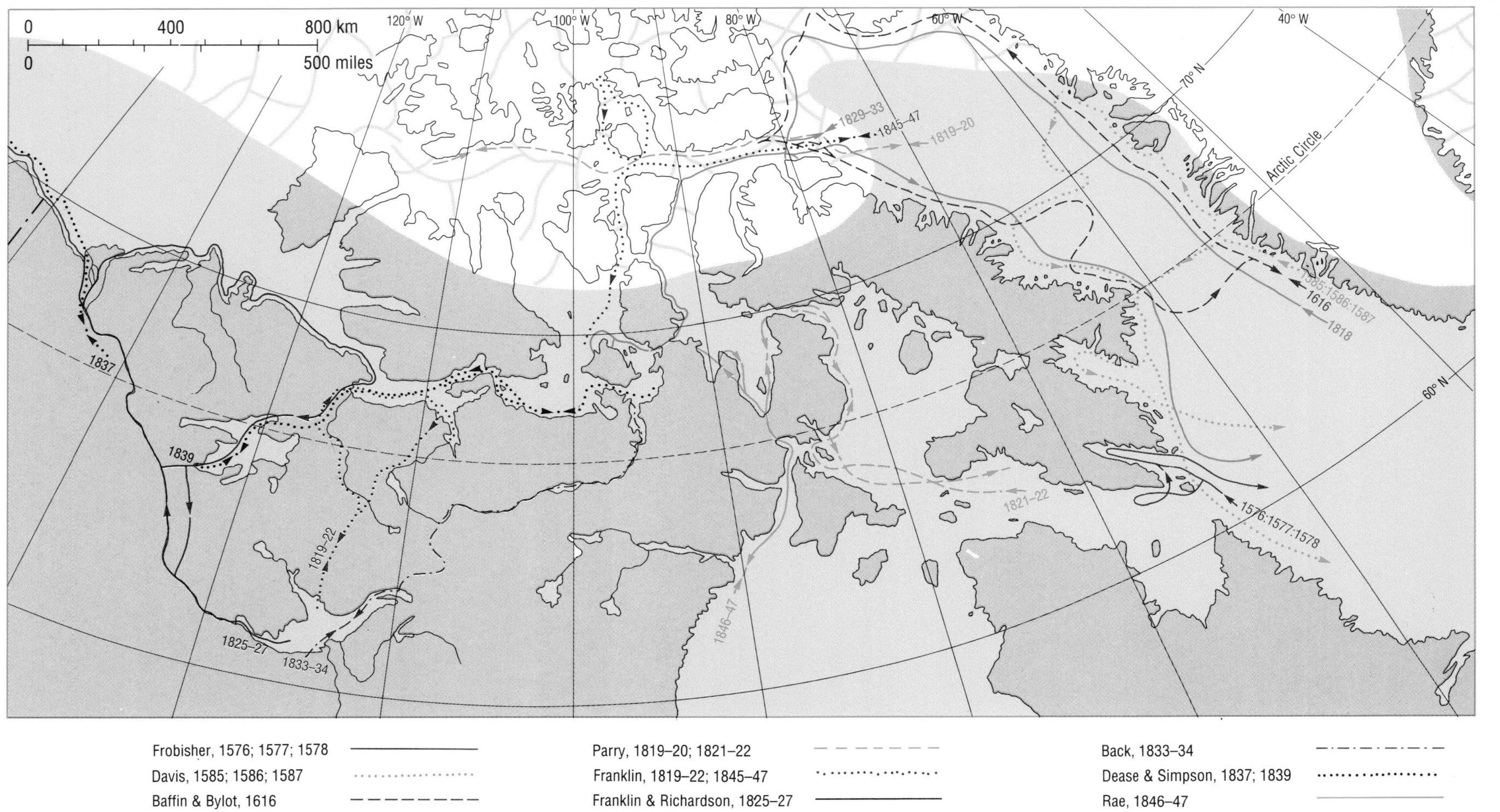

Figure 4.34 (a) Routes explored during the search for a North-West Passage between 1576 and 1847. As comparison with Figure 4.10 shows, many of the explorers (along with their vessels and patrons) have been honoured by the naming of a strait, sound or bay.

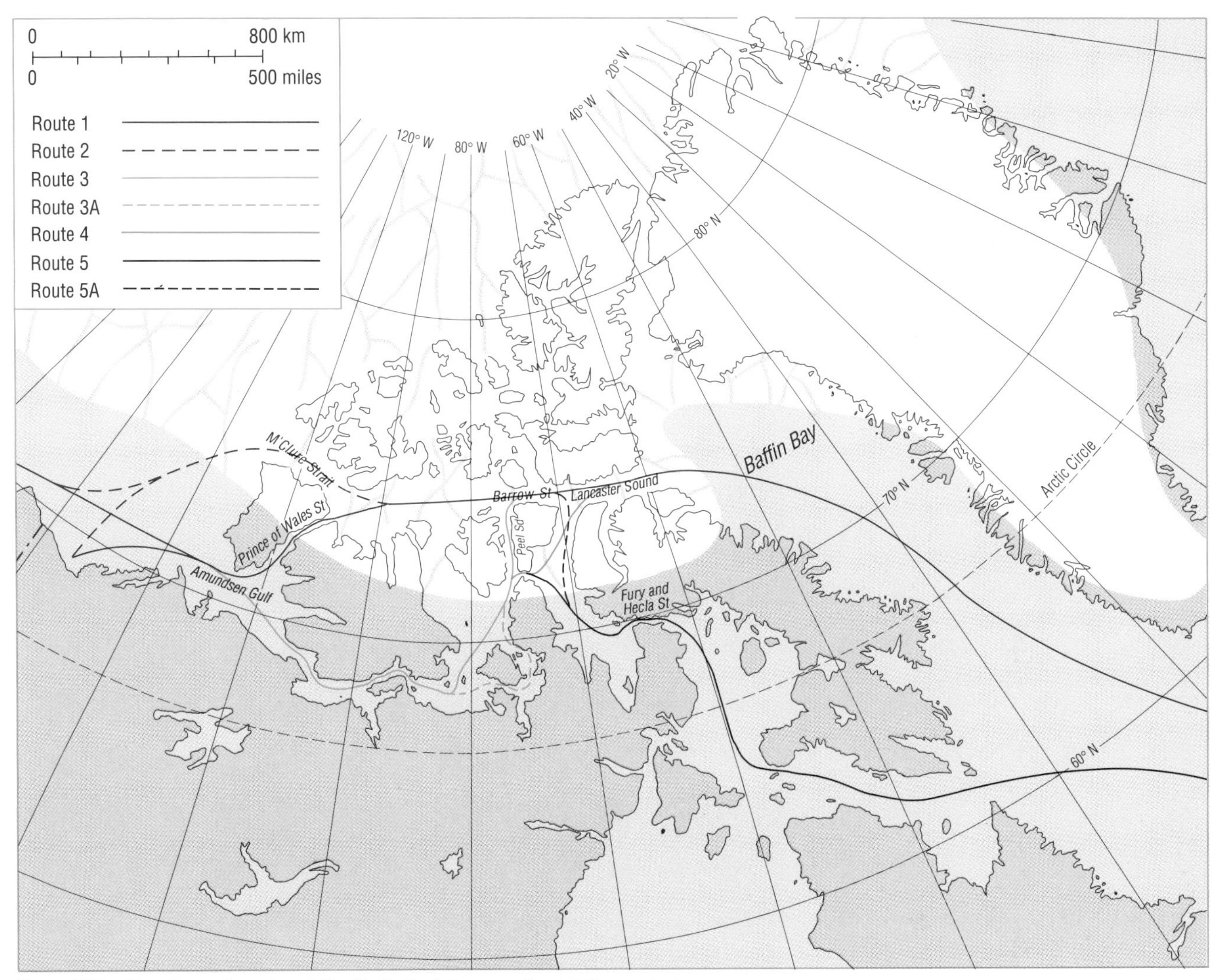

Figure 4.34 (b) The main routes of the North-West Passage. For details, see text.

Figure 4.35 The tragic end of Franklin's expedition, as depicted by W. Thomas Smith in his painting, *They forged the last link with their lives*, 1895.

4.4.1 THE *MANHATTAN* INCIDENT

In late summer, 1969, the ice-strengthened tanker *Manhattan* was making her way up the Davis Strait into Baffin Bay, with a swarm of US Coastguard helicopters in attendance. In a feasibility study by the US petroleum company Humble Oil, to see if crude oil could be transported by tanker from Prudhoe Bay to markets on the east coast of the United States, the *Manhattan* was to test her strength against the pack-ice of the North-West Passage (Figure 4.36). The US Government had not sought permission for the voyage from the Canadian Government. The route planned corresponds to route 1 on Figure 4.34(b), and the United States' view was that there was no justifiable reason for the Canadians to take exception to the journey as only waters of the *high seas* would be traversed.

Nevertheless, the Canadians were outraged. As discussed in Section 4.2.2, they considered the waters of the Arctic Archipelago to be Canadian, and resented what they saw as the less than 'humble' behaviour on behalf of the United States. Furthermore, they feared the implications of an established 'high seas' route through the islands.

The *Manhattan* succeeded in completing her voyage through the North-West Passage. However, she was defeated by the multi-year ice of the M'Clure Strait and, having been three times freed by the *John A. MacDonald* (Canada's then newest and strongest ice-breaker), she turned about and continued via the Prince of Wales Strait.

Figure 4.36 The *Manhattan* tests her strength against a large ice island in the North-West Passage.

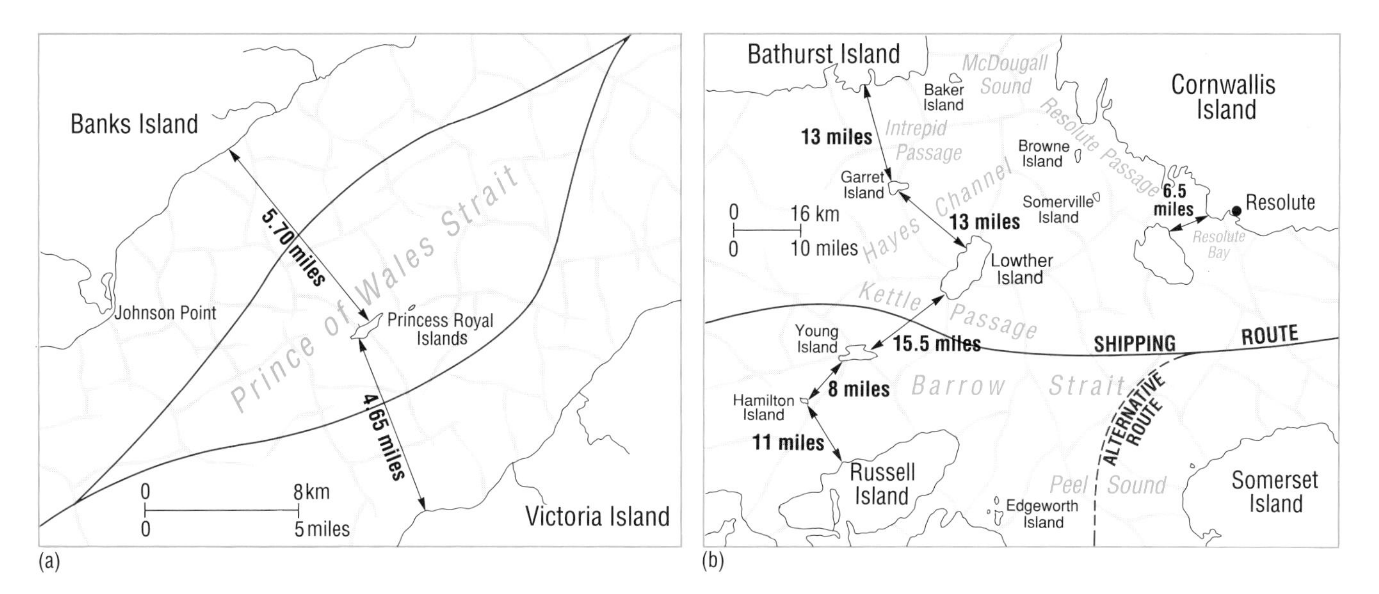

Figure 4.37 Geography of (a) the Prince of Wales Strait and (b) the western part of the Barrow Strait.

QUESTION 4.9 (a) By reference to Figures 4.37(a) and 4.34(b), can you see why this turn of events would have dampened any jubilation over the completed voyage?

(b) Can you also see why Canada then made plans to increase the width of its territorial sea from three to twelve miles?

Canada brought in the legislation to increase the width of the territorial sea in April 1970, at the same time as it brought in the Arctic Waters Pollution Prevention Act referred to earlier. It was not by chance that these measures were being enacted as the *Manhattan* was once again entering Arctic waters (Figure 4.38); the plan was that she should test her capability against the ice as far west as Resolute Bay (see Figure 4.37(b)). This time, the Canadian Government was prepared. The Arctic Waters Pollution Prevention Act would not come into force until 1972, but the *Manhattan* had to comply with a long list of anti-pollution regulations, and had to be accompanied by an ice-breaker. As the United States did not have any ice-breakers, Humble Oil had to ask Canada to supply one. In the event, the Canadian ice-breaker, the *Louis St Laurent*, was certainly needed. Having stopped overnight, the *Manhattan* was unable to get underway again, and the *Louis St Laurent* had to break up the pack-ice around her hull to enable her to reverse out.

The US Government refused to recognize both the increase in the width of Canada's territorial sea and the Arctic Waters Pollution Prevention Act, and took particular exception to the large area of Arctic waters which the Act was intended to control (*cf.* Figures 4.15 and 4.16). The conflict remained unresolved and when, in August 1985, the US Coastguard vessel *Polar Sea* was directed to travel from Thule, on the west coast of Greenland, to the Chukchi Sea where she was to carry out oceanographic work, the US Government refused to ask Canada's permission. The result was that although Canada cooperated in the voyage itself, straight baselines were established around the Arctic Archipelago shortly afterwards (*cf.* Section 4.2.2).

(a)

(b)

Figure 4.38 (a) The *Manhattan* incident was more than a conflict between Canada and an individual oil company (from the *Toronto Star*).

(b) That Canadians regarded the presence of the *Manhattan* in Arctic waters as a threat is shown clearly by this cartoon in the *Toronto Star* of 9 April, 1970.

In January 1988, the United States and Canada signed an Arctic cooperation agreement to the effect that the United States would seek permission from Canada to enter the North-West Passage, while still refusing to concede Canadian sovereignty over it. The strength of the agreement was tested shortly afterwards when the *Polar Star* (sister ship of the *Polar Sea*) failed to break through particularly thick ice off the coast of Alaska: the United States obtained permission from Canada before retreating through the North-West Passage.

You may have wondered why, when discussing the *Manhattan* voyage, we have tended to refer to 'the United States' when it might have seemed more appropriate to say 'the Humble Oil Corporation'. The point is that the US Government was firmly behind the *Manhattan* 'feasibility study', for strong economic reasons. At the time, it was thought that the Prudhoe Bay oilfield would have been so extensive that it could have provided enough crude oil to supply both the east and west coasts of the United States. In the event, it contained only enough oil to warrant transportation to the west coast, via the trans-Alaska Pipeline to Valdez. Of course, it is not only crude oil which is being supplied, but also *jobs*, as refining takes place in the United States.

The United States' behaviour was not a climb-down in the face of superior military strength, because—as Canadian Prime Minister Mulroney put it—Canada had 'not so much as a canoe' to defend her ice-bound territory. The concern over the North-West Passage brought this lack of appropriate naval power home to Canadians. In 1988, the

Canadian Government announced that it would be building a Polar Class 8 ice-breaker, capable of breaking ice eight feet thick at a continuous speed of three knots, so enabling Canada to exercise a reasonable degree of control over surface navigation in the North-West Passage (at least as far as Route 1 through the Prince of Wales Strait is concerned). Such a vessel could escort cargo ships, enforce pollution prevention laws and regulations, and respond to distress calls; furthermore, by being designed to accommodate a weapons system, it could provide a year-round quasi-military presence in the vicinity of the Arctic Archipelago, thereby strengthening Canada's assertion of sovereignty over those waters. However, in mid-1990, after it had been commissioned and laid down, the super-icebreaker was cancelled. At about the same time, plans to acquire a fleet of British- and French-built nuclear-powered submarines, capable of operating under the Arctic ice throughout the year, were also abandoned. As a result of all this, Canada's ability to demonstrate sovereignty over the archipelagic waters was weaker in 1990 than it had been in 1970.

4.4.2 CONSOLIDATION OF CANADA'S TITLE TO THE ARCHIPELAGIC WATERS

We noted in Section 4.2.2 that Canada's claim to historic title of the waters of the Arctic Archipelago is not strong. However, the evidence put forward in support of an 'historic' claim—exercise of state authority, long usage, general toleration of other states, and the vital interests of the coastal state—may be used to *consolidate* title to waters which have been claimed on some other basis. When the various factors are merely being used to support or confirm title acquired by another method, and are not themselves the basis of the claim, there is no special burden of proof on the claimant state, and the legal requirements are not so stringent as for a claim depending primarily on historic usage.

So, how good a position is Canada in to consolidate its title to the waters enclosed by the straight baselines (Figure 4.15)? In particular, how strong is Canada's position with respect to those waters at either end of the routes through the North-West Passage: Lancaster Sound and Barrow Strait in the east, and Amundsen Gulf in the west?

Despite the lack of a strong military presence, Canada has exercised legislative and administrative jurisdiction over the waters concerned for more than 80 years. As we have seen, it can legitimately claim to be protecting three 'vital interests': national security; the unusual Arctic marine environment; and the livelihood and culture of the Inuit.

Clearly, the strong possibility of increased traffic through the North-West Passage has implications both for national security and for the natural environment, and this strengthens Canada's case for the right to enact and enforce protective legislation. Many of the regulations introduced by Canada have been intended to protect the marine environment and the Inuit; for example, the 1970 Arctic Waters Pollution Prevention Act (*cf.* Section 4.2.2) states that Arctic waters are to be navigated only:

> '. . . in a manner that takes cognizance of Canada's responsibility for the welfare of the Eskimo and other inhabitants of the Canadian Arctic and the preservation of the peculiar ecological balance that now exists in the water, ice and land areas of the Canadian Arctic.'

Canada may even have a legal *obligation* to protect the Inuit, because it is a party to the International Covenant on Civil and Political Rights and has pledged not to deny its ethnic minority groups 'the right, in community with other members of their group, to enjoy their own culture . . . '.

Has there been a general toleration of other states towards Canada's activities with regard to the waters involved?

As discussed in Section 4.2.2, the United States objected strongly to the Arctic Waters Pollution Prevention Act. The United States is one of the states most affected by the legal status of Arctic waters and so this objection is significant. However, the World Conservation Strategy, adopted in 1980 by the International Union for Conservation of Nature, has designated Arctic waters as a priority for conservation; also, the Convention for the Protection of the World Cultural and Natural Heritage, to which Canada is party, obliges members to 'take appropriate legal, scientific, technical, administrative and financial measures' to protect 'natural areas of outstanding universal value from the point of view of science, conservation or natural beauty'. In other

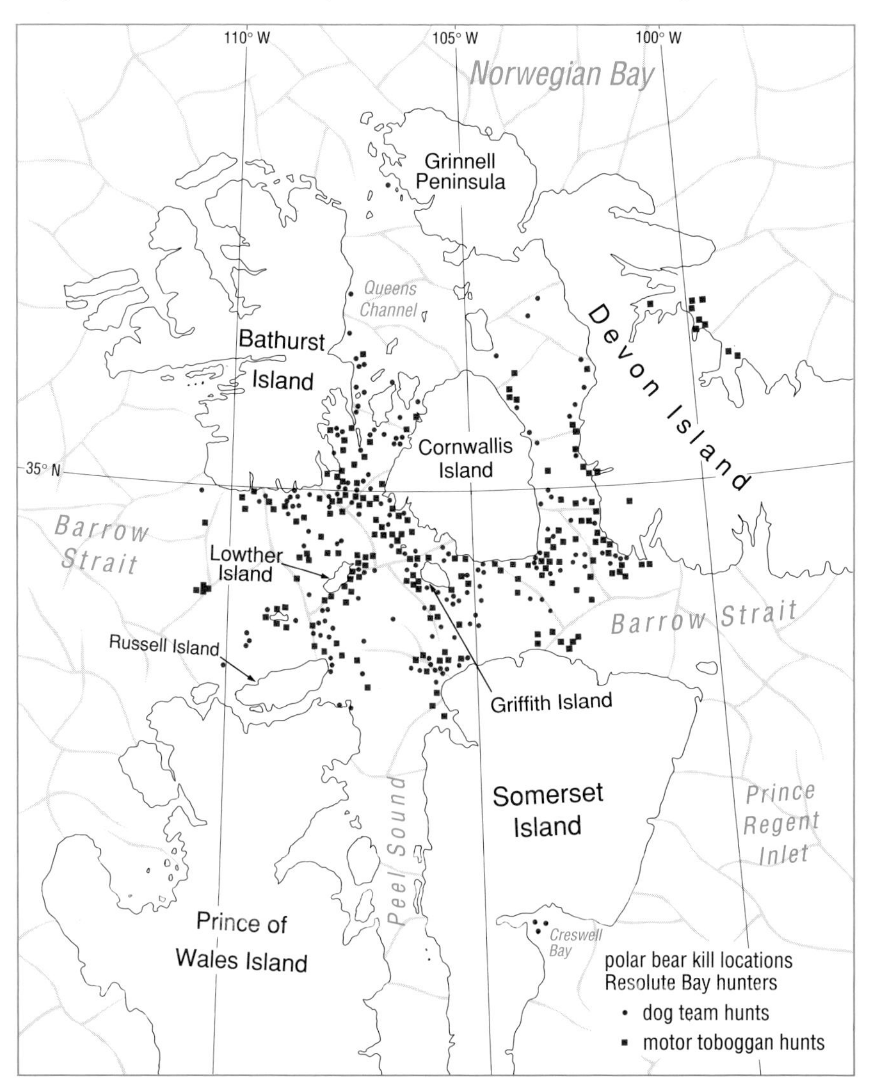

Figure 4.39 Locations of Inuit polar-bear kills around the Barrow Strait.

Article 234 Ice-covered areas

Coastal States have the right to adopt and enforce non-discriminatory laws and regulations for the prevention, reduction and control of marine pollution from vessels in ice-covered areas within the limits of the exclusive economic zone, where particularly severe climatic conditions and the presence of ice covering such areas for most of the year create obstructions or exceptional hazards to navigation, and pollution of the marine environment could cause major harm to or irreversible disturbance of the ecological balance. Such laws and regulations shall have due regard to navigation and the protection and preservation of the marine environment based on the best available scientific evidence.

words, Canada is considered by the world community to have a *duty* to take measures to protect Arctic waters.

Furthermore, the World Heritage Committee, established under the Unesco World Heritage Convention, has recommended that the whole of Lancaster Sound should become a World Heritage Area. Both Lancaster Sound and Barrow Strait are amongst the most biologically productive regions in the whole of the Arctic (in Lancaster Sound alone, the annual production of Arctic cod is estimated to be about 400 000 tonnes): they provide habitats for thousands of seabirds, marine mammals (*cf.* Figure 4.9 and Table 4.1) and polar bears, all of which are important to the Inuit (see Figure 4.39).

The most clear-cut demonstration of a 'general toleration of other states' towards Canada's wish to exercise full jurisdiction over the archipelagic waters is perhaps Article 234 of the 1982 Law of the Sea Convention. This Article, drafted by the Canadian delegation, is reproduced in the box.

The most important point here is that coastal states like Canada have the right to *apply and enforce* their national 'laws and regulations'. If one takes the view that the Convention (which Canada has signed) now represents part of customary international law, the fact that it has not yet come into force is irrelevant.

To summarize: Canada does have a case for consolidation of its title over the waters of the Arctic Archipelago, and has a particularly strong case as far as Lancaster Sound, Barrow Strait and Amundsen Gulf are concerned. However, it should be said that not all states accept that title to 'waters' *can* be consolidated, and of those states which do accept it, some do not regard 'vital interests' as a valid basis for consolidation of title.

4.5 THE ARCTIC AND CLIMATIC CHANGE

At the present time, global temperatures seem to be rising, and it is widely expected that over the next several decades they will continue to do so, at least partly because of the increased concentrations of CO_2 and other greenhouse gases in the atmosphere. We should briefly consider the Arctic in the context of climatic change, partly because global warming will have important impacts on the issues we have been discussing and partly because processes occurring at high latitudes play an important part in the way climatic changes come about.

4.5.1 THE GROWTH AND DECAY OF ICE-CAPS

The Earth's climate is determined both by the amount of solar radiation reaching its surface, and by changes on the Earth itself. The time-scale over which Ice Ages occur (tens of millions of years) strongly implicates changes at the Earth's surface—i.e. movements of the tectonic plates—as their main cause, and indeed major Ice Ages have only developed during periods of mountain building and continental uplift. However, mountain-building on its own does not seem to be sufficient cause for large ice-sheets to develop. The other prerequisite is that one, or both, polar regions are *thermally isolated.* During the warm (i.e. 'normal')

Mesozoic Period (between about 225 and 65 Ma ago), both polar regions were open oceans, and could be warmed by currents flowing from low latitudes; it is possible that at this time air temperatures could remain above freezing even during the dark winter months. By contrast, during the preceding Permo-Carboniferous Ice Age (270–310 Ma ago), the great continent of Gondwanaland was at the South Pole, preventing ocean currents from flowing sufficiently far south to affect southern polar regions. A similar situation exists at the South Pole at the present time; furthermore, the Arctic Ocean is almost completely isolated from the global heat-exchange system provided by ocean currents.

QUESTION 4.10 By reference to Figure 4.18, what is the main route by which water is exchanged between the Arctic Ocean and the world ocean? What is the main warm current affecting the Arctic region?

So, the ease with which heat can be exchanged between low and high latitudes may control whether or not there are large polar ice-caps. However, as we are now in an interglacial period, rather than a 'normal' warm period (see Section 4.1), in the context of present-day global warming it is probably more helpful to consider what brings about changes in ice cover *during* an Ice Age.

Figure 4.40(a) shows the variation in the volume of ice in ice-caps and glaciers over the last 600 000 years or so (it is the same curve as that in Figure 4.2). Figure 4.40(b) shows the variation in intensity of summer sunshine at high northern latitudes over the same period. The amount of solar radiation reaching the Earth varies periodically as a result of cyclical changes in the configuration of the Earth in its orbit. These cyclical changes are known as **Milankovitch cycles**; there are three different cycles, superimposed on one another, having periods of about 22 000 years, 40 000 years and 110 000 years.

From Figure 4.40, which of the three Milankovitch cycles would you say is having the most influence on the size of the ice-caps?

It is the 110 000-year cycle. This results from the Earth's orbit about the Sun becoming alternately longer and thinner (i.e. more elliptical) and more nearly circular. The more elliptical the orbit, the more the Earth–Sun distance changes between **perihelion** (when the Earth is closest to the Sun) and **aphelion** (when it is furthest away). The effect of a very elliptical orbit is to intensify the seasons in one hemisphere and moderate them in the other.

The implication of Figure 4.40(a) is that it is the *intensity* of the seasons which controls the growth and decay of ice-sheets. Furthermore, if you compare the two curves in (a) and (b) you will see that ice-sheets decline in periods when northern summer sunshine levels are rising (despite the fact that northern winter sunshine levels must be *falling*). Correspondingly, ice-sheets advance during periods of relatively mild northern winters but cool northern summers.

You may be wondering how the effect of Milankovitch cycles on seasonal variations in the *Northern Hemisphere* can be transmitted to the Southern Hemisphere, because while Figure 4.40(a) is for global ice cover, Figure 4.40(b) is for the Northern Hemisphere only. Various theories have been put forward, but it seems likely that the answer may

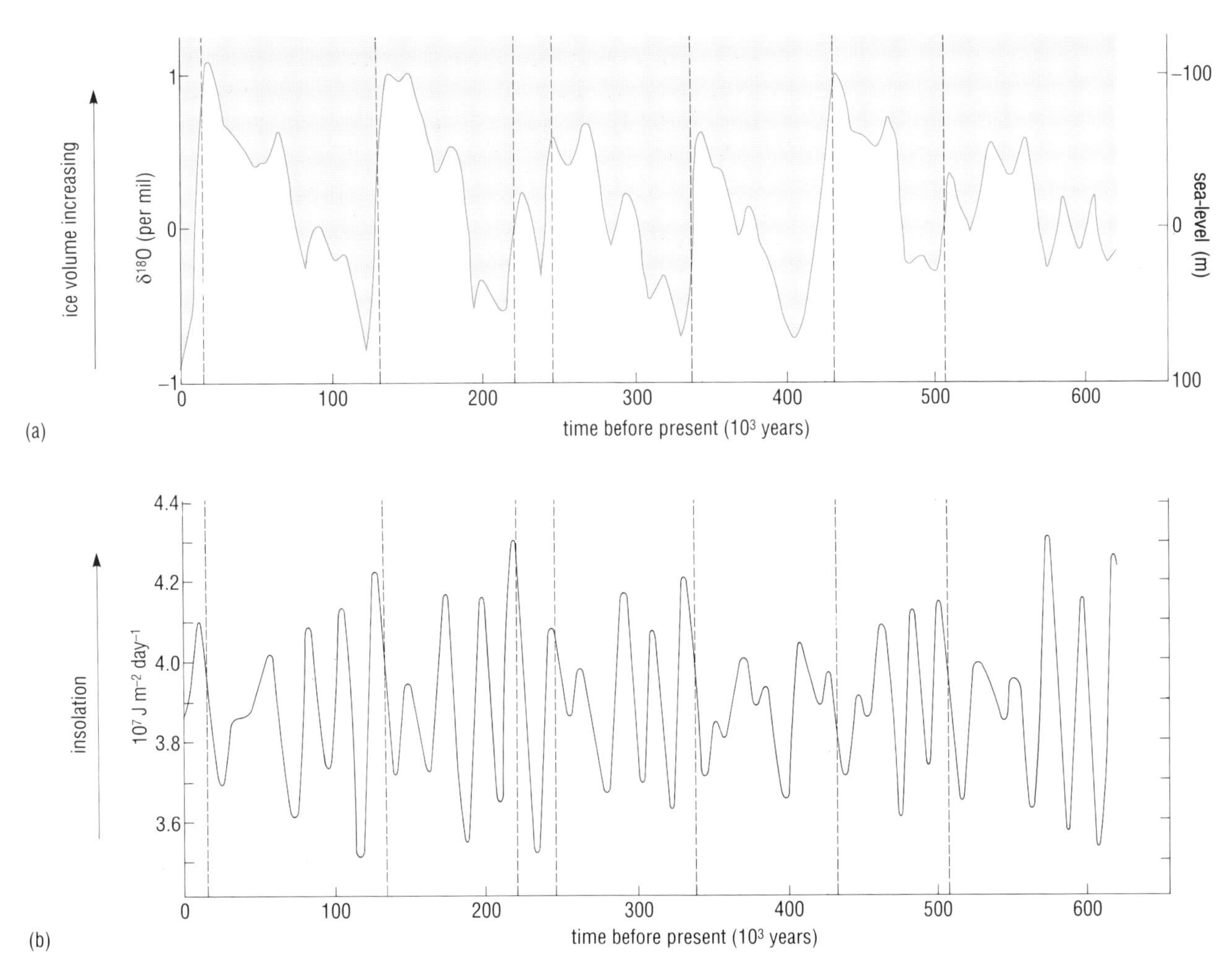

Figure 4.40 (a) The variation in the amount of ice in Northern Hemisphere ice-caps over the past 600 000 years (*cf.* Figure 4.2). (b) The variation of the intensity of summer sunshine at high northern latitudes, over the same period. The variation in insolation caused by changes in the intensity of the seasons is different in the Southern Hemisphere; because of the tilt of the Earth's axis, it is not simply a mirror image of that in the Northern Hemisphere.

lie with the ocean—either the rates at which it takes up CO_2, or the style of **thermohaline circulation**, or a combination of the two.

From Figure 4.40(a), what can you say about the rate at which glacial conditions become established compared with the rate at which they decline?

Figure 4.40(a) shows that ice-caps grow gradually as interglacial turns into glacial; then, after about 110 000 years, they decline rapidly. One explanation for this involves the changing albedo of the Earth's surface as the ice-sheets grow and retreat. For snow and ice, the **albedo**—the percentage of incoming solar radiation that is reflected—can be as high as 80–90%; by contrast, for land and vegetation, which tend to be darker and rough, it can be as low as 10%; for sea at high latitudes, where solar radiation has a shallow angle of incidence, it is usually about 15%. The high albedo of ice and snow cover means that once formed, ice tends to persist; the corollary of this is that if the Earth's surface is exposed by melting, the rate of absorption of solar radiation increases dramatically, and this causes yet more ice to melt.

In the context of global warming, can you now suggest why climatologists think that warming will occur at a higher rate around the rim of the Arctic than at lower latitudes?

Because it will not simply be the sea-ice cover that begins to melt but the snow cover on land. As we have just seen, the feedback loop between ice and albedo has a particularly strong effect as far as land is concerned. Some predictions suggest that over the next 70–100 years the Arctic region will warm by 3–5 °C, while the rest of the globe will only warm by 1°C or less.

4.5.2 IMPLICATIONS FOR MARITIME TRAFFIC

It is not easy to predict in detail the effect of climatic warming on moving sea-ice. First-year ice will not grow so thick, and neither will multi-year ice, and newly opened leads will refreeze more slowly. However, pressure ridges will form more easily and so the average ice thickness could stay about the same, despite the reduced thickness of undeformed ice.

It is likely that the major changes will occur first in those areas which at present experience only seasonal ice cover, i.e. the Arctic's marginal seas and much of its rim (*cf.* Figure 4.5). Figure 4.41 shows the major sea routes around the Arctic, including the various versions of the North-West Passage and the route known as the **Northern Sea Route** (the old North-East Passage). Along both of these routes, the ice cover is seasonal, and all the ice is locally grown first-year ice. Much of this ice is fast ice which grows through the winter and breaks up or melts in the short summer. And for fast ice it *is* possible to make some predictions. The maximum thickness reached by fast ice can be estimated simply by knowing the number of days in which winter air temperatures fall below 1.8 °C (the oceanic heat flux can be ignored, because fast ice occurs in shallow water). On this basis, it has been estimated that within a few decades the fast-ice thickness in the North-West Passage and the Northern Sea Route will be reduced by 0.5 m, and the navigation season along these routes will be extended by two months.

The amelioration of conditions along the North-West Passage and the Northern Sea Route will make them much more attractive as trade routes: the Soviet Union's existing plan to develop the Northern Sea Route may be more readily implemented; oil may be shipped from the Beaufort Sea to markets in eastern Canada and the Far East; and the Arctic Pilot Project (Section 4.3.2) may be revived. Hydrocarbons and other supplies may be shipped to Japan and the countries of the Pacific rim via the North-West Passage, rather than via the Panama Canal as at present—a saving in distance of about 3000 miles.

However, it is unlikely that a trans-polar route will be opened up. For one thing, as mentioned above, a thick ice cover is inherently stable. For another, the generally high stability of surface waters suppresses vertical mixing and turnover which could bring warmer (and more saline) water to the surface.

What aspect of climate, other than temperature, could be affected by 'greenhouse warming' which, as implied above, is likely to progress at different rates in different latitude belts?

Wind patterns, which in turn affect current patterns. These add another dimension to the problem of how the Arctic ice cover could change: as illustrated by Question 4.7, the thickness and distribution of ice is greatly influenced by winds and currents.

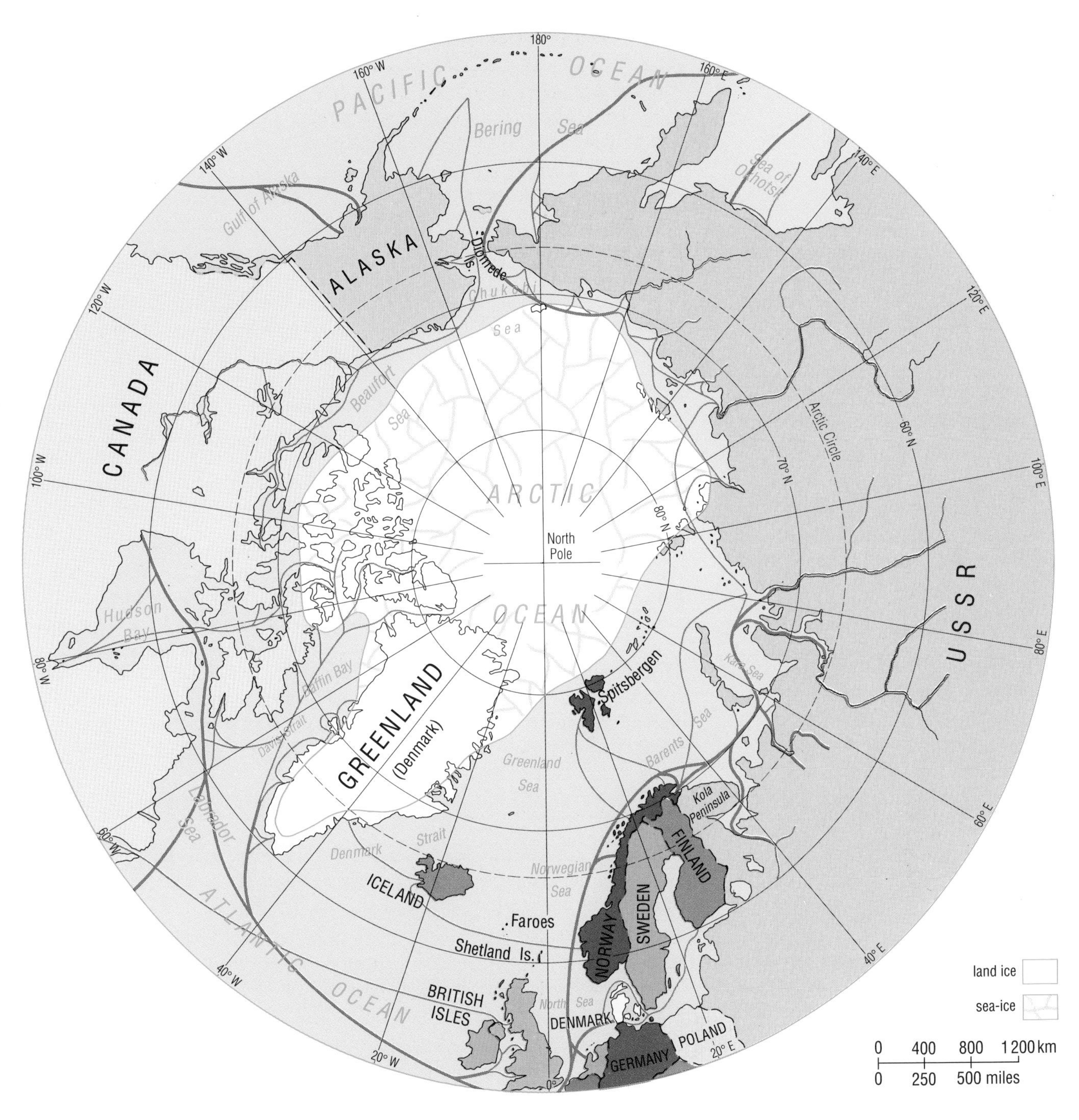

Figure 4.41 Present-day navigation routes through and around the Arctic Ocean. Thick lines indicate major routes; thin lines less important ones. The route known as the Northern Sea Route goes between Novaya Zemlya and the Bering Strait. The white region indicates the extent of summer ice cover.

In the event of global warming, should we expect there to be fewer or more icebergs calving from the Greenland glaciers?

This is difficult to answer. It may well be that for some time there will be an increased rate of iceberg production. We can be sure, however, that icebergs will melt much more quickly than at present so that few—perhaps none at all—will reach the coast of Newfoundland (*cf.* Figure 4.18).

4.5.3 LEGAL IMPLICATIONS

At first sight, it is perhaps surprising that a legal regime that is widely accepted might change in any way. But it is in the nature of laws to evolve as circumstances change. The present situation with respect to the waters of the Canadian Arctic Archipelago is that they are enclosed by the straight baselines declared by Canada in 1985 and are therefore internal waters. On the basis of customary law, as represented by the 1951 *Anglo-Norwegian Fisheries Case*, there is no clear right of passage through these waters.

Under what circumstances (not so far considered in this Chapter) could international shipping exercise the right to undertake 'continuous and expeditious transit' through the North-West Passage?

If it were regarded as a *strait* used for *international* shipping. If it *were* to become accepted as an international strait, then under Article 38 of the 1982 Convention, which is becoming widely accepted as representing customary law, foreign vessels would have the right of 'transit passage', described in Section 3.1.2.

There are three criteria which need to be satisfied if a strait, which in places is less than twice the width of the territorial sea (i.e. one in which there is no strip of high seas down the middle), is to be classified as an international strait. First, the strait must join two parts of the high seas (or two areas of exclusive economic zone) or part of the high seas and the exclusive economic zone of a foreign state. Secondly, the strait must be used for international navigation. And, thirdly, there must not be an alternative route of equal convenience seaward of any island forming the strait.

At present, does the North-West Passage satisfy these criteria?

It certainly satisfies the first criterion: the eastern end of the Passage leads to Baffin Bay, Davis Strait, the Labrador Sea and the Atlantic Ocean, while the western end leads to the Beaufort Sea, the Chukchi Sea, the Bering Strait and the Pacific Ocean. It also satisfies the third criterion, as waters to the north of the various North-West Passage routes are almost by definition more likely to be icebound, and for more of the year.

However, the second criterion, of use by international shipping, is *not* satisfied. Since 1969, there have been about 50 complete transits of the North-West Passage, of which only about 10% have been non-Canadian vessels. There have also been of the order of 200 partial transits, and most have been by Canadian vessels involved with resource extraction from the Arctic islands. Historically, the North-West Passage has *not* been a useful route for international maritime traffic, and this is how the requirement of 'international navigation' has been interpreted; *potential* use for international navigation is not relevant.

But what if global warming leads to Arctic waters in general, and the channels of the North-West Passage in particular, becoming more easily navigable? Will the North-West Passage *then* become an international strait?

It depends. The amount of traffic which is regarded as sufficient to constitute 'use for international navigation' is generally quite substantial. However, given the remoteness of the region, the lack of alternative routes of comparable convenience, and the difficult conditions for navigation (which would still obtain in the Arctic, at least for some decades), a pattern of limited international shipping, developed over a relatively short period, might be held sufficient to make it international. Of course, such 'internationalization' could only happen if Canada allowed foreign ships passage without exercising effective controls over the ships and the waters in question.

4.6 THE POLAR MEDITERRANEAN: WHAT DOES THE FUTURE HOLD?

As the 20th century becomes the 21st, it is inevitable that the Arctic Ocean and its surroundings will change. Its strategic and military importance may wane, but its economic importance will almost certainly grow.

As discussed in Section 4.3, the Arctic is rich in hydrocarbons and other resources; as technology improves and supplies diminish elsewhere, the Arctic environment and the traditional lifestyle of the indigenous people will come under increasing pressure. However, while industrialization erodes traditional lifestyles, it also brings the means to resist unwelcome change. Indigenous Arctic peoples are increasingly coming together (particularly since East–West relations have improved) and, at the same time, are attracting the support of pressure groups in developed countries.

Nevertheless, what constitutes a threat is not always obvious: for example, increased industrial activity and associated shipping could cause a significant increase in noise levels in an otherwise unusually quiet environment. This could severely disturb wildlife, particularly marine mammals. Another possible problem is the continual breaking-up of the pack-ice by ice-breakers—Inuit winter hunting techniques depend on being able to travel for long distances across the pack-ice. Research on topics such as these has already been initiated (e.g. in connection with the Arctic Pilot Project), but there is a danger that changes will occur too fast for researchers to be able to provide meaningful advice.

Effective protection of the Arctic marine environment will become possible only if the countries around its rim cooperate. Indeed, Article 123 of the 1982 Law of the Sea Convention lays upon the circum-Arctic countries the duty to do just this (see box). Progress *is* being made in this area. There is already an international agreement restricting dumping of waste in the Arctic Ocean; and, in February 1989, Finland urged the seven countries bordering the Arctic Ocean to draw up a treaty to limit trans-boundary pollution, protect wildlife and ban dumping of radioactive waste. The treaty would also allow for participation by other countries to the south, which contribute significantly to the pollution.

Article 123 Cooperation of States bordering enclosed or semi-enclosed seas

States bordering enclosed or semi-enclosed seas should cooperate with each other in the exercise of their rights and duties under this Convention. To this end they shall endeavour, directly or through an appropriate regional organization:

(a) To coordinate the management, conservation, exploration and exploitation of the living resources of the sea.

(b) To coordinate the implementation of their rights and duties with respect to the protection and preservation of the marine environment.

(c) To coordinate their scientific research policies and undertake where appropriate joint programmes of scientific research in the area.

(d) To invite, as appropriate, other interested States or international organizations to cooperate with them in furtherance of the provisions of this Article.

Concern about atmospheric pollution (of which the Arctic receives a disproportionate amount; *cf.* Figure 4.33) and global warming has stimulated a surge of interest in the Arctic. Like the Antarctic continent, the Arctic Ocean is an important heat sink; furthermore, as discussed above, if and when definitive signs of greenhouse warming occur, they will be observed first in the Arctic. For all these reasons, it is important that international scientific research in the Arctic be coordinated (Article 123 urges countries to cooperate in this respect; see box), and recent years have indeed seen a significant increase in the number of international scientific projects. With the coming of *perestroika*, it has become possible for eastern and western countries to work together, and for a non-governmental International Arctic Science Committee (IASC) to be established. The founding members of the IASC are Canada, Denmark, Finland, Iceland, Norway, Sweden, the United States and the Soviet Union, but other countries with Arctic research programmes may seek membership. The great strength of the IASC should be its ability to draw on the resources of all its members in investigating a region where logistical problems are often overwhelming; these resources include the United States' satellite programme, Canadian research vessels and a large fleet of powerful Soviet ice-breakers.

4.7 SUMMARY OF CHAPTER 4

1 The Earth is presently in an interglacial period of the Pleistocene Ice Age; the most recent glacial maximum was $\sim$25 000 years ago. The pattern of human settlement around the Arctic Ocean has been greatly affected by climatic fluctuations and the associated changes in ice cover and sea-level. At the present day, a number of indigenous peoples live around the Arctic rim; those of northern Canada and Greenland are the Inuit. The lifestyle of many Inuit is intimately related to marine resources and the natural Arctic environment.

2 According to the doctrine known as the sector theory, polar regions may be divided up between the surrounding states by lines of longitude. While lines of longitude have been used to define territory in the Arctic, the actual regions claimed have always been *land*; the same is true of claims of sovereignty made by virtue of discovery.

3 Although Canada has a fairly good case for claiming the Arctic Archipelago as Canadian territory on the basis of contiguity to the mainland, such a claim cannot be extended to sea areas outside the territorial sea. Furthermore, while some of the states bordering the Arctic Ocean have effectively used the sector theory to claim territory on the Antarctic continent, or to claim land territory, none has consistently supported the application of the sector theory with respect to the Arctic Ocean, much of which, in any case, qualifies as high seas under the Law of the Sea Conventions. Thus, the application of the sector theory in the Arctic cannot be justified on the basis of contiguity nor on the basis of customary law.

3 Canada has claimed that the waters of the Arctic Archipelago are Canadian internal waters on an historical basis. For various reasons, including protests by other states, and the fact that Canada itself has not consistently behaved as if the archipelagic waters *were* internal, Canada's claim that these waters are historic internal waters is not very strong.

4 However, Canada can put forward good arguments in support of the straight baselines which it established around the Arctic Archipelago in

1985. Its case is strengthened by the fact that for much of the year the islands are joined together by ice; furthermore, the Inuit's extensive use of the sea-ice for hunting provides a clear example of an 'economic interest', the reality and importance of which is 'clearly evidenced by long usage'. According to customary law (*Anglo-Norwegian Fisheries Case*), waters newly closed off by straight baselines are internal waters with no right of innocent passage.

5 The Arctic continental shelf is rich in resources, notably hydrocarbons. However, exploitation of hydrocarbons is fraught with difficulty because of the severe environmental conditions, particularly the multi-year pack-ice which drifts under the influence of winds and currents. Drilling rigs may be mounted on bases (including artificial islands) designed to deflect the force of the ice; alternatively, they may be mounted on mobile platforms, successful use of which depends on pack-ice movement being accurately forecast; increasingly, this is being done using remote-sensing techniques. At present, few oilfields in the Beaufort Sea are commercially viable; whether others will become so depends on the economics of transportation out of the Arctic.

6 Oil pollution incidents in the Arctic are potentially very damaging as ice hinders the dispersal of oil, and under low ambient temperatures oil takes much longer than usual to break down; sea-bed blow-outs are a particular hazard. The ecological damage done by an oil spill would depend crucially on where and when the spill occurred.

7 In some senses, the Arctic can rightly be regarded as 'fragile'. A low diversity of species means that food webs are simple, so that damage to a given link in the web could be very serious for the ecosystem as a whole. Furthermore, the low Arctic temperatures mean that growth and reproduction rates are very slow, so that populations affected by an oil spill, for example, would take a long time to recover.

8 The Arctic pack-ice contains algae able to photosynthesize at low light levels. However, by cutting down light levels in the water column, and contributing to the production of a low-salinity surface layer, the sea-ice generally has a negative effect on marine productivity. Large concentrations of organisms may nevertheless be found in the vicinity of polynyas, or near the ice edge during the spring bloom.

9 In addition to the pollution which results from local activities such as resource extraction and fishing, the Arctic Ocean receives pollution from far away, brought in by ocean currents, and in the atmosphere, where it leads to the phenomenon of 'Arctic haze'.

10 Trade with the Orient was the original incentive for Europeans to find a way through Arctic regions to the Pacific; the westerly route through the Arctic Archipelago became known as the North-West Passage. Today, it is the wish to transport oil from the Beaufort Sea to the eastern seaboard of America which provides the United States with the incentive to establish an international seaway through the North-West Passage. Following the entry into the Passage of the ice-strengthened oil tanker *Manhattan*, Canada increased its territorial sea width from three to twelve miles, and brought in the Arctic Waters Pollution Prevention Act. The present situation is that the United States does not acknowledge Canadian sovereignty over the North-West Passage (despite the establishment of the straight baselines) but nevertheless agrees to ask Canada's permission to enter it.

11 Although it does not have a strong military presence in the Arctic, Canada is in a good position to consolidate its title to the waters enclosed by the straight baselines, particularly Lancaster Sound and Barrow Strait at the eastern end of the North-West Passage, and Amundsen Gulf in the west. Canada's case for consolidation of title is helped by the need, accepted by other states, to protect the marine environment and the lifestyle of the Inuit.

12 It seems that Ice Ages are brought about by one or both polar regions being thermally isolated, as a result of the configuration of the continental masses. Climatic fluctuations *within* Ice Ages are probably related to fluctuations in the amount of solar radiation reaching high northern latitudes, as determined by the 110 000-year Milankovitch cycle. In the context of enhanced global warming as the result of the greenhouse effect, it is expected that, because of the ice–albedo feedback loop, warming will occur faster in polar regions than at lower latitudes. Climatic warming will have considerable significance for the viability of the North-West Passage (and other Arctic seaways) as trade routes, and this could affect the legal status of the waters concerned. For scientific research in the Arctic to be successful, it is vital that the states involved cooperate.

Now try the following questions to consolidate your understanding of this Chapter.

QUESTION 4.11 In 1907, Senator Poirier of Canada proposed that Canada, Russia, the United States, Norway and Sweden should each be allocated a sector of the Arctic (Section 4.2.1).

(a) Why was the United States entitled to a sector?

(b) What country was (without any explanation) omitted from Senator Poirier's list?

QUESTION 4.12 We have seen that contiguity cannot be used as a justification for the sector theory. Explain why contiguity is, nevertheless, the basis of most, if not all, claims for maritime territory.

QUESTION 4.13 In 1953, there was a dispute between France and the United Kingdom over two small groups of islands in the English Channel: the Ecrehos, which are located 3.9 miles from the British island of Jersey and 6.6 miles from the French coast; and the Minquiers which are 9.8 miles from Jersey and 16.2 miles from the French coast. Both parties claimed to have an ancient and original title to the islands; during the argument, France invoked the proximity and geographic dependency of the islands with the coast, while the United Kingdom referred to the natural unity of the Channel Islands archipelago. On the basis of what you read in Section 4.2.1, do you think the International Court would have come down in favour of the United Kingdom or France? Why?

QUESTION 4.14 (a) The continental shelf around Alaska contributes 65% of the total US legal continental shelf area (*cf.* Figure 3.1). What geographical feature is largely responsible for this large area of US continental shelf?

(b) According to the 1982 Law of the Sea Convention, in which maritime zones would the Bering Sea oilfields off Alaska (Figure 4.17) be located? (i) The US territorial sea. (ii) The US exclusive economic zone. (iii) The US legal continental shelf. (iv) The high seas.

(c) What are the implications of your answer to (b) so far as the profits of the US offshore oil industry are concerned?

QUESTION 4.15 (a) In discussion of likely pollutants of the Arctic environment, we mentioned *sound* (Section 4.6). What other type of energy produced by human activities could also be particularly disruptive in polar regions?

(b) Bearing in mind what you read in Section 4.5, can you suggest why oil spills are regarded by some scientists as a threat to the stability of the Arctic pack-ice?

QUESTION 4.16 (a) What might the significance of 'Arctic haze' be so far as global warming is concerned?

(b) One of the predicted consequences of global warming is that while sea-ice begins to melt, land ice—e.g. the glaciers of Greenland and Scandinavia—may well expand in volume. Why is this?

CHAPTER 5 THE GALÁPAGOS: ISLANDS OF VARIETY AND CHANGE

'The archipelago is a little world within itself, or rather a satellite attached to America, whence it has derived a few stray colonists, and has received the general character of its indigenous productions. Considering the small size of these islands, we feel the more astonished at the number of their aboriginal beings, and at their confined range. Seeing every height crowned with its crater, and the boundaries of most of the lava-streams still distinct, we are led to believe that within a period geologically recent the unbroken ocean was here spread out. Hence both in space and time, we seem to be brought somewhat near to that great fact—that mystery of mysteries—the first appearance of new things on this earth.'

Charles Darwin, 1845

Darwin wrote this description of the Galápagos Islands in his *Journal of Researches into the Natural History and Geology of the Countries visited during the Voyage of HMS Beagle around the World*. The words convey his intense interest not only in living creatures but in the whole of the natural world. He realized that the unusual flora and fauna of the archipelago were the result of its short geological history and its setting in the eastern equatorial Pacific, far from land. In this Case Study we look at how processes under and within the oceans have had—and continue to have—a profound effect on the Galápagos Islands, their plants and animals, and their human population.

5.1 THE TECTONIC SETTING

Geologically, the Galápagos archipelago is very young, probably between 3 and 5 million years (Ma) old. Its islands and islets are the tips of the tallest **seamounts** that together form the Galápagos Plateau (Figure 5.1(a)). They are still active volcanically, especially the westernmost islands—Isabela and Fernandina (Figure 5.1(b))—and are believed to lie at present over the mantle '**hot spot**' which generated the Plateau. The geographic trends of certain chains of islands and seamounts—notably the Line Islands and the Hawaiian–Emperor Chain—appear to be related to the motion of the Pacific Plate over more or less immobile hot spots in a fairly straightforward manner (Figure 5.2). The same is not true of the Galápagos, which lies just south of the Galápagos **spreading axis** (Figure 5.3). This spreading axis appears to have been initiated about 25–30 Ma ago along the line of a pre-existing **fracture zone**, splitting the oceanic plate on the east side of the East Pacific Rise into the Cocos Plate to the north and the Nazca Plate to the south of the new axis (Figure 5.4). It is possible that the initiation of this spreading axis was related to the onset of hot-spot activity, and certain that the axis lay almost directly over the hot spot for about the first 20 Ma of its existence. The extra submarine volcanism over the hot spot created an aseismic ridge on both the Cocos and Nazca Plates—the Cocos Ridge and the Carnegie Ridge respectively.

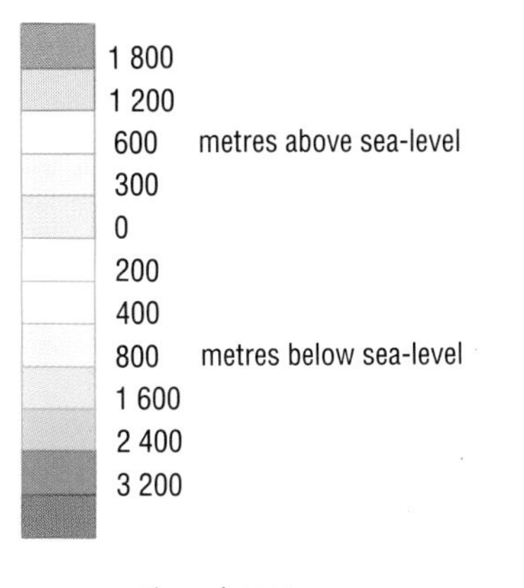

* active volcanoes

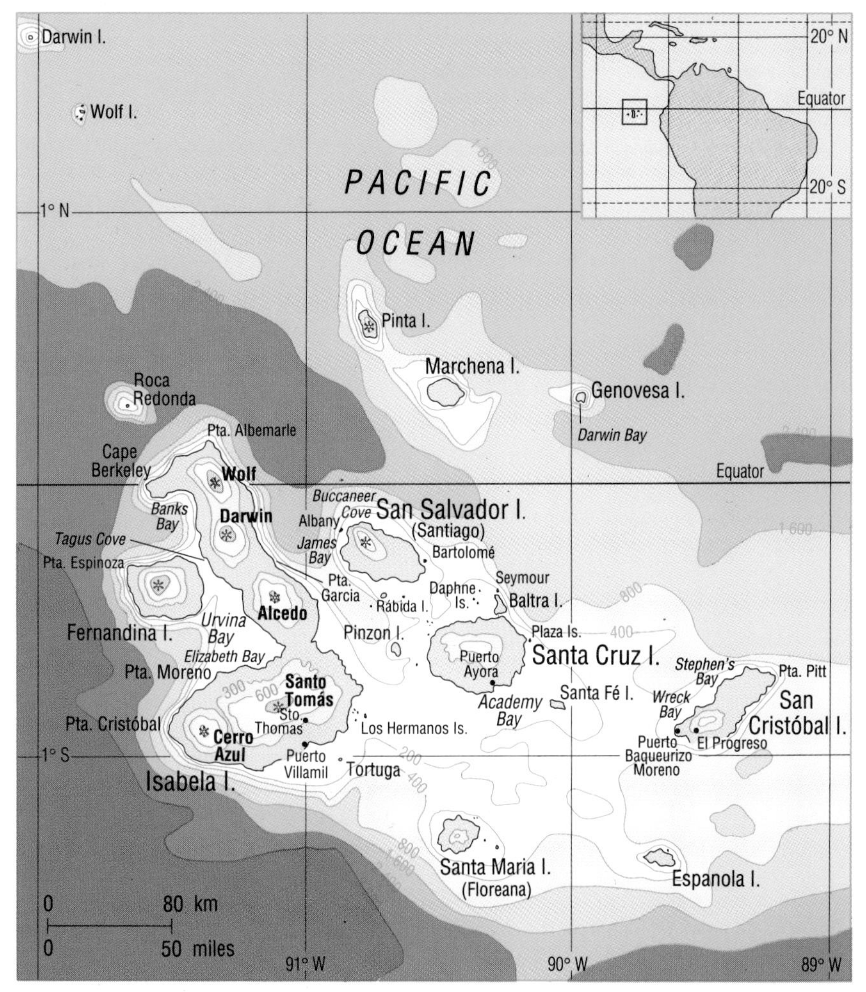

Figure 5.1 (a) The Galápagos archipelago in relation to the Galápagos Plateau. Red stars indicate volcanic centres that have been active during the 20th century. Contours are at intervals of 200 m above sea-level (yellows, browns and greens) and of 200 m, 400 m and 800 m below sea-level. Western and southern aspects of the submarine plateau are steep, while to the east and north-east (where the Cocos and Carnegie Ridges merge, see text) water depths increase gradually.

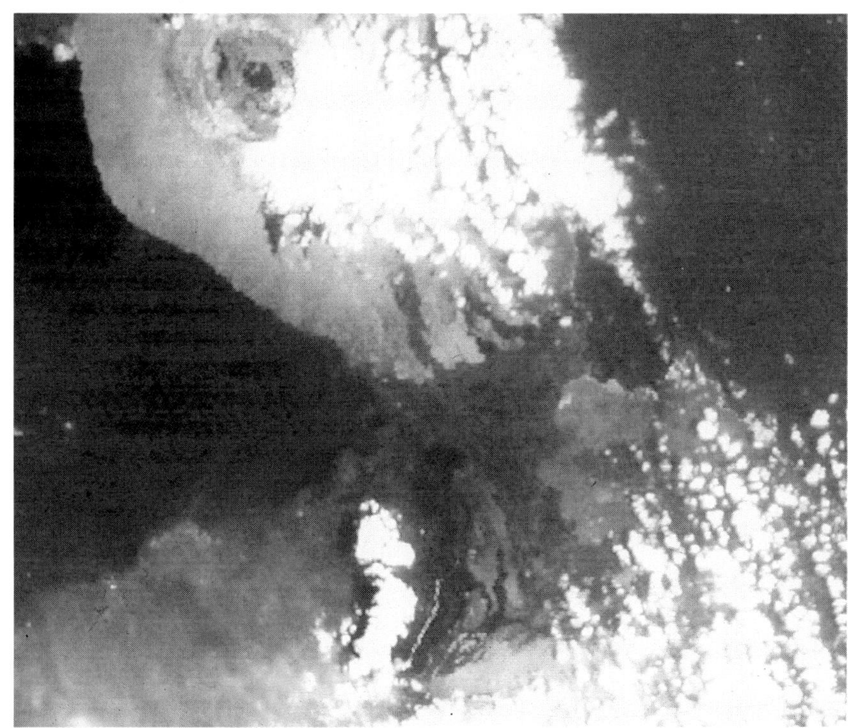

(b) *Landsat* image of part of Isabela Island, in November 1979. An eruption was in progress on the volcano Cerro Negro, producing a lava flow, the red glow from which can be seen extending northwards from the summit towards the coast.

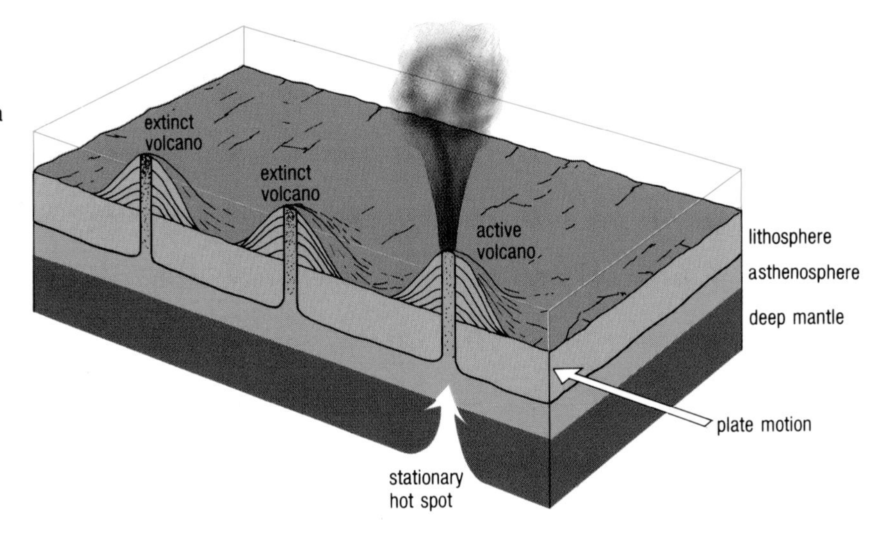

Figure 5.2 Schematic diagram (not to scale) illustrating how volcanic island chains may be formed by an oceanic plate moving over a mantle plume or 'hot spot'. If volcanism is less episodic, a continuous submarine ridge, known as an aseismic ridge, may be formed instead of discrete islands or seamounts.

Figure 5.3 The tectonic setting of the Galápagos Islands, at the western end of the Carnegie Ridge. The position of the propagating axis tip at 95.5° W (see text; *cf.* Figure 5.6) is indicated by the red box; the red ring on the Carnegie Ridge indicates the position of the hydrothermal vent field mentioned in Section 5.6.3.

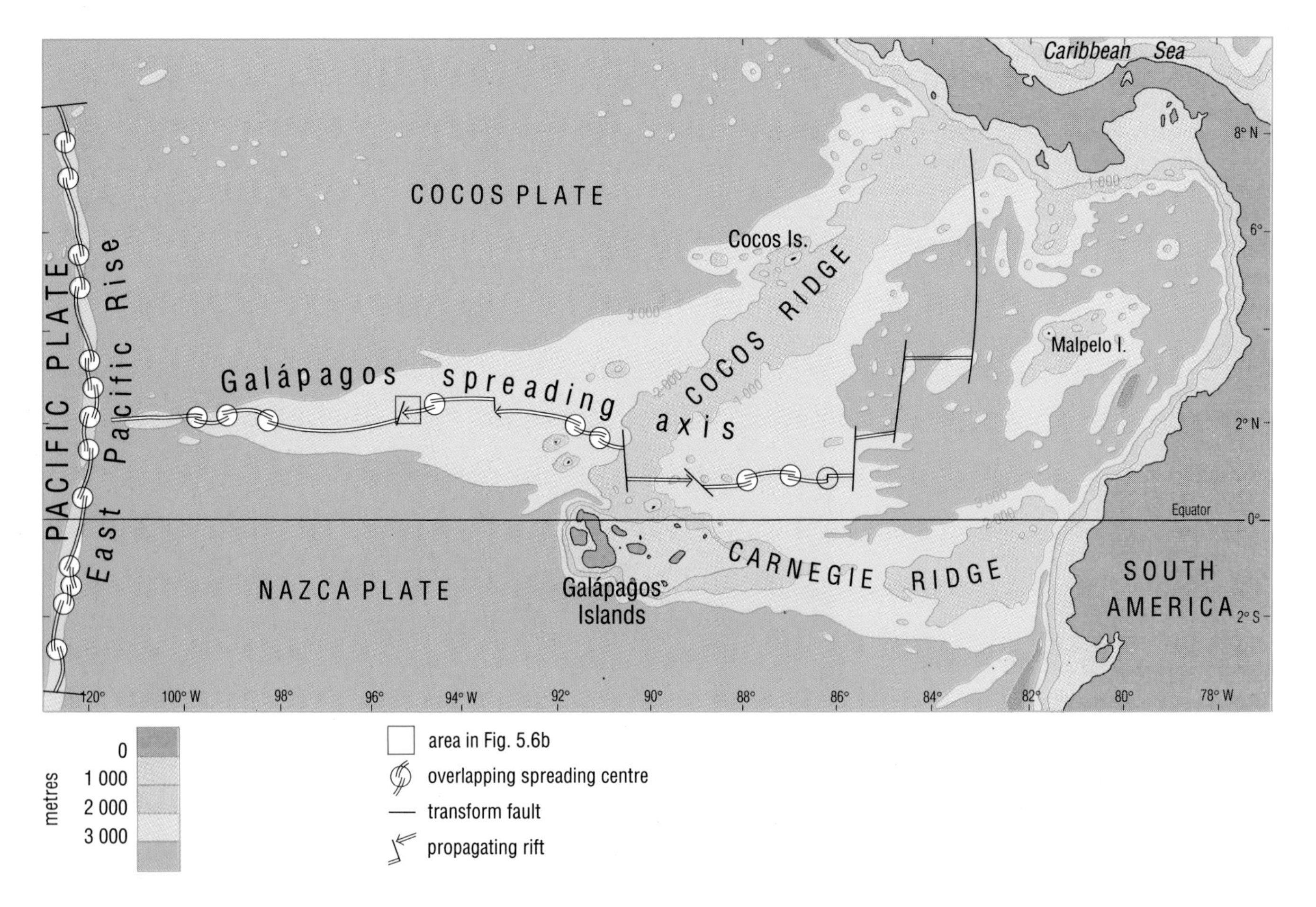

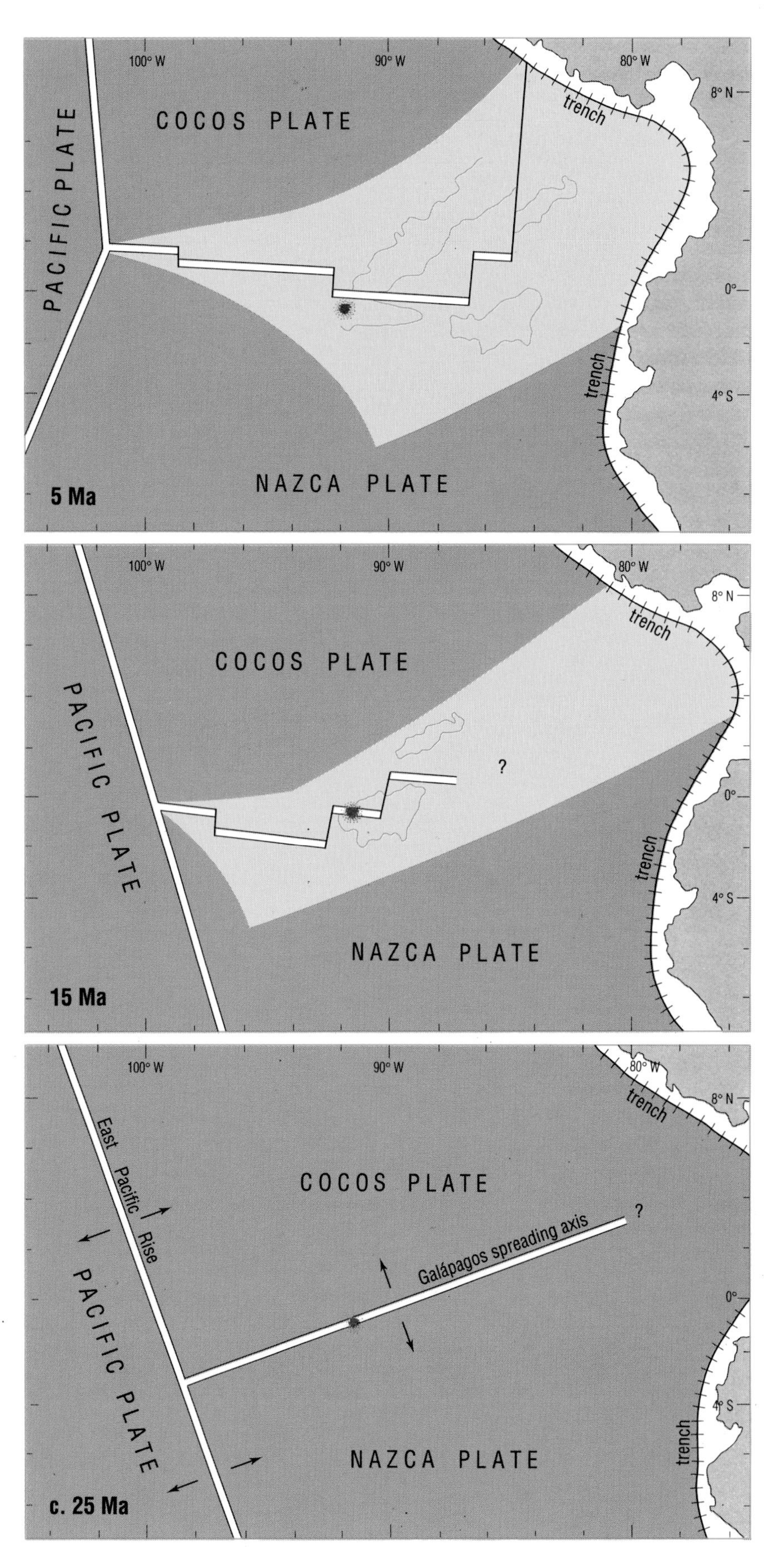

Figure 5.4 The development of the Galápagos spreading axis and its motion relative to the hot spot (shown in red). Oceanic crust generated by spreading at the Galápagos spreading axis is shown in orange; oceanic crust generated at the East Pacific Rise is shown in brown. The approximate positions of the Cocos and Carnegie Ridges are indicated by the blue contour. See Figure 5.3 for the present-day situation.

QUESTION 5.1 Figure 5.3 shows that these ridges form a 'v' shape, converging on the Galápagos spreading axis. Can you think why these ridges are not at right angles to the spreading axis?

By about 5 Ma ago the Galápagos spreading axis had drifted north of the hot spot (Figure 5.4), so at this time hot-spot volcanism ceased to add to the Cocos Ridge and from then on was confined essentially to the young end of the Carnegie Ridge only.

What is the youngest part of the Carnegie Ridge? And what is happening to the oldest part?

The youngest part of the Carnegie Ridge is the Galápagos Plateau (*cf.* Figure 5.3) and the oldest part is being **subducted** at a trench off the coast of Ecuador (generally known as the Peru–Chile trench). The far end of the Cocos Ridge is being subducted beneath Central America in a similar fashion.

The details of the disposition of the Galápagos Islands cannot be entirely explained by the simple hypothesis shown in Figure 5.2. Individual volcanoes are not arranged in a single line along the Carnegie Ridge in order of increasing age; instead, the islands cover a broad area, and adjacent volcanoes have overlapping age ranges with no obvious trend parallel to the hot-spot trace.

Volcanoes on several of the Galápagos Islands have erupted during the past century (Figure 5.1(a)). Geochemical studies of the basalts and related volcanic rocks occurring on the islands suggest that magmas generated from two distinct sources—partial melting of the mantle beneath the spreading axis and partial melting of mantle material brought from much greater depths by the rising hot-spot plume—have been mixed in varying proportions from place to place and from time to time as the islands have developed. Evidently the hot spot is a somewhat diffuse feature and is complex in its surface expression.

The interplay between the two magma sources as expressed in the chemistry of the erupted lavas may be at least partly related to the style of tectonic evolution of the Galápagos spreading axis. For much of its length, the axis has no **transform faults**. Instead, it is broken into segments terminated by **overlapping spreading centres**, where portions of the axis are starved of magma, but the tips of ridge segments that are well-supplied can propagate and eventually truncate the ends of adjacent segments (Figure 5.5).

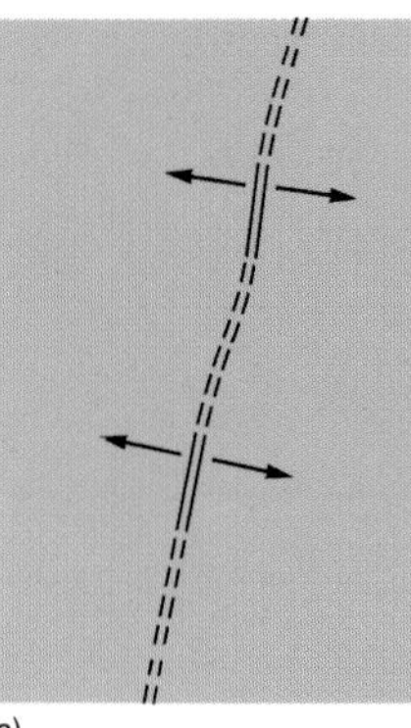

(a)

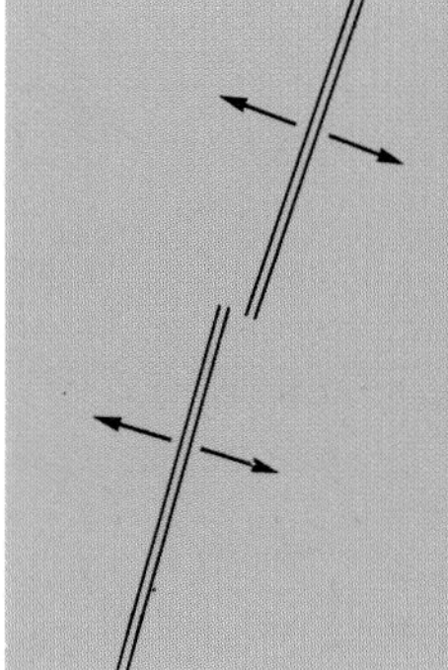

(b)

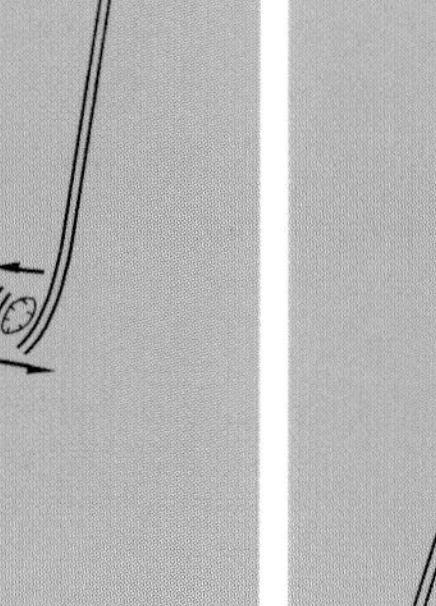

(c)

(d)

(e)

Figure 5.5 Model for the formation and evolution of an overlapping spreading centre. Double lines show the axial crest; lines are continuous over active segments of the axis, which are well-supplied with magma, and broken over lengths where the magma supply has failed.

(a) Start of the cycle: activity has been subdued for some time in the gaps between active segments.

(b) New supplies of magma cause one or (in this case) both segments to propagate.

(c) Misalignment between propagating tips causes them to overlap, and the stress field encourages them to curve in towards one another.

(d) The more vigorous of the propagating tips truncates the other one. At this stage, the underlying magma chambers may have merged.

(e) The old overlap tip is abandoned, but remains as a topographic high

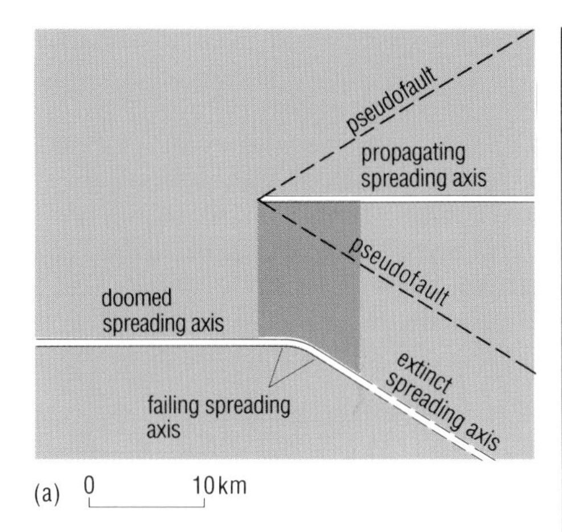

Figure 5.6 (a) Simplified sketch of the 95.5° W propagator on the Galápagos spreading axis. The darker area is the transform zone, a region of shearing and rotation of the crust. The two 'pseudofaults' are the boundaries between crust created at the failing spreading axis and the propagating spreading axis.

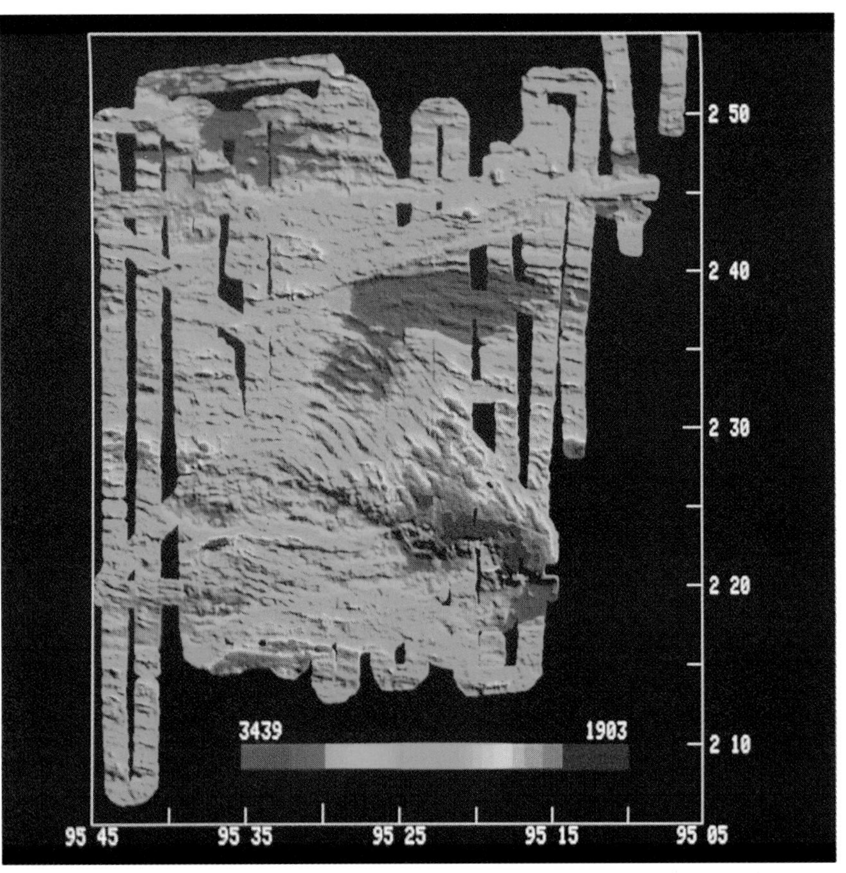

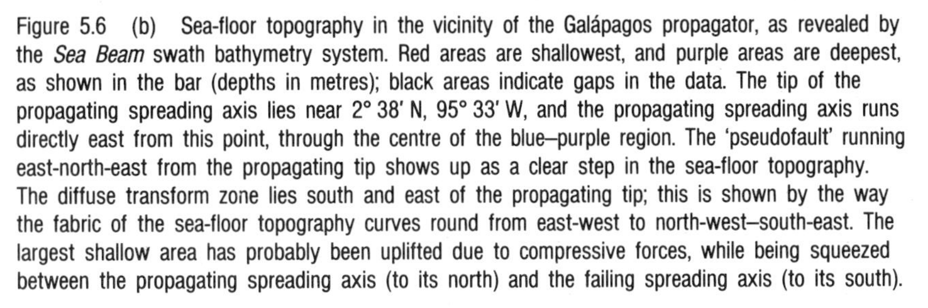
Figure 5.6 (b) Sea-floor topography in the vicinity of the Galápagos propagator, as revealed by the *Sea Beam* swath bathymetry system. Red areas are shallowest, and purple areas are deepest, as shown in the bar (depths in metres); black areas indicate gaps in the data. The tip of the propagating spreading axis lies near 2° 38′ N, 95° 33′ W, and the propagating spreading axis runs directly east from this point, through the centre of the blue–purple region. The 'pseudofault' running east-north-east from the propagating tip shows up as a clear step in the sea-floor topography. The diffuse transform zone lies south and east of the propagating tip; this is shown by the way the fabric of the sea-floor topography curves round from east-west to north-west–south-east. The largest shallow area has probably been uplifted due to compressive forces, while being squeezed between the propagating spreading axis (to its north) and the failing spreading axis (to its south).

Clearly the waxing and waning of magma supply along the axis could influence the volcanism on the islands. Quite apart from overlapping spreading centres, at 95.5° W the Galápagos spreading axis contains one of the best-studied examples of another class of ridge offset, a *propagating rift* (Figure 5.6(a)). In this case, two lengths of ridge are offset by 13 km (*cf.* the 2–15 km offset typical of overlapping spreading centres), and one of these is propagating at the expense of the other. Instead of having a discrete transform fault across the offset, there is a block of ocean floor in the overlap zone between the propagating spreading axis and the failing spreading axis. This acts as a diffuse transform zone and is subject to considerable amounts of stress and rotation. Marine scientists have begun to unravel the tectonic and volcanic history of the 95.5° W propagator by means of detailed bathymetric studies using side-scan sonar and narrow-beam echo-sounding (Figure 5.6(b)). Further detailed surveying remains to be done around the islands, to learn how much this part of the ocean floor has been affected by similar processes.

5.2 THE OCEANOGRAPHIC SETTING

The most striking aspect of the Galápagos climate—both above and below sea-level—is its variability. You might suppose that, because they lie on the Equator, the islands experience year-round hot sunshine and are constantly bathed by warm waters. In fact, their position in the eastern equatorial Pacific means that conditions around the islands change dramatically as winds and ocean currents vary seasonally in strength and direction.

The most powerful ocean current affecting the equatorial Pacific is the **South Equatorial Current** (Figure 5.7(a)). This is fed mainly by the Peru Current, which carries water chiefly from the South Pacific **sub-tropical gyre**.

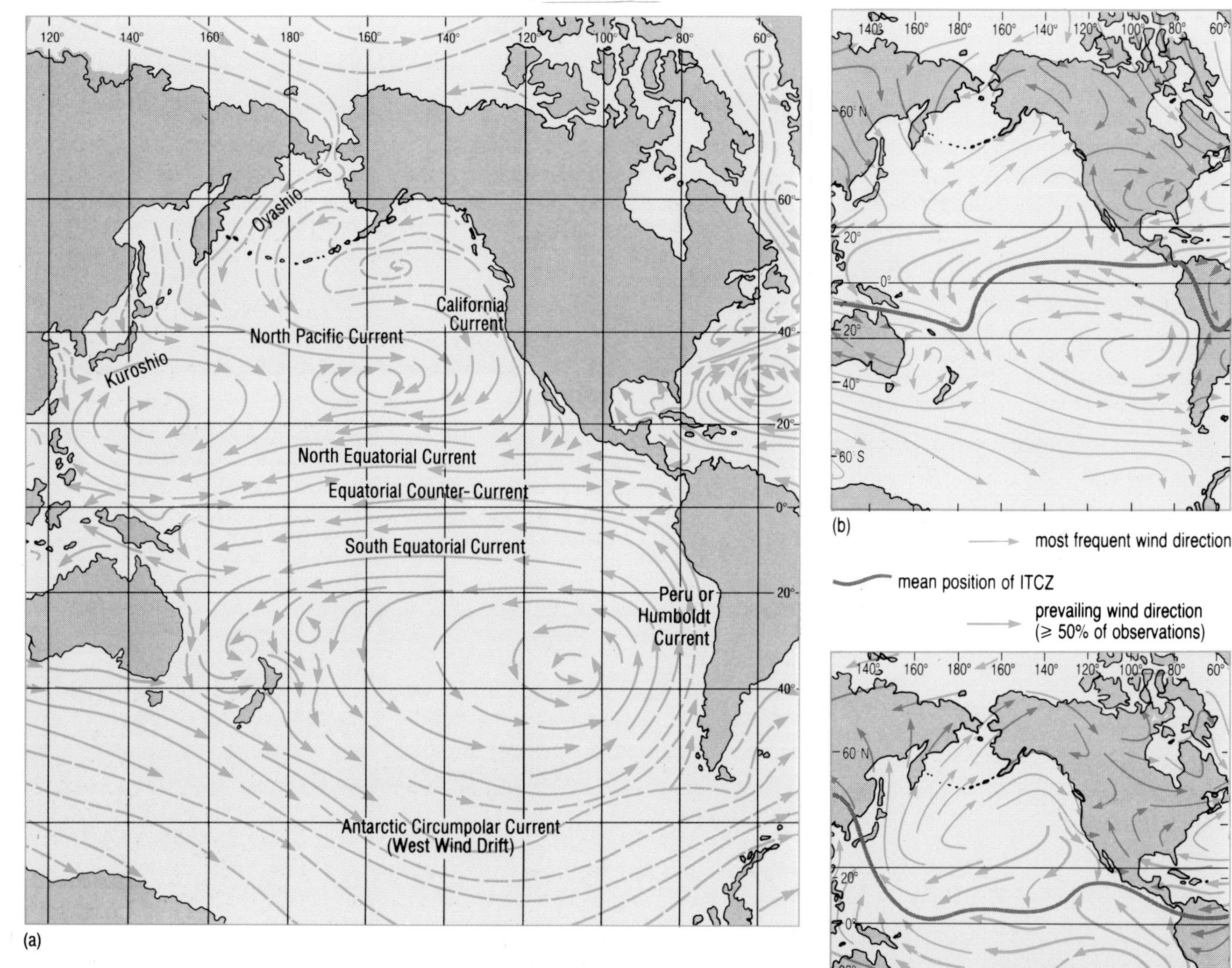

Figure 5.7 (a) The surface current system of the Pacific Ocean, showing average conditions for the Southern Hemisphere summer. Solid lines show warm currents; broken lines show cool currents. For seasonal differences in the Galápagos region, see Figure 5.8.

(b) Prevailing surface winds in the Pacific, and the position of the Intertropical Convergence Zone (ITCZ) for January (southern summer/northern winter).

(c) Prevailing surface winds in the Pacific, and the position of the ITCZ for July (southern winter/northern summer).

This water has been circulating beneath a sub-tropical high-pressure region, where evaporation rates are high and precipitation rates are low. It will therefore be both fairly warm (20–24 °C) and saline (salinity ~35).

The strengths of the Peru Current and the South Equatorial Current vary according to the strength of the South-East Trade Winds. These blow persistently between June and September, so the Peru Current and the South Equatorial Current are best developed then. At this time of year, the **Intertropical Convergence Zone (ITCZ)**—the zone of light winds (Doldrums) which corresponds roughly to the region of highest sea-surface temperatures—is at its most northerly position: between 8° N and 17° N in the eastern Pacific (Figure 5.7(c)).

During this season (the Southern Hemisphere winter), water in the Peru Current is cooled as a result of the South-East Trades blowing obliquely along the edge of the South American land-mass.

How is this?

The winds cause offshore movement of surface waters, which allows **upwelling** of cooler sub-surface water along the coast; and upwelling also occurs along the Equator, within the South Equatorial Current. As a result of these effects, water reaching the Galápagos during the Southern Hemisphere winter is relatively cool for equatorial latitudes—generally less than 24 °C, and sometimes as low as 18 °C.

The two current patterns associated with stronger and weaker South-East Trade Winds are shown in Figure 5.8. Figure 5.8(a) shows that during the Southern Hemisphere winter, the South Equatorial Current consists not only of water from the Peru Current but also of water which has turned south from the **Equatorial Counter-Current** (also sometimes referred to as the North Equatorial Counter-Current). The mauve zone in Figure 5.8 is the **Equatorial Front** (or Convergence). As the Southern Hemisphere summer develops, the South-East Trade Winds weaken and the ITCZ moves to its southerly position—although, in the

Figure 5.8 (a) Currents in the eastern equatorial Pacific during the Southern Hemisphere winter.

(b) Currents in the same region during the Southern Hemisphere summer.

Mauve represents warm tropical water; blue is cool subtropical and equatorial water; the grey zone is the Equatorial Front (see also Figure 5.9). The Galápagos archipelago lies just below the Equator at about 91° W. Note: off northern South America, the Peru Current is made up of the relatively warmer and more saline Peru Oceanic Current and the relatively cooler and less saline Peru Coastal Current. (The alternative name of the Humboldt Current strictly speaking only applies to the Peru Coastal Current.)

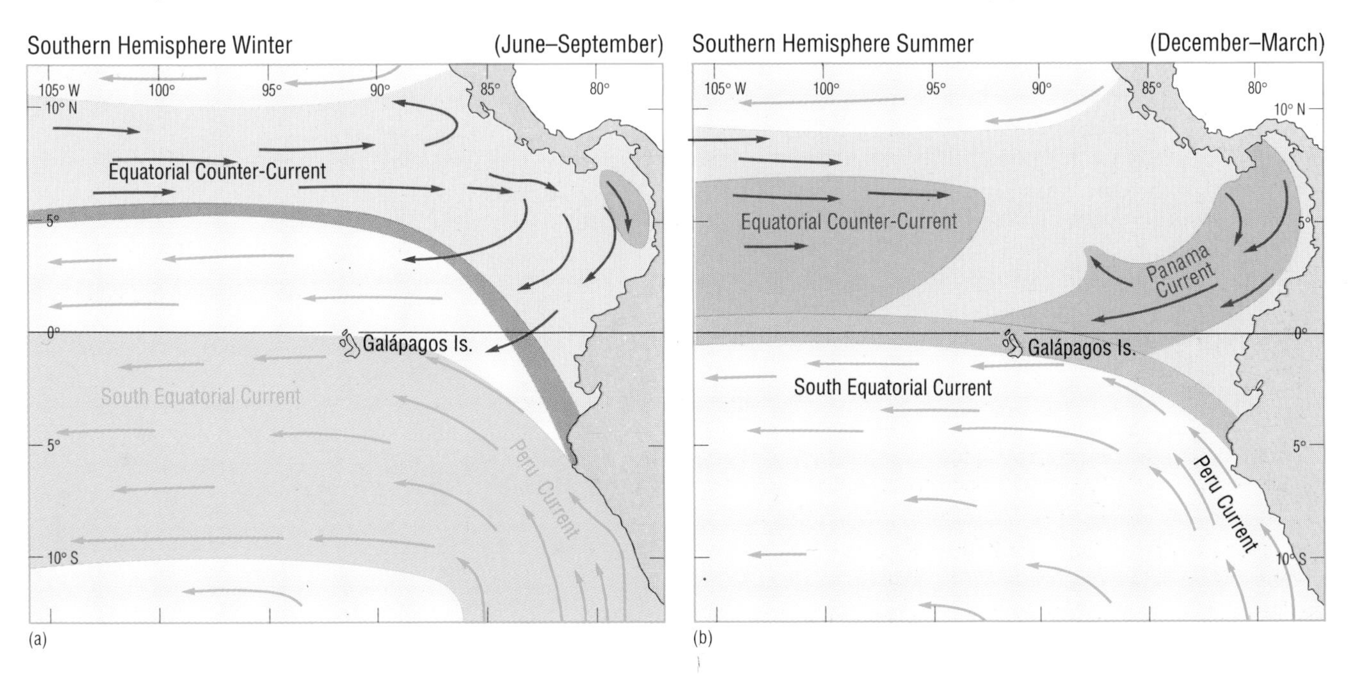

eastern Pacific, it usually remains in the Northern Hemisphere (Figure 5.7(b)). The South Equatorial Current has now diminished and the boundary between it and the Equatorial Counter-Current—the Equatorial Front—has shifted southwards by about 4° of latitude. The water in the Equatorial Counter-Current now flows through the tropical zone of high temperatures and high rainfall for a number of months and so is warmed and reduced in salinity by the time it reaches the eastern equatorial Pacific. The Galápagos Islands are now also affected by warm, low-salinity tropical waters which flow south from the Panama Bight as the Panama Current.

As well as being the boundary between the Equatorial Counter-Current and the South Equatorial Current, the Equatorial Front is the boundary between warmer and less saline tropical waters to the north and cooler and more saline equatorial and subtropical waters to the south. In the vicinity of the Galápagos archipelago, this boundary is known as the *Galápagos Front*. The long-term mean position and average temperature–salinity characteristics of the Front are summarized in Figure 5.9, although clearly its position changes with the seasons (Figure 5.8(a) and (b)). The strength of the Front (i.e. the sharpness of the change in water characteristics across it) also varies with the seasons.

By reference to Figure 5.8 and from what you have read so far, would you expect the Front to be better developed in the Southern Hemisphere winter or the Southern Hemisphere summer?

During the Southern Hemisphere winter, when the waters of the South Equatorial Current are coolest. From May to November, the change in temperature across the Front can be up to 5 °C in 50 km. From January to March, southern waters are sufficiently warm for the Front effectively to disappear as far as sea-surface temperature is concerned.

Figure 5.9 (a) Mean temperature and (b) mean salinity concentrations along 95° W, based on historic hydrographic data. The arrow indicates the mean position of the Galápagos Front.

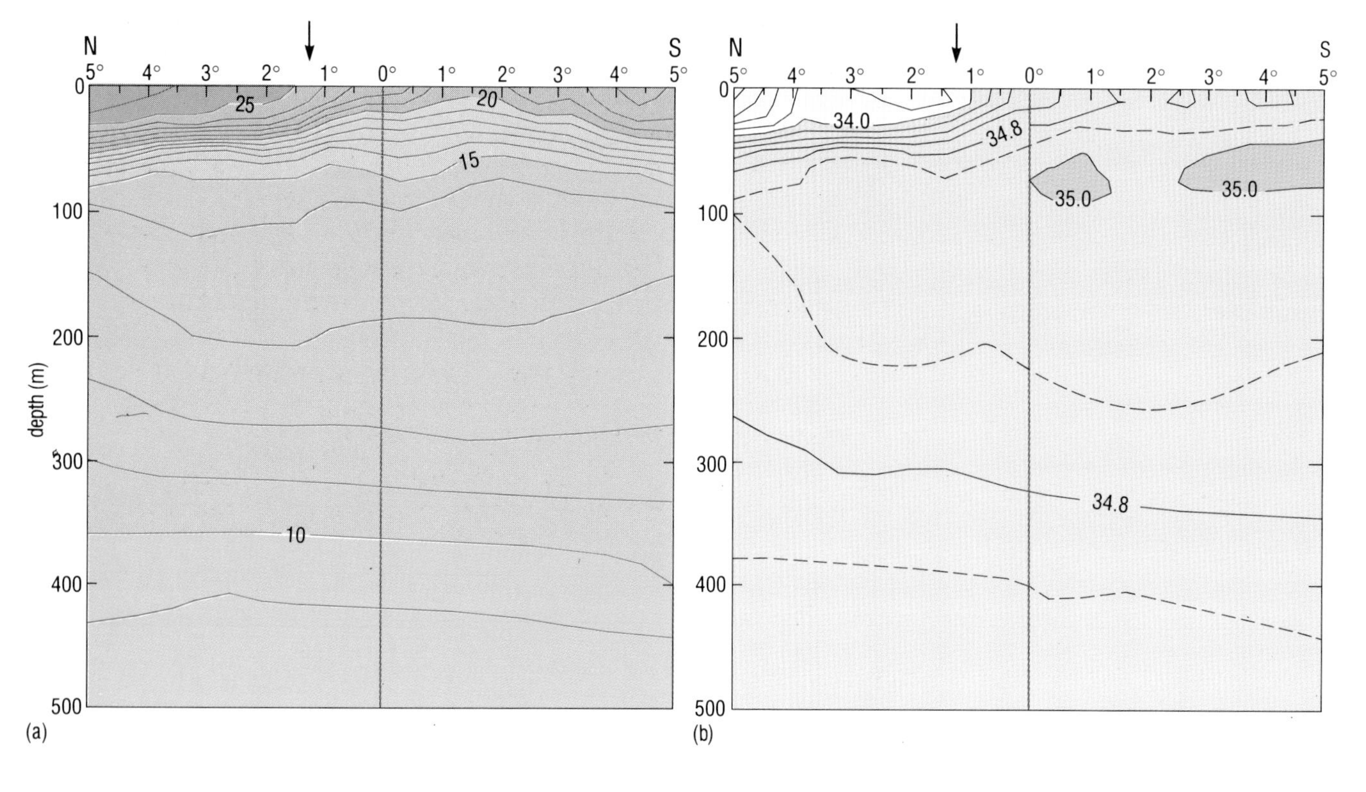

Would it still be possible to detect the boundary between the two water masses?

Yes, by using salinity (and perhaps other indicators such as nutrient content or oxygen concentration). The salinity gradient across the Front can be as much as 1 (part per thousand) in 50 km.

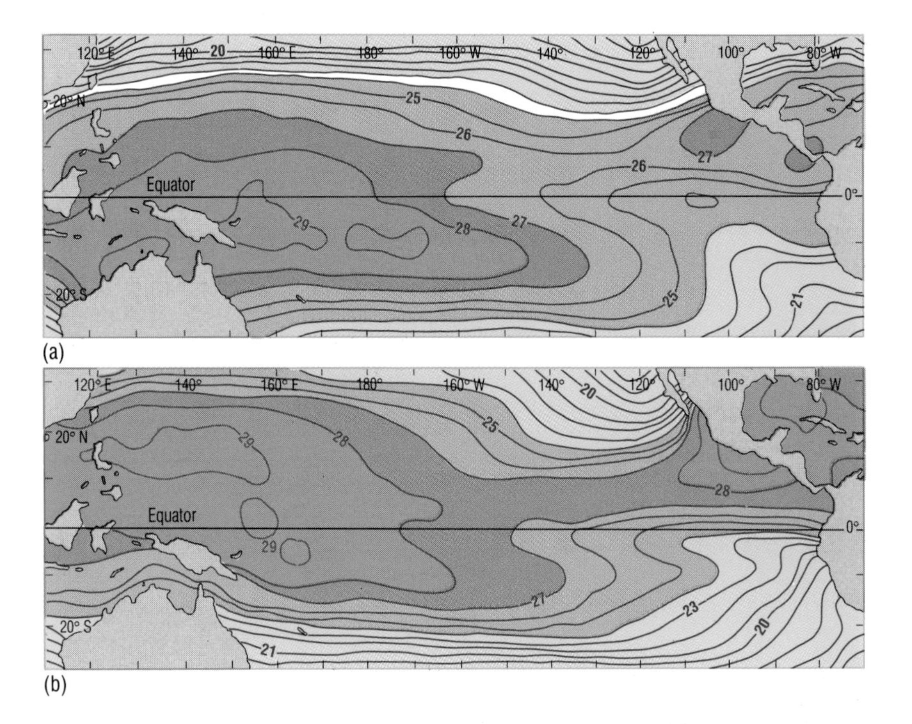

Figure 5.10 Mean sea-surface temperature contours (1 °C intervals) in the tropical Pacific for (a) January and (b) July.

At this point, it is appropriate to place Figures 5.8 and 5.9 in a broader context. Figure 5.10 shows average sea-surface temperatures in the tropical Pacific for the two main seasons of the year. Salient points to note are:

1 In the western half of the Pacific, there is a more or less permanent 'pool' of warm surface water straddling the Equator. This warm-water 'pool' shifts slightly northwards in the Southern Hemisphere summer (January) and southwards in the Southern Hemisphere winter (July).

2 In the eastern half of the Pacific, sea-surface temperatures are generally greater to the north of the Equator than to south of the Equator.

3 In the Southern Hemisphere winter (July) when the South-East Trades are normally strong, equatorial waters in the eastern Pacific are actually cooler than those off Central America.

4 In the eastern Pacific, the north–south gradient in surface temperature is steeper during the Southern Hemisphere winter (July).

QUESTION 5.2 Below are given descriptions of the atmosphere and ocean in the vicinity of the Galápagos during the two seasons of the year. Read these descriptions carefully and decide which set of conditions characterizes the season of stronger South-East Trades (June–September, i.e. Southern Hemisphere winter) and which typifies the season of weaker South-East Trades (December–March, i.e. summer).

1 During this time of year, local winds are gentle and the sea is calm. Skies are clear except for large cumulus clouds which produce tropical thunderstorms. This is called the wet season.

2 During this time of year, the sky is often overcast with low stratus clouds and a persistent misty drip known as *garua*. There are strong local winds and the seas are rough. This is (somewhat perversely, perhaps) known as the dry season.

The large cumulus and cumulonimbus clouds and thunderstorms which occur during December–March are a result of the vigorous upward convection of moist air which occurs, especially in the ITCZ, over a warm sea-surface—and convection can become vigorous enough to generate tropical cyclones when sea-surface temperatures exceed 29 °C. By contrast, dry weather tends to occur when the sea-surface is cold and cools the overlying air, so that convection is suppressed. However, the layer of cool air is only a few hundred metres thick; above this is a layer of warm air. In other words, there is a temperature inversion, of which low stratus clouds and misty weather are a typical indication.

Vegetation on the islands tends to be vertically 'stratified': desert at sea-level gives way to forest at about the 'dry season' cloud base (i.e. at the temperature inversion, which tilts to windward—it is lower in the east than in the west); and at higher levels there is marshy grassland.

In times past, when Galápagos islanders noticed the waters warming and storm clouds gathering near Christmas-time, they would say 'El Niño is coming'. El Niño is the Christ Child: warm conditions are identified with The Child because, bringing much-needed rain, they are a blessing. The best-known El Niño tradition is that of Ecuador and Peru. Every year when the ITCZ moves south, the South-East Trades falter, and the current system off South America changes: not only does the Peru Current weaken, but upwelling is reduced and the warm waters flowing south from the Panama Bight (Figure 5.8(b)) bring mild weather and welcome rain.

As far as the local inhabitants of western South America and the Galápagos are concerned, the annual 'El Niño', or warm, wet season, is a short-lived version of what happens during the large-scale climatic perturbations which have taken its name. At intervals of 2–10 years, the eastern tropical Pacific is influenced by exceptionally warm tropical waters, and experiences intense wet-season conditions for an extended period. These perturbations are sometimes called El Niño events, but are more properly termed El Niño–Southern Oscillation (ENSO) events; and they appear to be associated with anomalous climatic conditions in other parts of the world. ENSO events are discussed in Section 5.4.

Another important influence on life on and around the Islands is that of a sub-surface current, the Pacific **Equatorial Undercurrent** (also known as the Cromwell Current). This horizontal ribbon of water flows eastwards in a meandering fashion along the Equator, with its main flow in the thermocline, which in the vicinity of the Galápagos is at a depth of about 100 m. On reaching the steep submarine slopes on the western side of the Galápagos Plateau, the Equatorial Undercurrent is split up into filaments and deflected upwards, coming to the surface in some places (Figure 5.11).

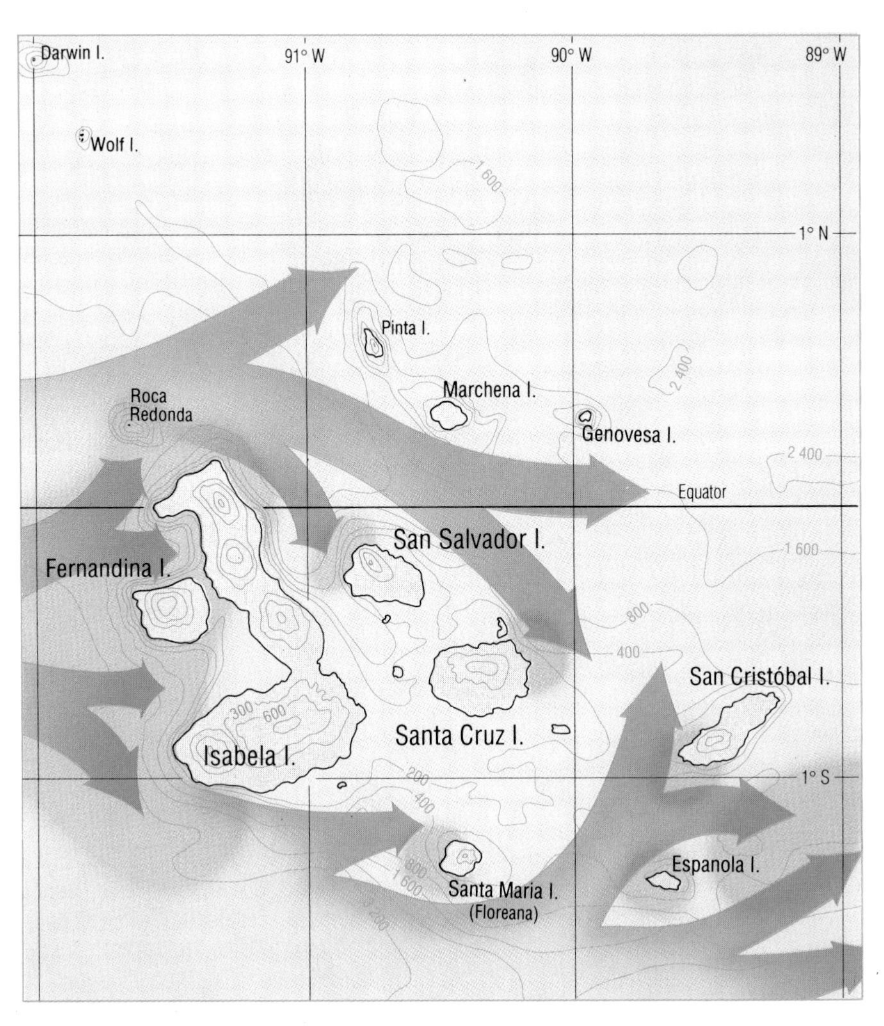

Figure 5.11 Most common flow paths of the Equatorial Undercurrent through the Galápagos archipelago, and locations of the most persistent upwellings (green areas).

Why are these upwellings so important for life on and around the Galápagos?

Because they bring nutrients up into the sunlit **photic zone**, where they support increased populations of phytoplankton, which in turn support zooplankton and a wide variety of higher organisms. Also, because the upwellings are cool, they can cause surface temperatures to be considerably less than 24 °C, even in summer; and this allows organisms to live around the Galápagos which are not normally seen in the vicinity of the Equator (see Section 5.3).

However, as noted above, the Cromwell Current does not flow straight along the Equator, but meanders along it, being diverted not only by the presence of islands, but also by the effect of northerly or southerly winds. For much of the time it passes to the north or south of the Galápagos and there is little coastal upwelling. Figure 5.11 shows the position of the most persistent of the upwellings but even these occur only intermittently and—more important—unpredictably.

5.3 LIFE AROUND THE GALÁPAGOS

The Galápagos Islands, now famous for their peculiar flora and fauna, were visited by Charles Darwin for five weeks in 1835. Over the next twenty years or so, with the help of extensive discussion with colleagues, he formulated his thoughts about the evolution of distinct species. In *The Origin of Species*, eventually published in 1859, he set down his belief that new species develop through organisms being isolated under particular environmental conditions; and, furthermore, that the species which ultimately thrive in a given region will depend not only on which organisms find their way there, but also on the *other* organisms that are already established. The reason that the Galápagos are still an inspiration to scientists is that they form a natural laboratory in which the results of such evolutionary processes can be seen more clearly than elsewhere.

For mid-oceanic islands like the Galápagos, natural colonization of both terrestrial *and* near-shore marine organisms can occur only through the agency of winds or ocean currents. By looking at the Galápagos in relation to their oceanographic setting (Figure 5.8), we can get some insight into how they acquired their unique range of flora and fauna.

The main routes whereby organisms have reached the Galápagos are the South Equatorial Current and the Peru Current from the south, and the Panama Current from the north (Figure 5.8). Many organisms must have arrived on rafts of branches or chunks of soil and vegetation. The chance of such a raft reaching the islands is extremely small, but it has been estimated that all the species of reptiles now living on the Galápagos could be accounted for by only five successful landings every million years. Reptiles are well represented on the islands: indeed, their most famous inhabitants are the terrestrial and marine iguanas, and the great tortoises or *galápagos* after which the islands were named. Reptiles can generally survive transportation by sea better than other terrestrial vertebrates; they can go for long periods without food or water, and their eggs are more resistant to desiccation than are those of amphibians, which are absent from the Islands. Apart from introduced species, the only terrestrial mammals are bats, blown in on the wind, and several species of rice rat.

The majority of the marine organisms of the Galápagos have come from the coast of Central and South America, between about 16° N and 5° S. This region, known as the Panamic Province, has a tropical floral and faunal assemblage which is distinct from that of the western Pacific and—as the seaway between North and South America closed in the late Pliocene, about 3 Ma ago—that of the Atlantic.

By which current would these organisms probably have been carried to the Galápagos?

At first sight, the most likely carrier is the Panama Current. However, organisms are generally transported as seeds, eggs or larvae, and the times when the influence of the Panama Current on the Galápagos is strongest (the Southern Hemisphere summer) are likely to correspond to periods of low reproductive activity along the coast of Central America.

For this reason, it is believed that these tropical organisms reached the Galápagos indirectly, via the coast of Ecuador and Peru and the South Equatorial Current.

A small proportion of Galápagos species originated in the western Pacific, and they probably came via the Cocos and Clipperton Islands which lie directly in the path of the eastward-flowing Equatorial Counter-Current. This current is therefore a way through what biogeographers call the *East Pacific Barrier*—the expanse of deep water and westward currents underlying the prevailing easterly Trade Winds which effectively prevents organisms from travelling from west to east across the Pacific. A possible 'tunnel' through the East Pacific Barrier is the eastward-flowing Equatorial Undercurrent, but larvae of species from the tropical western Pacific would be unlikely to survive in the low-temperature waters of this current.

The prevalence of Central/South American organisms around the Galápagos is clearly shown by Table 5.1 where the column for the Panamic Province is printed in bold.

How would most of the relatively small number of temperate species (see Table 5.1) have reached the Galápagos Islands?

In the Peru Current, probably via staging posts up the western coast of South America.

Table 5.1 also gives some indication of the large numbers of *endemic* species—plants or animals which have been so long isolated in conditions different from those in which they originated that they have evolved sufficiently to be classified as new species. However, perhaps the most notable aspect of Table 5.1 is that the diversity—the total number of species—is high for an oceanic archipelago of small islands. This is partly the result of the variety of plants and animals brought by the different current systems; it is also related to the wide range of habitats

Table 5.1 Summary of the biogeographic affinities of various groups of marine organisms in the Galápagos Islands. This Table is a compilation of data collected at different times. As research continues, so we can expect the numbers of identified species to increase.

Type of organism	Total number of species	Endemic to the archipelago	Endemic to one island	Western Pacific/ cosmopolitan	**Panamic**	Temperate	Atlantic
Algae	333	116 (35)*	3 (<1)	33 (10)	**133 (40)**	13 (4)	46 (14)
Scleractinia (stony corals)							
reef-building	13	0	0	9 (69)	**4 (31)**	0	0
non-reef-building	31	10 (30)	0	11 (33)	**6 (19)**	?	4 (13)
Brachyura (true crabs)	117	25 (22)	4 (3)	9 (8)	**82 (70)**	11 (19)	8 (7)
Mollusca							
gastropods	399	140 (35)	11 (2)	9 (2)	**235 (59)**	?	4 (1)
chitons	11	7 (64)	0	—	**2 (18)**	1 (10)	—
Echinoderms							
urchins	24	4 (17)	1 (4)	1 (4)	**17 (71)**	1 (4)	?
sea-stars	28	3 (11)	2 (7)	6 (21)	**14 (58)**	1 (3)	2 (7)
sea-cucumbers	30	1 (3)	1 (3)	9 (30)	**19 (63)**	?	?
Fish	306	51 (17)	7 (2)	43 (14)	**177 (58)**	21 (7)	4 (1)

*Numbers in parentheses give the percentages of the total number of species (as given in the first column of data).

to be found around the islands' shores (ranging from coral reefs, rocky shores and vertical rock walls to sandy beaches, coastal lagoons and mangrove swamps). Another important factor seems to be that each island appears to have evolved its own set of species, which have been exchanged between islands to give a complex mosaic of diversity. In general, the diversity of both plants and animals, terrestrial and marine, is greatest on the larger islands.

Whether a plant or animal carried to an island of the Galápagos becomes established there depends on whether there is an ecological niche for it, both in terms of the other organisms already established and of the environmental conditions. Thus, the great tortoises became the main herbivores, not only because there were large areas of suitable habitat, but also because there was no competition (e.g. no deer or other ruminants).

Suggest *six* physical factors that contribute to the environment so far as marine organisms are concerned.

Perhaps the most obvious factors are water temperature, salinity and nutrient content; also important for some organisms are the light regime/degree of cloud cover, the type of sea-floor/coastline, and local current systems.

Figure 5.12 shows the distribution around the Galápagos Islands of two groups of organisms which need contrasting environmental conditions to flourish. These are:

1 Hermatypic (i.e. reef-building) corals.

2 Macrophytic algae, i.e. seaweeds such as sea-lettuce *Ulva* and *Sargassum* weed.

QUESTION 5.3 (a) By comparing Figures 5.11 and 5.12, how can you infer that water temperature and nutrient content seem to be the controlling influences on whether corals or macrophytic algae colonize a given stretch of coastline?

(b) What is the significance of light levels, so far as reef-building corals are concerned?

We know that reef-building corals flourish only where seawater temperatures are generally above 18 °C, and there is only one large coral reef structure off the Galápagos (this coral was killed during the 1982–83 ENSO event—see Section 5.4). In the cooler-water areas, Galápagos coral colonies tend to form isolated, monospecific stands; even along the colder coasts subject to upwelling, there are 'refuges'—protected shallow inlets where corals can survive.

Luxuriant growths of macrophytic algae are found only where nutrient concentrations are high, but other factors also play a part. For example, upwelling around the Galápagos occurs where the submarine topography is steep (Figure 5.11), and the offshore volcanic rocks and islets often have near-vertical submarine cliffs. Under these conditions, only a limited amount of light falls directly onto organisms growing on the rock surface; and after upwelling events, phytoplankton blooms can severely cut down light penetration. By contrast, the more gently sloping eastern aspects of the Galápagos Plateau are orientated to receive more light.

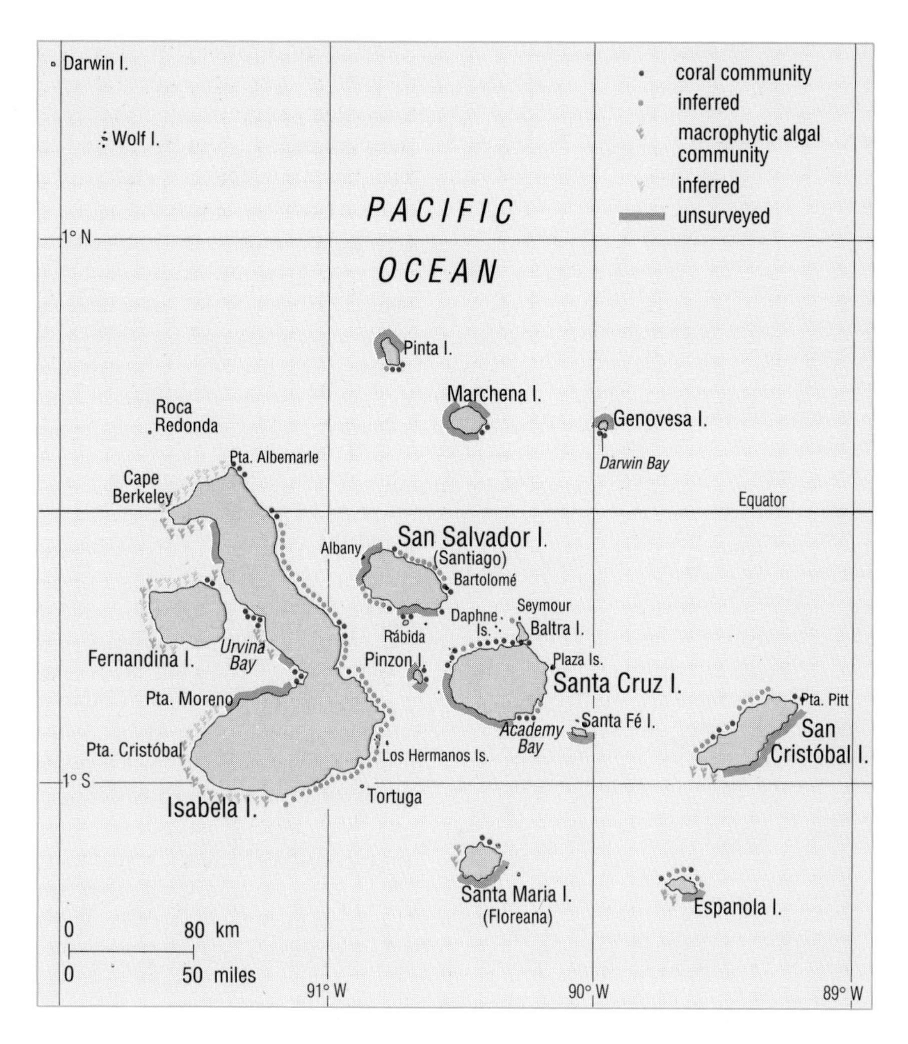

Figure 5.12 Summary map of observed and inferred distribution patterns of reef-coral communities and macrophytic algal communities around the Galápagos.

Cloud cover does not seem to have a significant influence on coral-reef growth around the Galápagos. The clouds which form as the moist air in the South-East Trades is forced upwards over the islands' steep volcanic topography are aligned along the eastward-facing coasts. These same prevailing winds also bring large waves and swell to the south-easterly aspects of the islands, and reef-building coral communities are dominated by the more robust species of branching and massive corals.

Corals cannot tolerate very low salinities, but under normal conditions, freshwater run-off from the Galápagos is very small, even in the wet season, and changes in salinity are brought about through changes in the current pattern.

Which current brings the lowest salinity water?

The Panama Current, which affects the archipelago in the early part of the year. However, even when the Panama Current has a strong influence on surface salinities, they fall only to around 31.5–32. Although low for seawater, this is significantly above the minimum tolerance limit for reef-building corals; this can be between 27 and 19, depending on the type of coral and the period of reduced salinity.

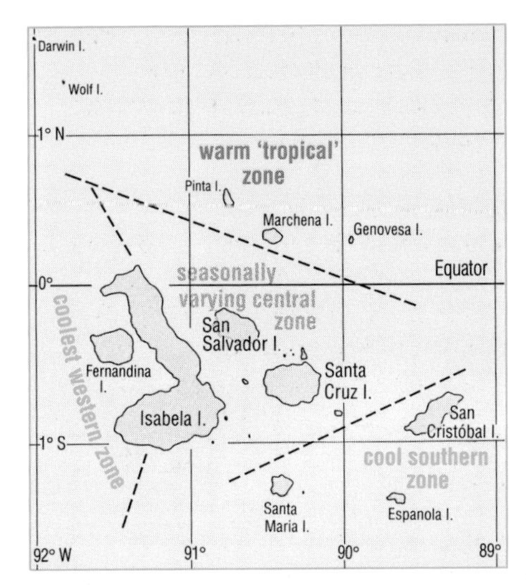

Figure 5.13 Zones of sea-surface temperature for the waters around the Galápagos.

The fate of a coral colony in the Galápagos may be dramatically affected by geological processes. The Galápagos Plateau forms part of the bulge in the oceanic crust associated with the magmatic activity of the Galápagos spreading axis (Section 5.1). Corals may be killed by hot magma which can stream down into the sea during a volcanic eruption, or by landslides associated with volcanic activity. They may also be affected by tectonic processes—rifting and faulting—with blocks of crust being moved up and down relative to one another. There have been several examples of coral reefs being lifted—either instantaneously or in several stages—several metres above sea-level. Such events are recorded by masses of bleached dead coral (see Section 5.5) but other, less dramatic shifts, either up or down, must presumably affect coral growth.

The temperature of surface waters around the archipelago is not solely determined by whether or not there has been recent upwelling in the vicinity.

What else determines water temperatures around the Galápagos?

The season of the year or, more precisely, the position of the Equatorial Front (Figure 5.8). The combined effects of upwelling and the position of the Equatorial Front mean that the waters of the archipelago can effectively be divided into four different zones so far as sea-surface temperature is concerned (Figure 5.13).

Warm 'tropical' zone

During the Southern Hemisphere summer, the Front lies south of the two most northerly islands, Darwin and Wolf, but the influence of warm tropical waters can be felt around the islands of Pinta, Marchena and Genovesa as well. These five islands to the north of the Equator are therefore in the warmest zone and have the most tropical species in their marine flora and fauna. Reef-building corals proliferate, and support numerous reef fish species from the western Pacific, such as the black surgeon fish, the Moorish idol and the sunset wrasse; there are also moray eels which shelter in cracks and caves, and many pelagic fish species (Figure 5.14).

(a) (b) (c)

Figure 5.14 Tropical fauna of the more northerly Galápagos islands: (a) Guinea fowl puffer fish; (b) moray eel; (c) Moorish idols.

By contrast, many tropical algae are missing from these northern islands; most of these are species which typically occur in warm, shallow, protected waters with a sandy or rocky substrate, a type of environment that is not widespread around the Galápagos.

Coolest western zone

Not surprisingly, the western zone (Figure 5.13) is the most productive. Around the rocky shoreline, there is a wide variety of invertebrate species, and although reef-building corals are scarce (*cf.* Figure 5.12) there are abundant 'soft' corals, including the beautiful black corals which, as they may be easily carved into trinkets, are threatened by the growing tourist industry. There is an abundant inshore fish fauna, largely derived from the temperate regions off South America; this includes the harlequin wrasse and damsel fish.

The high productivity of upwelling water supports a variety of land-based animals. Marine iguanas graze on the luxuriant growths of red and green algae on subtidal rocks, and notable inshore feeders include the flightless cormorant which lives mainly on small benthic fish; the brown pelican; the blue-footed booby; and the Galápagos penguin, which lives mainly on schools of fish fry (Figure 5.15). These birds breed only where plentiful food is within swimming distance.

Galápagos penguins now form a distinct species endemic to the islands; but how would penguins have reached the Galápagos in the first place?

Being a Southern Hemisphere cold-water group, they would have reached the Galápagos via the coast of South America in the cold waters of the Peru Current, which also transports water from the Southern Ocean (Figure 5.7(a)).

The great abundance of pelagic organisms—krill, squid and various cold-water fish—which occur in the upwelled water around the islands, supports huge populations of seabirds (e.g. wedge-rumped and white-vented storm petrels, red-footed boobies, and frigate birds); as well as dolphins, sharks, pilot whales, and baleen whales such as the sei, fin and humpback. Sperm whales are also abundant—indeed, it was their populations which the early whalers came to exploit.

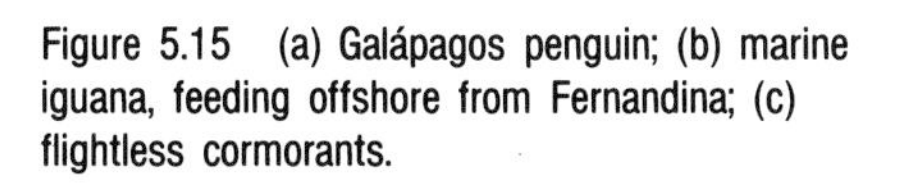

Figure 5.15 (a) Galápagos penguin; (b) marine iguana, feeding offshore from Fernandina; (c) flightless cormorants.

(a)

(b)

(c)

The whales feed in the productive waters which have upwelled within the northern part of the South Equatorial Current just south of the Equator. This upwelling zone is the **Equatorial Divergence** which can be clearly seen in Figure 5.9(a), centred on 1°30′ S. This upwelling is at its most intense when the South Equatorial Current is strongest (i.e. in the Southern Hemisphere winter) and so is fairly predictable. The same cannot be said of the more localized upwelling along the western coasts of Fernandina and Isabela.

Why is this?

In the discussion of Figure 5.11 you read that upwelling associated with the Cromwell Current occurs '... only intermittently and—more important—unpredictably'. There may be six months or even a year when the western waters are not especially productive, and then suddenly they are cold, and green with algae. Populations of boobies, albatrosses and other birds converge to feed on the shoals of small fish attracted by the plankton, and penguins, cormorants and pelicans begin a frenzy of courtship and reproduction.

Cool southern zone

The south-eastern islands lie furthest from the Equatorial Front and, in addition to Cromwell Current upwelling (Figure 5.11), are strongly influenced by the cool waters of the Peru Current and the South Equatorial Current. In many ways, the flora and fauna of the surrounding waters resemble those of the western region.

Seasonally varying central zone

The waters of the central region are characterized by moderate temperature fluctuations, and there is a mixture of faunas. Most species originated in the Panamic Province, along the coasts of Peru and Chile or even Mexico, and there are also a number which have become distinct endemic species.

Figure 5.16 Two endemic Galápagos species.

(a) The black coral *Antipathes galapagensis*; the living polyps are yellow, but the skeleton is black. A sea-lion swims past in the background.

(b) Galápagos batfish *Ogrocephalus darwini*, photographed at night.

(a)

(b)

In later Sections we will see how life around the Galápagos has been affected both by natural changes and by human activities, and we will also look at what is being done to protect these islands. Let us end this Section with Figure 5.16 which shows just two of the strange and beautiful endemic marine species to be found around the shores of the Galápagos.

5.4 THE GALÁPAGOS AND EL NIÑO

The warm wet conditions associated with 'El Niño' come to the Galápagos almost every year; only rarely does 'El Niño' fail to return. However, as mentioned in Section 5.2, every 2–10 years there are also El Niño *events*, large-scale climatic perturbations which are also known as **El Niño–Southern Oscillation (ENSO) events**.

5.4.1 EL NIÑO–SOUTHERN OSCILLATION (ENSO) EVENTS

Climatic conditions in the tropical Pacific are closely related to the difference in pressure between the atmospheric low over Indonesia and the atmospheric high in the south-eastern Pacific (Figure 5.17). When the pressure difference between these two is greater than usual, the South-East Trades are particularly strong and the ITCZ is unusually far north; when the pressure difference is lower than usual, the South-East Trades are weak, and the ITCZ is unusually far south. During ENSO events, the pressure difference between Indonesia and the South-East Pacific is exceptionally low, and the South-East Trades become very weak—in the western Pacific they may even reverse.

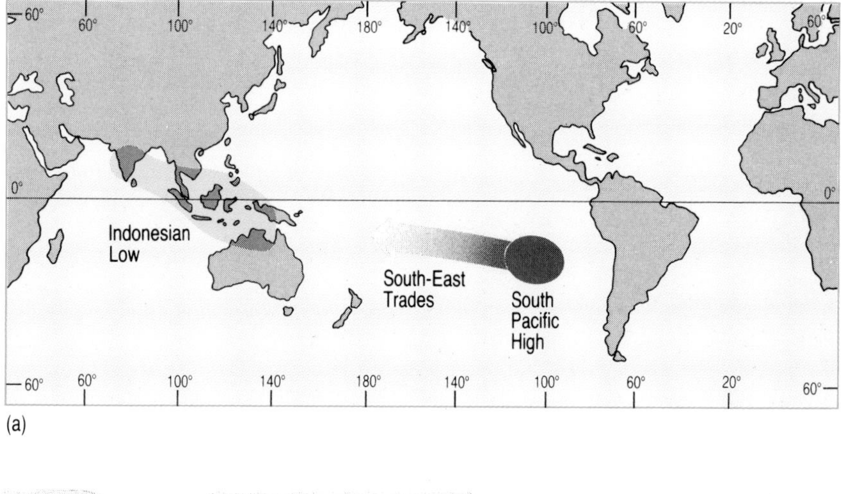

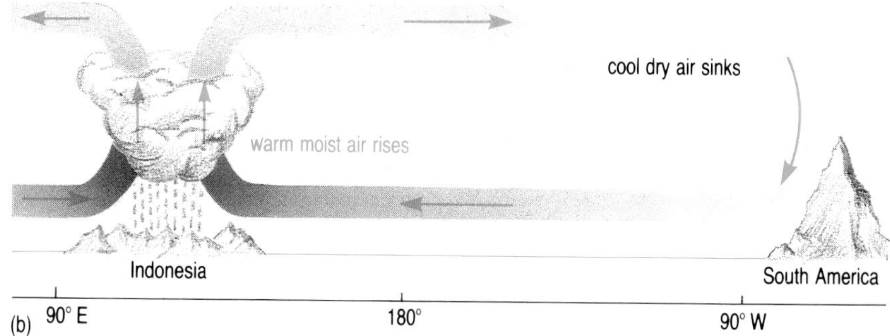

Figure 5.17 (a) Schematic map showing the position of the Indonesian Low and South Pacific High.

(b) The zonal atmospheric circulation between the Indonesian Low and the South Pacific High.

Figure 5.18 summarizes the main differences between conditions in the Pacific basin during a normal year and conditions during an ENSO event. During an ENSO event, not only does the region of precipitation associated with the ITCZ shift southwards, but the region of vigorous convection usually over Indonesia (*cf.* Figure 5.17(b)) also moves into the central equatorial Pacific, bringing rain to regions that are normally arid.

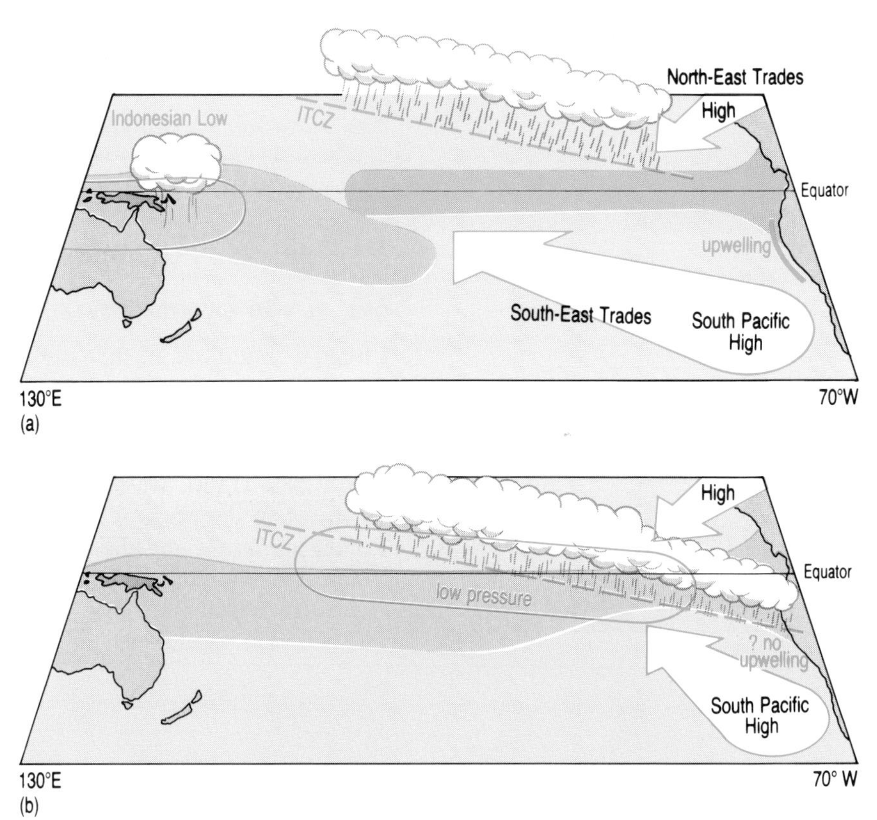

Figure 5.18 Schematic diagrams showing conditions in the Pacific (a) in a normal year, and (b) during an ENSO event. The pink area indicates regions where the sea-surface temperature is higher than about 28 °C; the orange area in (a) indicates regions that are normally dry.

The spread of warm water eastwards across the Pacific, illustrated in Figure 5.18, may be explained as follows. Under normal conditions, strong South-East Trade Winds cause the sea-surface along the Equator to slope up towards the west; this upward slope of the sea-surface is, of necessity, accompanied by a downward slope of the **thermocline** so that, on the western side of the ocean, the warm **mixed surface layer** is thick, while in the east it is very thin (Figure 5.19(a)). During an ENSO event, the South-East Trades relax, the sea-surface slope collapses and both it and the thermocline become near-horizontal, and a considerable volume of warm water flows eastwards across the Pacific (Figure 5.19(b)). The resulting changes of sea-level in the eastern and western Pacific are an exaggerated version of the seasonal changes that occur normally. Figure 5.20 shows that sea-level falls more than usual in the western Pacific and rises more than usual in the eastern Pacific. Although no ENSO event can be described as 'typical', it is probably fair to say that the features illustrated in Figures 5.17 to 5.20 seem to occur in most of them.

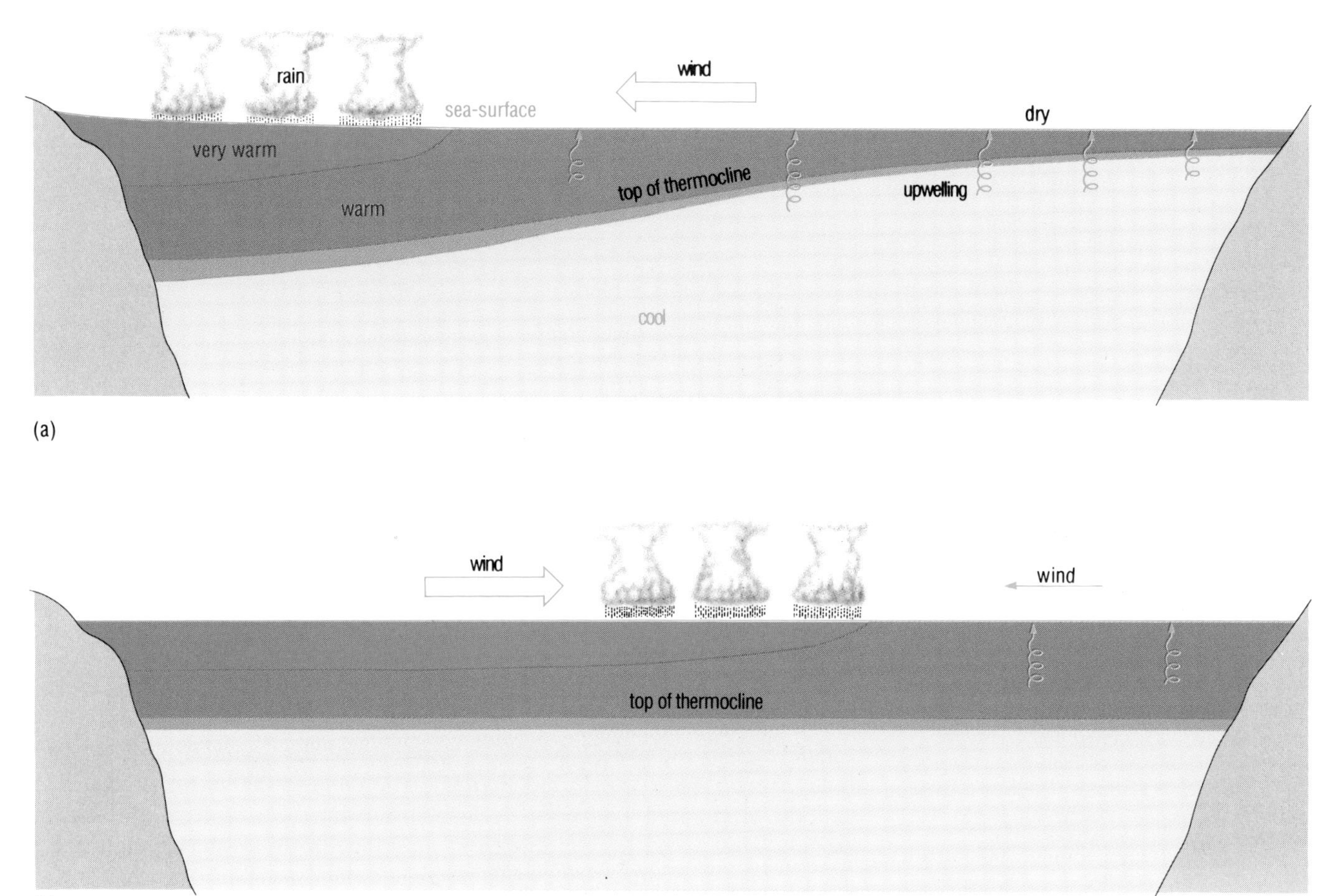

Figure 5.19 Cross-section along the Equator in the Pacific.

(a) In a normal year, the sea-surface slopes up to the west, where sea-surface temperatures are highest and rainfall is heaviest; the thermocline is deep and fairly diffuse in the west and shallow and sharply defined in the east; and upwelling brings cool nutrient-rich water from below the warm surface layer.

(b) During an ENSO event, when Trade Winds are weak (even reversed in the west), rainfall is heaviest in the central–east Pacific; warm surface water spreads eastward, so that both sea-surface and thermocline are near-horizontal; and weakened upwelling brings warm nutrient-poor water from within the warm surface layer.

From the foregoing, you might infer that ENSO events are well understood, and that they occur in a predictable way, with changes in the distribution of atmospheric pressure over the Pacific resulting in changes in the wind pattern, which in turn cause changes in the current pattern. However, the links between atmospheric pressure patterns, winds and currents are not as simple as might at first appear. Changes in the atmosphere can result from changes in the ocean—specifically, changes in the distribution of sea-surface temperature affect atmospheric pressure and winds (*cf.* Question 5.2). To complicate things further, the ocean takes much longer to react to changes in the atmosphere than *vice versa*; the feedback loops between the ocean and the atmosphere are very complex.

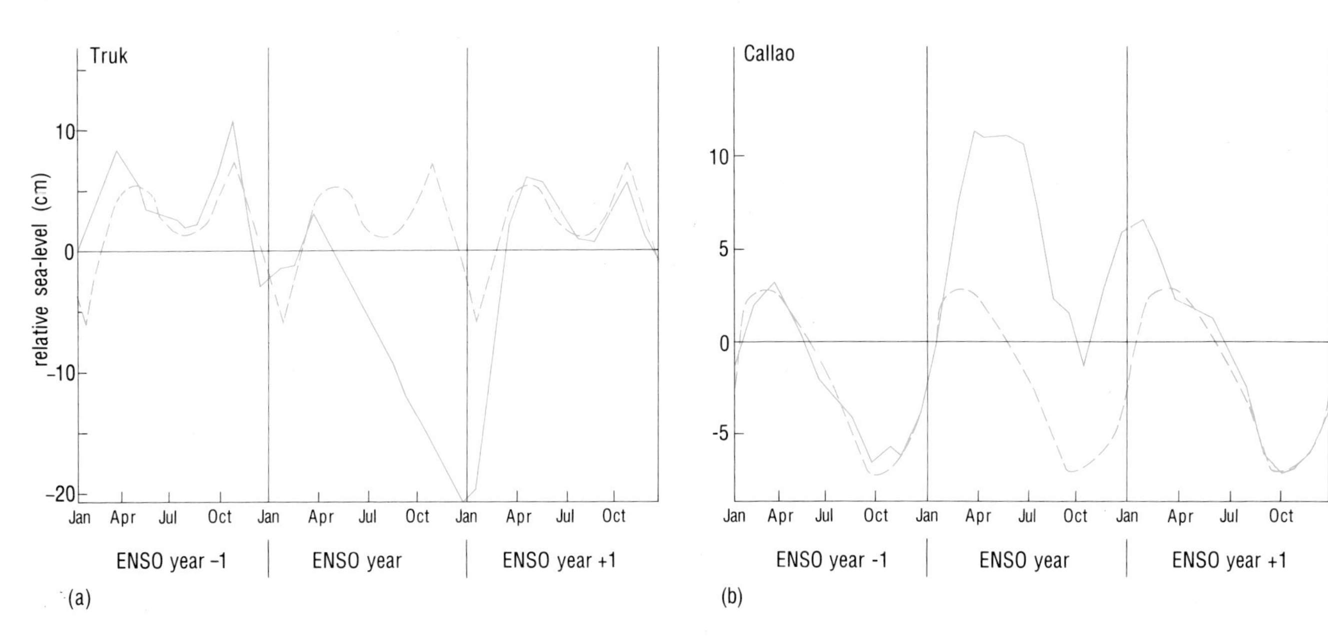

Figure 5.20 Sea-level curves for (a) Truk in the Caroline Islands (152° E, 7° N) and (b) Callao on the coast of Peru (77° W, 12° S). The solid curves are composite ENSO curves for comparison with the dashed curves, which show the seasonal changes in sea-level that occur in normal years. Note that while in the western Pacific the sharp fall in sea-level during ENSO events contrasts somewhat with the normal twice-yearly rise and fall in sea-level, in the eastern Pacific the ENSO sea-level change is an exaggerated version of what happens in a normal year.

There is, in fact, a continual see-sawing in the relative strengths of the South Pacific High and the Indonesian Low—this is known as the **Southern Oscillation**. A useful indicator of the state of the Southern Oscillation is the difference in atmospheric pressure between Tahiti and Darwin (Australia). The departure of this difference from the long-term mean difference is known as the *Southern Oscillation Index*.

Figure 5.21(a) (opposite) shows the variation of the Southern Oscillation Index from 1968 to 1983, and Figure 5.21(b) shows the variation in the easterly wind speed *anomaly* (i.e. departure from the long-term average easterly wind speed for the month in question) in the western equatorial Pacific. Negative values do not necessarily indicate westerly winds; mostly they merely indicate less strong easterly winds.

QUESTION 5.4 ENSO events occurred in 1972–73, 1976–77 and 1982–83. They are indicated by the arrows in Figure 5.21.

(a) Do ENSO events occur every time the Southern Oscillation Index goes negative?

(b) Were the two ENSO events covered by Figure 5.21(b) *immediately preceded* by a period in which the South-East Trades were normal, unusually strong or unusually weak? Why might this be relevant to the scale of an ENSO event?

(c) From the limited information in Figure 5.21(a) and (b), would you expect the 1982–83 ENSO event to have been either unusually severe or relatively mild?

Fluctuations of the kind depicted in Figure 5.21 suggest that ENSO events are best regarded not as isolated 'happenings', but rather as extreme conditions that recur sporadically at intervals of 2 to 10 years within a somewhat more regular warm–cold cycle. It is not yet certain, however, what it is that puts the cycle into an 'ENSO mode'. One possibility is that ENSO events are set off by monsoonal reversals over the Indian Ocean and western Pacific; another is that they result from imbalances between the Indonesian Low–South Pacific High atmospheric system and other systems based on low pressure centres over central

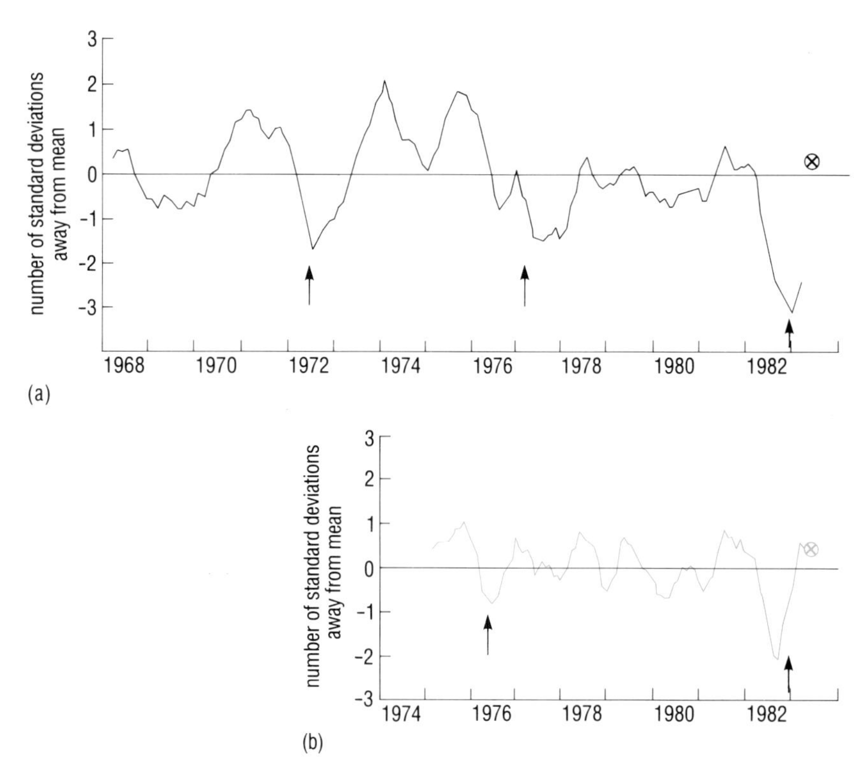

Figure 5.21 (a) The variation of the Southern Oscillation Index between 1969 and 1983. The two stations used here are Tahiti and Darwin, and the Index is expressed in terms of the departure from the normal difference in sea-level atmospheric pressure between the two stations, divided by the standard deviation for the appropriate month. Negative values mean pressure differences smaller than normal.

(b) The average easterly wind speed anomaly over the western equatorial Pacific, from 5° S to 5° N, in the area between 135° E and 170° W. The value shown is the difference from the normal easterly wind speed, divided by the standard deviation for the appropriate month. Negative values indicate easterly wind speeds lower than normal, not necessarily westerly winds. (The wind speed used was actually that at about 1 500 m above sea-level, as that enabled additional data from low-level cloud movements observed by satellite to be included.)

In both diagrams, the cross indicates the value for August 1983; arrows indicate ENSO events.

Africa and the Amazon basin; periodic bursts of wind (atmospheric waves) and internal ocean dynamics may also play a part.

It is also difficult to predict how an ENSO event might develop (*cf.* Question 5.4(c)). Some are more severe than others: the 1982–83 ENSO event was more extreme than events previously studied. In addition, it was better and more fully observed than any previous event, partly because data collected by remote-sensing satellites provided information about sea-surface temperature, wind and precipitation, atmospheric water vapour, and sea-surface topography, over a huge area in which observations would normally be few and far between. Also, by chance, at the time there were a number of ship-based observational programmes running in the equatorial Pacific.

5.4.2 THE ENSO EVENT OF 1982–83

As noted above, the 1982–83 event was probably the most severe since meteorological records began. It was also one of the longest: most ENSO events last little more than a year, from about January of one year to February or March of the following year (Figure 5.20). This event began in May–June 1982 and came to an end in September–October 1983, nearly a year-and-a-half after it began.

Despite the exceptionally high sea-surface temperatures experienced across the central and eastern Pacific during 1982–83, it was not generally the warmth of the water *per se* which caused havoc to ecosystems and human communities; it was to a large extent the unusual weather accompanying the warm water. As the zone of upward convection of moist air which normally extends south-eastwards from New Guinea moved eastwards, equatorial islands which normally experience little rain from one year to the next were battered for days on end by torrential rainfall. Some of the Gilbert Islands were affected, and also the Line Islands. At Christmas Island, which is 2° north of the Equator and just west of the international date-line (180°), 2 279 mm of rain fell between August 1982 and January 1983—more than ten times the average rainfall for these months and double the annual rainfall; the 17 million bird population all but disappeared. In Section 5.4.3, we will see how the bird populations of the Galápagos also suffered.

The region of vigorous convection and high precipitation moved eastwards across the Pacific over the course of about a year. Associated with it was a region in which near-surface winds had an unusually strong westerly component. This region of westerly winds appeared in the western Pacific in June 1982 and then moved eastwards along the Equator, increasing in extent and strength. During late 1982–early 1983, these westerlies shifted southwards to lie between the Equator and ~10° S.

It has been proposed that the unusual wind directions were linked to the unusually high sea-surface temperature as follows. A warm sea-surface warms the overlying atmosphere, and convection cells result, with warm, moist air flowing in at the bottom. On the Equator, air flows into these cells along an east–west axis rather than equally from all directions, because of the rotation of the Earth. In other words, the convection cells are elliptical in plan view rather than circular. So, in equatorial regions where the prevailing wind direction is easterly, unusually strong convection can give rise to less strong easterly winds, or even westerly winds. In 1982, the increase in sea-surface temperatures along the Equator allowed the ITCZ to move further south than usual, and westerly winds were particularly strong.

What other manifestation of atmospheric convection would you expect to be related to the region of high sea-surface temperature?

Tropical storms, also known as hurricanes or cyclones. These powerful convective events, which remove huge amounts of heat from the sea in the form of latent heat of evaporation, seem to be generated most readily when the temperature of the sea-surface is higher than about 29 °C. During the ENSO event, therefore, the region of cyclone generation, normally in the western Pacific, moved eastwards. Since 1940, there had only been one cyclone east of the date-line but during 1982–83 cyclones occurred *east of 140° W*. French Polynesia was hit by five hurricanes between December 1982 and April 1983. In Tahiti alone, 25 000 people were made homeless by cyclone damage.

With the eastward shift of the zone of convection and precipitation, the western Pacific, from the Philippines to the Hawaiian Islands, was visited by drought and crop failure. Australia was also affected; here, the worst drought of the century was accompanied by vast dust storms as the dry soil blew away, and by catastrophic bush fires. Stock and wildlife died on a massive scale.

On the eastern side of the equatorial Pacific, the people were suffering from the opposite problem. The anomalously warm conditions that spread along the coast of South America in September and October brought with them the type of climatic effects normally experienced during the El Niño season, but much more severe than usual. For nine months, northern Peru and Ecuador were affected by storms and torrential rain, which led to catastrophic flooding and mud-slides. Many people died, buried in mud or drowned; others succumbed to diseases such as typhoid which reached epidemic proportions in the wet conditions.

Offshore, patterns of primary productivity were drastically altered, and the Peruvian anchoveta fishery collapsed.

By reference to Figure 5.19, can you suggest why this happened, given that longshore winds continued to blow and coastal upwelling did not cease?

As shown in Figure 5.19(b), during ENSO events the thermocline in the eastern equatorial Pacific is unusually diffuse and, rather than coming near to the surface as it does normally, is depressed. As a result, when upwelling occurs, the water that is brought up to the photic zone is water that has been depleted in nutrients, and is not nutrient-rich water from below the thermocline.

The west coast of North America was also adversely affected: there was flooding from rains and melting snow, and mud-slides destroyed crops and homes. On the Atlantic side, Cuba and the area to the north of the Gulf of Mexico were affected by heavy rains, which caused the destruction of crops and property, and many deaths.

So, much of what occurred during the 1982–83 ENSO event can be explained in terms of the spread into the eastern Pacific of warm, wet conditions normally in the western Pacific. This can be related to the shift in position of the Indonesian Low and the ITCZ, associated with a change in the distribution of the temperature of the sea-surface. But from Figure 5.21 and the answer to Question 5.4(c) it appears that the 1982–83 ENSO event was *not* preceded by unusually strong South-East Trades piling up an extra-large reservoir of warm water in the western Pacific, to spread eastwards across the ocean on the collapse of the Trade Winds.

So, what *did* happen to make this ENSO event so severe? The answer is unclear, but it may be related to the timing of this ENSO in relation to the normal seasonal cycle (cf. Figure 5.20). We mentioned earlier that from mid-1982 westerly winds were unusually strong along the Equator in the western Pacific. Their influence extended eastwards as the year progressed, and they then shifted south to lie roughly along 10° S latitude. These changes in the wind field from the prevailing easterly regime resulted in major changes in the relative strengths of equatorial currents: the eastward-flowing Equatorial Counter-Current increased in strength by between 25 and 50 per cent, while the westward South

Equatorial Current became significantly weaker; and the Equatorial Undercurrent (Cromwell Current) could not be detected in the central Pacific from about mid-1982 to early 1983.

Figure 5.22 shows maps compiled using data from 35 sea-level monitoring stations on islands scattered across the tropical Pacific. Sea-level is a convenient measure for analyzing the response of the sea-surface, not only to changes in the wind field and in surface currents, but also to changes in the density structure of the upper ocean—specifically, the thickness of the warm low-density surface layer above the thermocline. In Figure 5.22, pink regions (positive anomalies) indicate sea-levels higher than usual, and the presence of unusually large amounts of warm, low density water; blue regions (negative anomalies) indicate sea-levels lower than usual. Taken together, maps (a)–(c) in Figure 5.22 show a large-scale rearrangement of the upper ocean in the tropical Pacific, with large volumes of warm water initially in the west being transferred to the east. The sea-surface topography in these maps also indicates how this rearrangement was brought about. For example, in Figure 5.22(a), the trough aligned along 10° N corresponds to the northern edge of the Equatorial Counter-Current, and shows that by July 1982 that current was stronger than usual: the faster the current, the greater the sea-surface slope across it, down towards 10° N. Signs of a similar 'trough' feature south of the Equator suggest increased flow also in a normally weak to non-existent South Equatorial Counter-Current.

By December 1982 (Figure 5.22(b)), much of the warm water previously in the western tropical Pacific had been transferred to the eastern tropical Pacific, and this helps to explain the temporary demise of the Equatorial Undercurrent. The current is driven by an *eastwards* pressure gradient caused by the sea-surface being higher in the western than the eastern equatorial Pacific (*cf.* Figure 5.19(a)). Figure 5.22(b) shows that the sea-surface was now lower than usual in the west and much higher than usual in the east.

By May 1983 (Figure 5.22(c)), after the southward shift of the belt of westerly winds, warm water had also flowed eastwards from much of the region south of the Equator. Throughout the intervening period, warm waters were also spreading north and south along the coasts of North and South America.

Even though the major disturbance in the wind field (from easterlies to westerlies) had disappeared by mid-1983, the sea-surface topography and currents were slow to return to normal, as Figure 5.22(d) shows: in December 1983, sea-levels were slightly higher than normal north of the Equator, but remained low to the south.

However, this is not the whole story. While the South Equatorial Current became weaker, and the Equatorial Counter-Currents became stronger, current flow alone cannot account for the *speed* with which the changes in the upper layer travelled from west to east. A clue is to be found in the elongate 'high' in the sea-surface, which can be seen in the vicinity of the Equator in Figure 5.22(a).

It is thought that when winds in the western Pacific became westerly, locally driving surface water eastwards, it piled up, causing a local rise in sea-level, and a deepening of the thermocline (Figure 5.23(a)); a group or 'packet' of such 'bulges' (giving rise to the elongate 'high' in

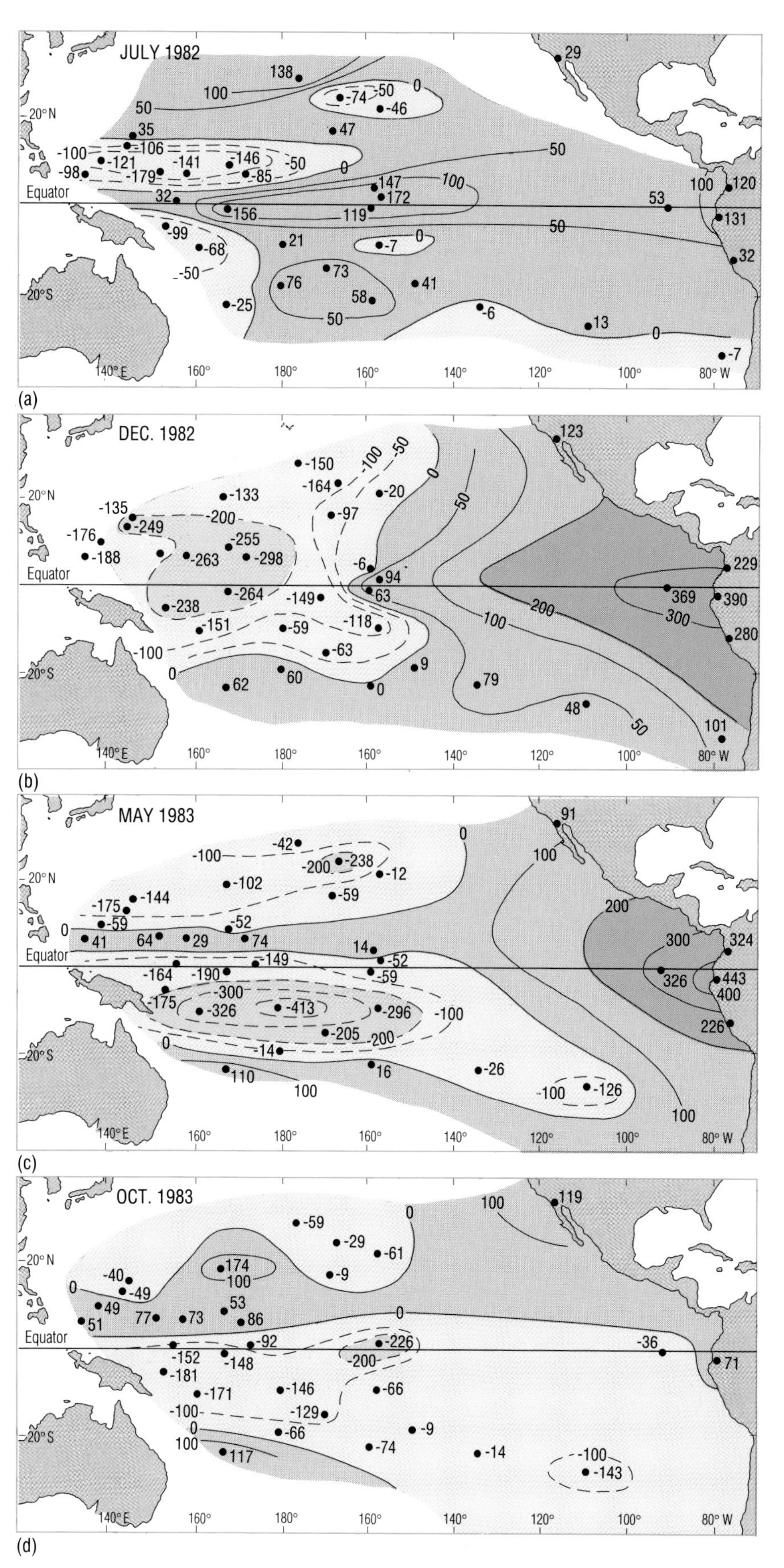

Figure 5.22 The anomaly in the monthly mean sea-level (in mm) for (a) July 1982, (b) December 1982, (c) May 1983, and (d) October 1983. Pink areas are higher than normal for the month in question; blue areas are lower than normal.

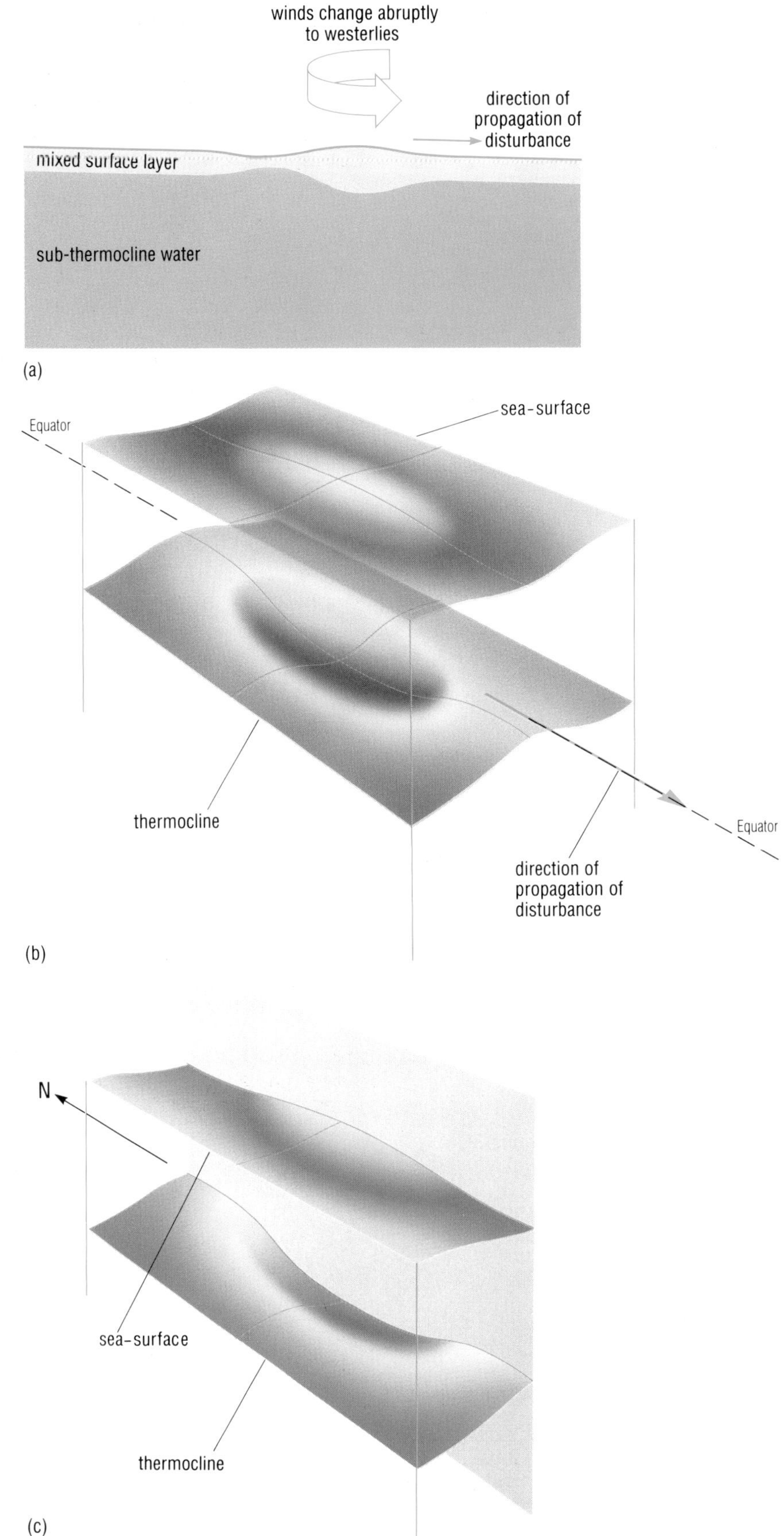

Figure 5.23 (a) Westerly winds in the western Pacific cause a local bulge in the mixed surface layer.

(b) An equatorial Kelvin wave travelling in the thermocline: one or more 'packets' of these Kelvin waves was mainly responsible for the rearrangement of the upper ocean which brought large amounts of warm water into the central and eastern Pacific. The disturbance extends several hundred kilometres either side of the Equator.

(c) A coastal Kelvin wave: as in the case of equatorial Kelvin waves, the depression of the thermocline is accompanied by an opposite (and much smaller) bulge in the sea-surface.

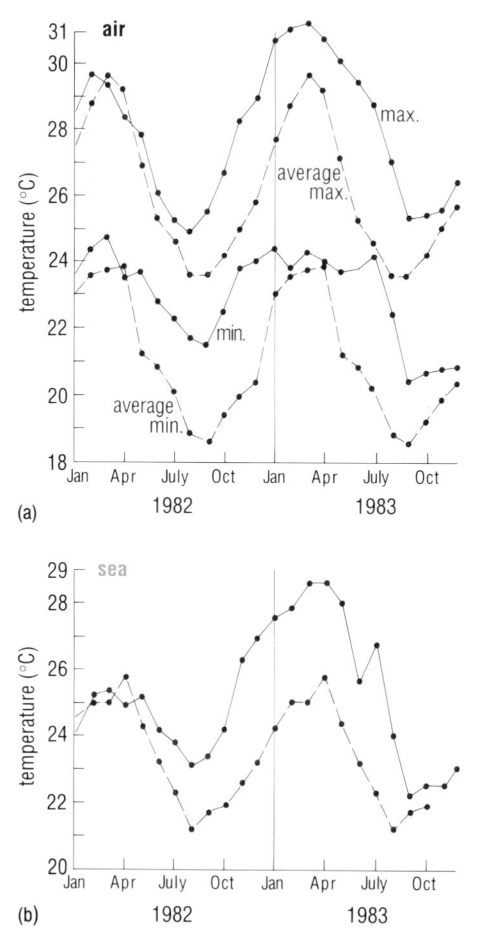

Figure 5.24 (a) Average monthly maximum and minimum air temperatures at the Charles Darwin Research Station on Santa Cruz during 1982–83 (solid curves), compared with the average maximum and minimum air temperatures for 1979–81 (dashed curves).

(b) Monthly average sea temperatures in Academy Bay during 1982–83 (solid curve) compared with the averages for 1979–81 (dashed curve).

Figure 5.22(a)) then propagated eastwards across the Pacific as the belt of westerly winds also moved east. This type of propagating disturbance of the upper ocean is known as a **Kelvin wave** and has its greatest amplitude (greatest change in sea-surface height and thermocline depth) on the Equator (Figure 5.23(b)); for this reason, the passage of these waves was clearly detectable at equatorial islands like the Galápagos. Kelvin waves travel fastest on the Equator, and within a few months the effect of the change in the winds in the western Pacific could be felt in the east. On reaching the coast of South America, each equatorial Kelvin wave split into two coastal Kelvin waves, one travelling northwards and one travelling southwards, each with their maximum amplitude at the coast (Figure 5.23(c)). It was through these coastal Kelvin waves, which caused the thermocline near the coast to be distorted downwards, that ecosystems were affected by the ENSO event even in sub-Arctic regions of the eastern Pacific.

5.4.3 THE 1982–83 ENSO EVENT IN AND AROUND THE GALÁPAGOS

Physical manifestations

Air and sea temperatures around the Galápagos began to rise above normal in the first part of 1982, and remained so virtually until the end of 1983 (Figure 5.24).

QUESTION 5.5 Rainfall data for the period of the 1982–83 ENSO event are shown in Figure 5.25, with data from normal years for comparison. From Figures 5.24 and 5.25, when might the Galápagos islanders have been sure that an ENSO event was underway?

In studying Figure 5.25, you may have noticed that the principal difference between wet and dry seasons in normal years is *not* the number of days on which it rains, but the total amount of rain that falls when it does rain—tropical storms shed a great deal more water than do stratus clouds.

During 1982–83, rainfall over the islands was about nine times the average, and some months had rain nearly every day. Warm air,

Figure 5.25 Histograms showing (a) monthly rainfall at the Charles Darwin Research Station, Santa Cruz, during 1982–83; (b) number of days with rain each month during 1982–83. In both cases, the dashed curve is the mean for 1979–81.

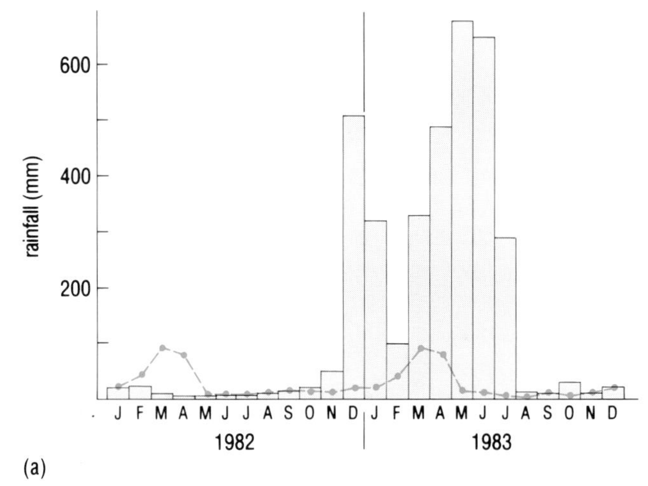

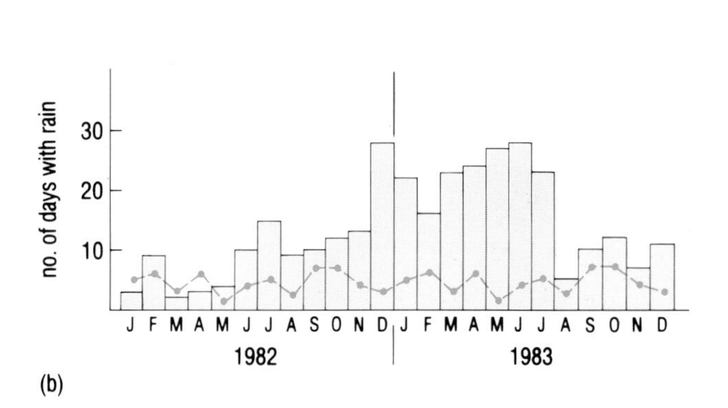

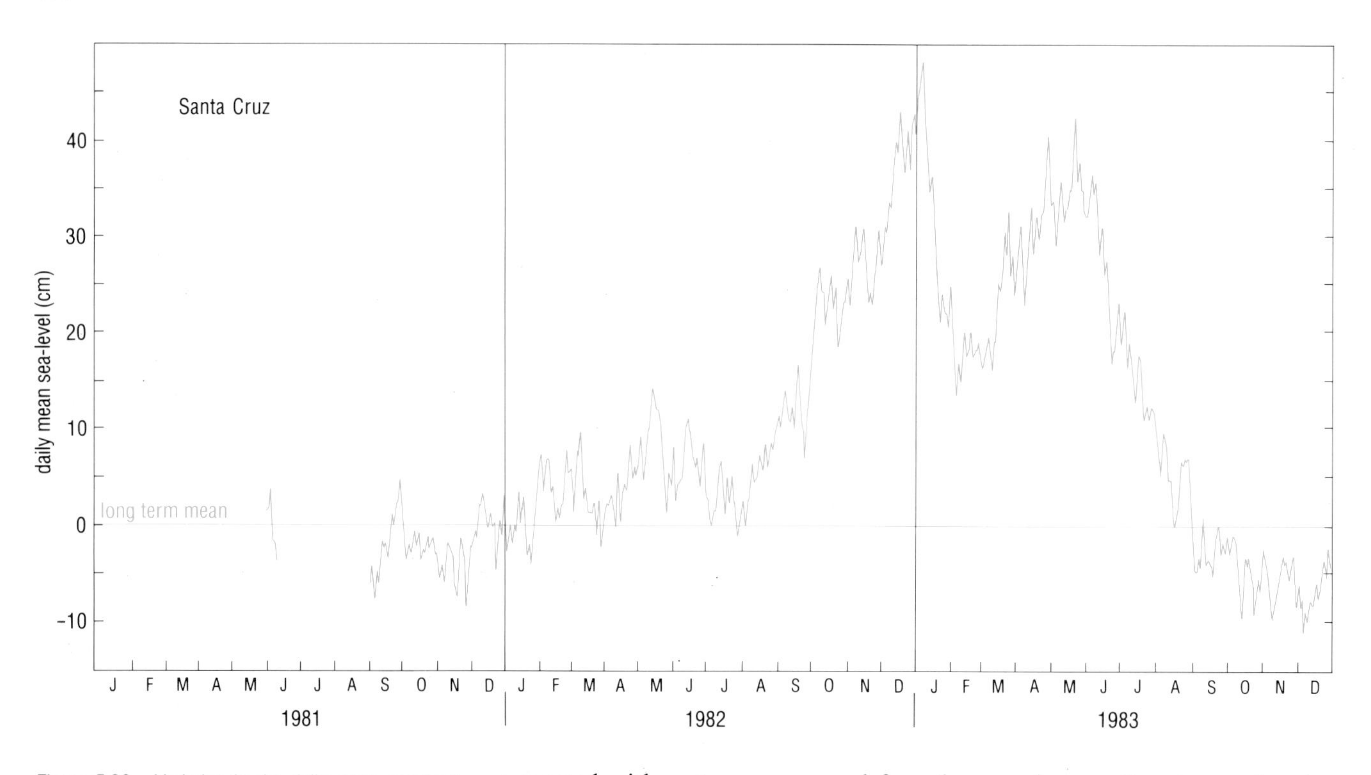

Figure 5.26 Variation in the daily mean sea-level (cm) at Santa Cruz, Galápagos, relative to the long-term mean.

saturated with water vapour and forced upwards over the steep volcanic topography, released its moisture as torrential rain. The normally arid islands were transformed by an explosive growth of vegetation; several volcanic craters became freshwater lakes; roads became quagmires or were washed away in flash floods. The havoc caused by the rain was compounded by damage to trees and power stations by lightning—a phenomenon rarely seen before by the islanders.

Figure 5.26 shows the variation in daily mean sea-level at Santa Cruz, from mid-1981 until the end of 1983.

What aspects of this sea-level record correlate with the precipitation records in Figure 5.25, and what might be the significance of such a correlation?

Both sets of data have a bimodal shape, with peaks towards the end of 1982 and again in about May 1983. The significance of this correlation is that the rise of sea-level signals the arrival of higher sea-surface temperatures in the surrounding ocean; and the higher the sea-surface temperature, the greater the incidence of tropical storms. (It is *not* likely that the increased rainfall itself contributed to the sea-level rise.)

Sea-level at the Galápagos was already above normal in mid-1982, as you would expect from Figure 5.22(a). The sharp rise commencing in August–September was almost certainly caused by the arrival of the Kelvin waves mentioned in the previous Section. The first of these waves (triggered by the onset of strong westerly winds in the western Pacific in mid-1982, as described earlier) took about a month-and-a-half to reach the Galápagos, travelling at some 3 m s^{-1}—this is characteristic of equatorial Kelvin waves travelling in the thermocline. Sea-level continued to rise for the rest of 1982 (Figure 5.26) as a result of the arrival of warm water from north of the Equator, advected east by the

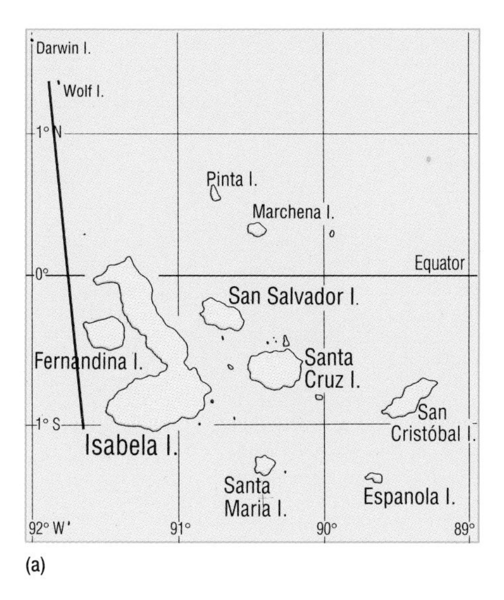

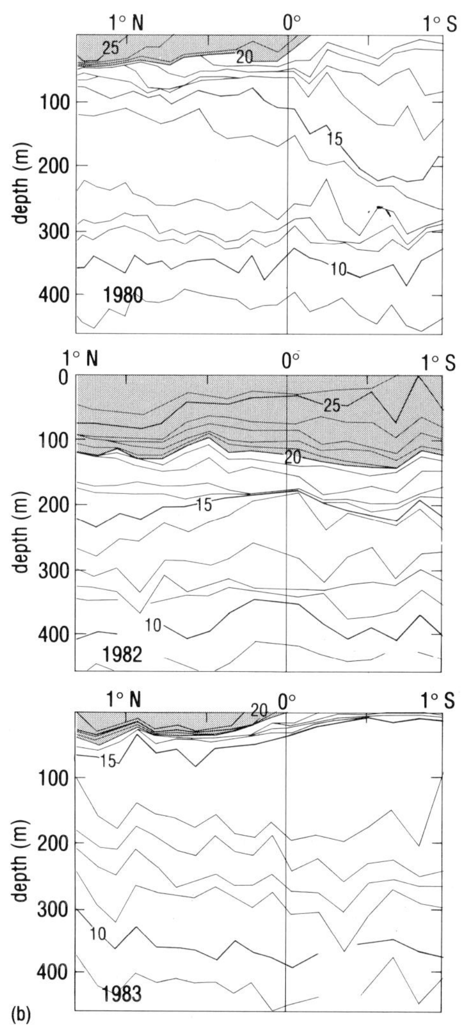

strengthened Equatorial Counter-Current(s); and probably also as a result of further Kelvin waves, generated as the zone of westerly winds moved east along the Equator. The second sea-level peak, in May 1983, was the result of warm water flowing eastwards from south of the Equator, after the belt of westerly winds had shifted to lie along 10° S earlier in the year (*cf.* Figure 5.22(c)).

The great change in the structure of the upper ocean in the vicinity of the Galápagos during the 1982–83 ENSO event inevitably affected the Galápagos Front, which separates warm and less saline tropical waters to the north from cooler, more saline waters to the south (Figures 5.8 and 5.9). Figure 5.27(b) shows vertical temperature distributions along the red line in Figure 5.27(a) for the month of November in 1980, 1982 and 1983.

What happened to the Front during the 1982–83 ENSO event?

It moved nearly to 1° S and all but disappeared. Water warmer than 20 °C extended down to an average depth of about 125 m all the way across the section (the 20 °C isotherm is commonly taken to mark the middle of the thermocline at low latitudes). By November of 1983, however, the Front was re-establishing itself. The sequence of events depicted in Figure 5.27(b) is very much what would be expected from the basin-wide changes summarized in Figure 5.22.

Could the heavy rain which fell on the Galápagos during late 1982–early 1983 (Figure 5.25) have affected sea-surface salinities?

Rainfall nine times the average might be expected to have some significant effect, and it seems highly likely that it did. Salinities as low as 32.5 were observed in Galápagos surface waters during this period, in a layer some 10–15 m thick, overlying water of more normal salinity (> 34.5), with a steep salinity gradient (**halocline**) between the two.

Could there be another reason for these low near-surface salinities?

It has been suggested that the influence of the Panama Current (Figure 5.8) extended unusually far to the south during the 1982–83 ENSO event. That would certainly have brought low-salinity water to the Galápagos, and there is some evidence that it may have happened at least sporadically, for some current-borne Panamic species turned up round the islands during 1982–83 (see later). However, flows during this period were predominantly towards the east.

Nutrients and primary productivity

As discussed in Section 5.3, many Galápagos animals—marine and shore-based—depend upon the high primary productivity of the

Figure 5.27 (a) The line along which the temperature data used to construct the cross-sections in (b) were collected. Measurement was done using expendable bathythermographs (XBTs).

(b) Temperature sections along the red line in (a) for November in the years 1980, 1982 and 1983.

Figure 5.28 The variation of (a) nitrate, (b) phosphate, and (c) silica at the surface and at 60 m depth in Academy Bay, during the course of the 1982–83 ENSO event. ($\mu M = 10^{-6}$ mol l^{-1}.)

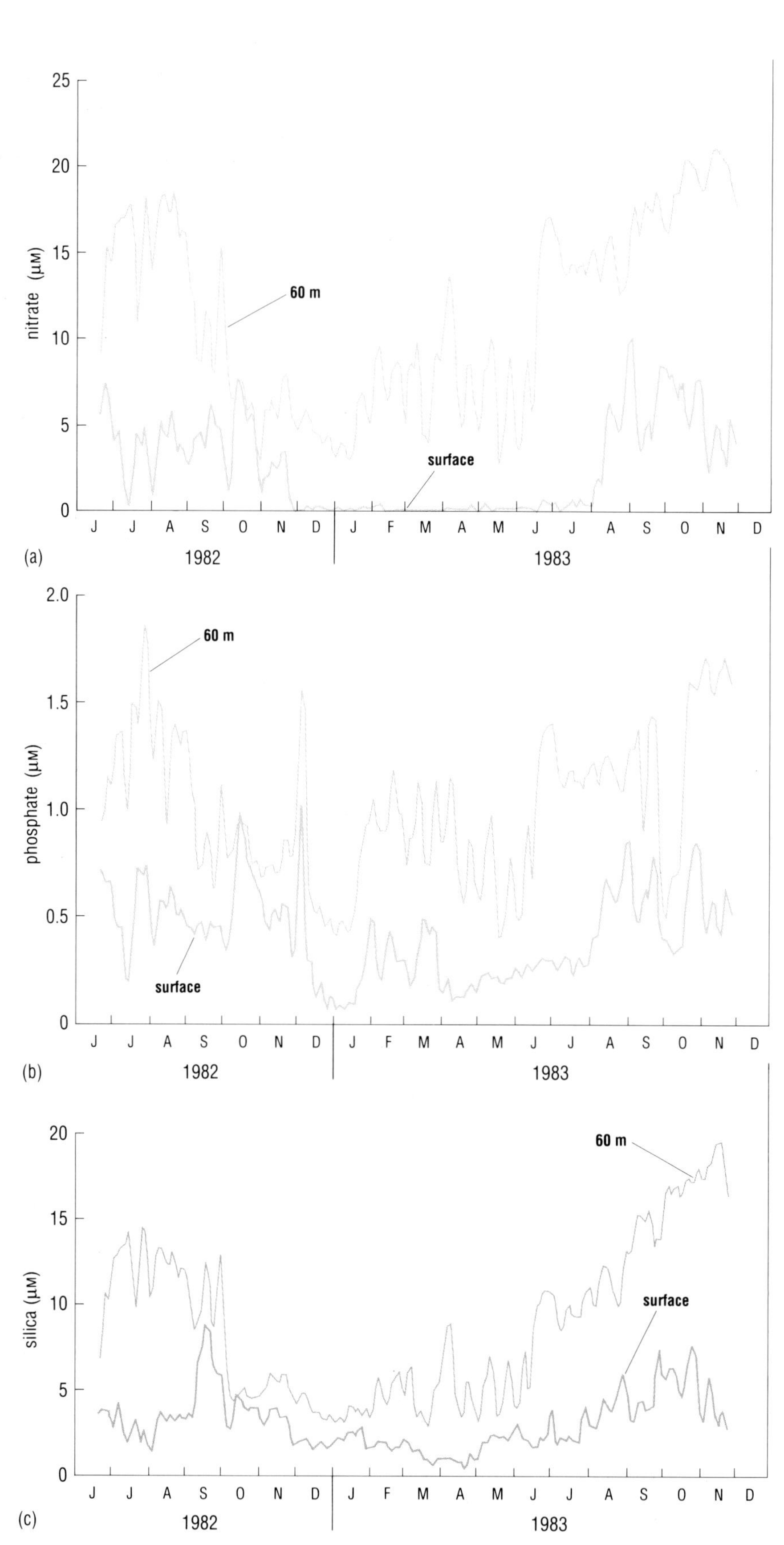

surrounding waters, supported by nutrients brought up into sunlit surface waters by upward mixing of sub-surface water. As mentioned in Section 5.4.2, however, the thermocline in the eastern equatorial Pacific deepens during ENSO events (Figures 5.19 and 5.27). Upwelling water is therefore mixed surface layer water which is both relatively warm and already depleted of nutrients by phytoplankton growth.

QUESTION 5.6 How would the low surface salinities described above, and the sharp halocline, have exacerbated this nutrient depletion?

A series of oceanographic measurements were made off Academy Bay, on the eastern side of Santa Cruz (Figure 5.1(a)) from June 1982 to December 1983. The researchers had been aware of the possibility of an ENSO event and, as their main aim was to study changes in phytoplankton abundances, which occur very quickly, they collected frequent water samples from the surface and a depth of 60 m using a small boat with an outboard motor. This simple programme yielded useful results.

Figure 5.28 shows how nitrate, phosphate and silica concentrations varied at the Academy Bay station over the course of 1982–83.

Which of these nutrients was most severely depleted in surface waters during the ENSO event?

Nitrate. Under normal (non-ENSO) conditions, nitrate concentrations in Galápagos waters are nearly always sufficient to support phytoplankton growth. There were signs of sporadic nitrate depletion in surface waters as early as July 1982, but severe depletion began in November 1982 and continued until August 1983. Phosphate and silica were also relatively depleted in surface waters but they were never completely used up: nitrate was the limiting nutrient.

Figure 5.29 is a record of the concentration of chlorophyll-*a* and of salinity in the surface waters of Academy Bay.

What is the most noticeable aspect of the chlorophyll record between November 1982 and June 1983?

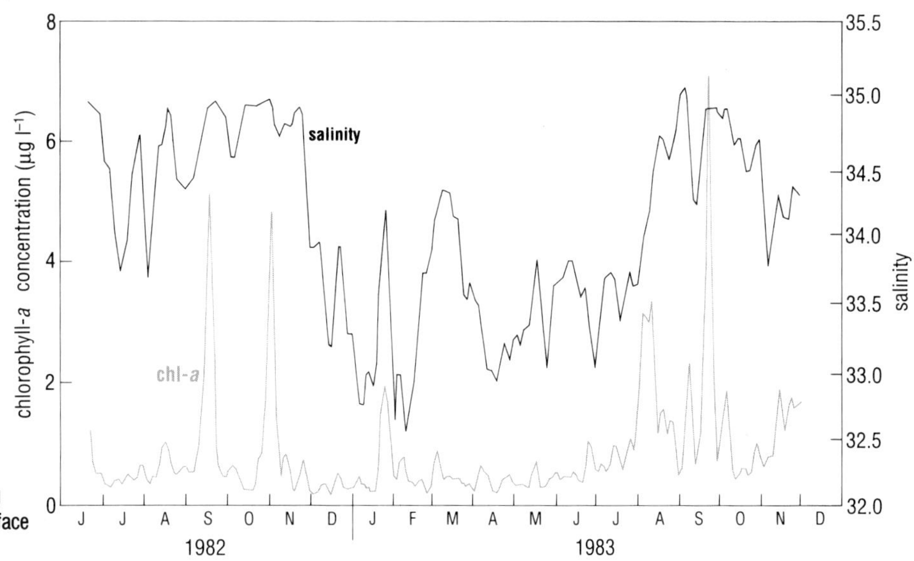

Figure 5.29 As for Figure 5.28 but showing chlorophyll-*a* concentration and salinity in surface waters.

There are no large peaks, corresponding to phytoplankton blooms, as there are before and after the event, although there is a smallish bloom in January 1982. The effect of the loss of the peaks is to reduce the average chlorophyll concentration down to about $0.4\,\mu g\,l^{-1}$, approximately a third of its normal value. Moreover, when phytoplankton populations contributing to the primary production were examined, it was found that the species composition was different from normal. Species which flourished during the ENSO event were better able to make efficient use of limited nutrient supplies than were the usual species.

All phytoplankton 'leak' soluble organic substances into surrounding waters, where they are utilized by bacteria (and by some zooplankton) which release inorganic nutrients (ammonia and phosphate) back into the water. Such recycling of dissolved nutrients from phytoplankton 'leakage' becomes important in otherwise nutrient-poor waters, such as those typical of tropical latitudes. Indeed, from the point of view of nutrient supply, the Galápagos experienced typical tropical conditions during the 1982–83 ENSO event: surface waters were depleted in dissolved nutrients (Figure 5.28), and phytoplankton had to depend in large measure on nutrients recycled by bacteria from the dissolved organic substances which they (the phytoplankton) had released.

The species composition of the phytoplankton in Galápagos waters changed from large-celled and diatom-dominated populations to small-celled *picoplankton* (diameter *c.* 10^{-6} m), which are less available to filter-feeding zooplankton. Of course, if nutrients *were* brought into the photic zone by upwelling, then phytoplankton blooms could occur.

The researchers who made the measurements shown in Figure 5.29 noticed that peaks in chlorophyll content nearly always occurred when salinity of surface waters was greater than 34.9. They reasoned that as the only water in the vicinity of the Galápagos with salinity greater than 34.9 was normally below 50 m depth, these peaks must represent phytoplankton blooms in water which had upwelled or been mixed upwards.

How might that have occurred?

To answer this question, let us take a more general look at primary productivity around the Galápagos during the peak of the ENSO event. The Coastal Zone Color Scanner (CZCS) images (Figure 5.30(a) to (c)) are for the three days indicated by the red lines in Figure 5.30(d)); and this shows near-surface wind and current data over the period, at a station on the Equator some 450 km west of the islands—sufficiently far away from the islands for regional patterns to have been relatively unaffected by local influences.

According to Figure 5.30(d), until late January the South-East Trades blew fairly strongly towards the north-west, i.e. they were at least temporarily normal and may have caused the surface current flow to change from eastwards to quite strongly westwards—typical of flow in the South Equatorial Current—at the turn of the year.

How might these conditions be related to what can be seen on the first CZCS image, for 1 February 1983 (Figure 5.30(a))?

(a)

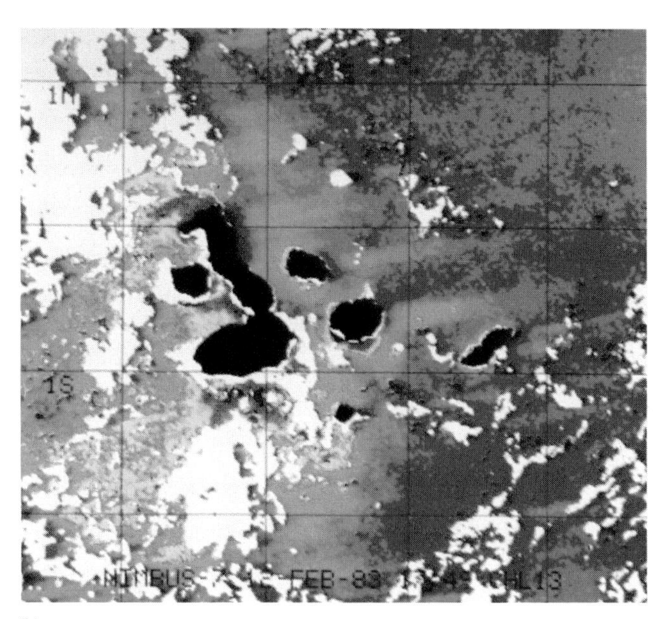

(b)

(c)

Figure 5.30 The distribution of phytoplankton pigment concentrations around the Galápagos, derived from *Nimbus-7* CZCS images acquired on overpasses of the islands (indicated by red lines on (d)) on (a) 1 February, (b) 12 February, and (c) 28 March, 1983. Major islands are shown black; clouds are white. (d) 'Stick diagrams' showing daily average direction and speed (in m s^{-1}) of the wind 3.5 m above the sea-surface, and current at 15 m depth, at 0° 95′ W.

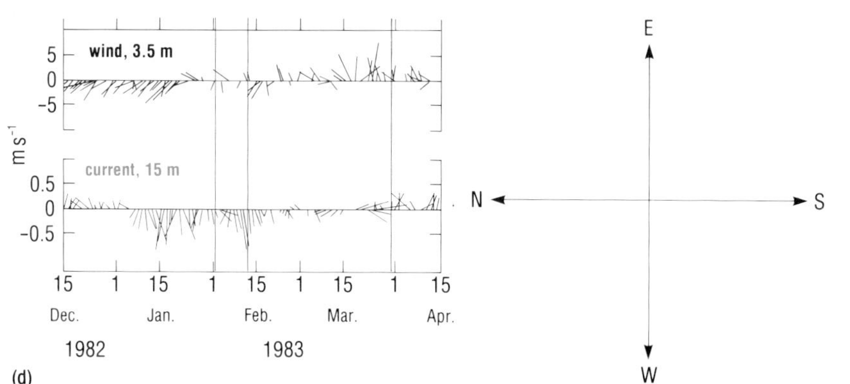

(d)

The most notable characteristic of the chlorophyll pattern is the large plume of relatively high chlorophyll concentration ($>1.5\,mg\,m^{-3}$) extending westward from Isabela; smaller plumes are visible to the west of the other islands in the archipelago. This pattern corresponds to what has been observed at this season of the year under normal conditions. It is thought to be the result partly of coastal upwelling driven by the South-East Trades, and partly by what is known as the *island-wake effect*: currents flowing around islands produce wakes of eddying currents, with vigorous vertical mixing, on the downstream side of the islands; and the South Equatorial Current was strong on 1 February (Figure 5.30(d)). The fact that the high-productivity plume is particularly strong to the west of Isabela might normally be attributable to the effect of upwelled Equatorial Undercurrent water, but we have seen that the current had become weak or non-existent at this time.

How had the picture changed two weeks later (Figure 5.30(b)), and can we account for the changes?

The plume of high pigment concentrations, with its sharp frontal regions, broke up, and the only regions with high concentrations were now on the northern and southern tips of Isabela and to the south of Santa Cruz. Although pigment concentrations were still higher on the western than the eastern side of the archipelago, the mean concentration for the entire archipelago had decreased from $0.30\,mg\,m^{-3}$ on 1 February to $0.17\,mg\,m^{-3}$ on 12 February. There is no satisfactory explanation for these changes. Although the winds had become variable in the intervening period, the near-surface current flow was still quite strongly westward (Figure 5.30(d)). It is possible that in this case the wind and current measurements were not in fact representative of conditions at the islands, 450 km to the east.

However, it is easier to account for the third image, for 28 March, 1983 (Figure 5.30(c)), on the basis of the available wind and current data. Westerly winds blew for most of March, and by the end of the month surface currents were again often eastwards. So it is not surprising that the distribution of chlorophyll pigment had now changed dramatically: although the mean pigment concentration was not significantly different from that on 12 February, the region of highly productive water was now on the eastern side of the archipelago; streamers of phytoplankton-rich waters extended towards the north-east and a large patch of productive water could be seen north-east of Santa Cruz.

Satellite images like those in Figure 5.30, combined with measurements in the water column such as those in Academy Bay discussed earlier, have helped to explain why the marine fauna, and some land-based animals and birds, suffered so severely during 1982–83.

The effects on Galápagos marine life

During the course of 1982–83, environmental conditions around the Galápagos were altered in a variety of ways, all of which affected the flora and fauna to a greater or lesser extent.

The high sea-levels, combined with heavier than normal swell from the north-west, physically altered some coastlines, particularly during periods of high tide. Sand was temporarily removed from beaches to reveal cobbles and boulders, and large subtidal boulders were overturned and tumbled up the beach-face. Large heads of massive corals (*Pavona* and *Porites*) were broken off in shallow subtidal waters and deposited above the normal high-tide line; the more delicate branching corals (*Pocillopora*) were torn loose and ground to beach rubble, along with many sea-urchins and starfish. The onslaught of breaking waves and boulders beat the shoreline vegetation back several metres. On some beaches, high tides breached berms and cut channels through to back lagoons; these lagoons, and others which became filled with rainwater, were abandoned by flamingos and other wading birds.

The rise in sea-level shifted shoreline zones landward. For example, in Academy Bay, there are two species of fiddler crab, one of which (*Uca galapagensis*), occupies a distinct high-tide band in muddy sediment around mangroves. As sea-level rose, these fiddler crabs migrated up the shore and were to be seen burrowing into a low section of the dirt road between the town of Puerto Ayora and the Charles Darwin Research Station. The second species of fiddler crab (*Uca helleri*) generally prefers to burrow into coarser sandy sediments in the middle of the intertidal zone. The unusually large waves shifted sandy sediments further up the shore, and so these crabs also migrated shoreward.

There were dramatic changes in the marine vegetation of the islands.

With reference to Figure 5.12, can you suggest which type of algae would be particularly adversely affected?

The macrophytic algae which flourish in nutrient-rich, cool water. In the central islands, the brown alga *Blossevillea galapagensis* which normally flourished in the surf zone (at the bottom of the intertidal zone) began to die back in early 1983; by April, none at all remained in Academy Bay. *Sargassum* weed similarly disappeared from the western coast of Isabela and around Fernandina.

Reef-building corals around the Galápagos were also adversely affected, as they were throughout the equatorial Pacific. By February 1983, it was clear that all the major coral reef species were becoming bleached (Figure 5.31(a)).

Bearing in mind the discussion about reef-building corals in Section 5.3, what could be the possible causes of coral decline?

Figure 5.31 (a) Bleached *Pavona* coral; (b) coral colonized by 'algal turf' of *Bryopsis*..

(a)

(b)

At first it was suspected that the corals were reacting to the extremely low salinity of surface waters brought about by the torrential rain. However, when it was seen that bleaching occurred down to depths of 30 m (below the influence of rainwater), another cause had to be sought.

It seems that the corals were probably stressed mainly by the increased water temperature. It has been discovered that corals have not only a minimum water temperature but also a maximum, which depends on the species. The symbiotic photosynthesizing algae (the **zooxanthellae** which through their metabolism supply the coral tissue with the major proportion of its nutrients) lose photosynthetic pigment and become less productive when water temperature exceeds 28 °C. The reduced nutrient supply from such zooxanthellae could lead to metabolic upset within coral tissue, resulting in atrophy or death. Eventually, the zooxanthellae either leave the coral polyps or are expelled from them in streams of mucus, and the corals become bleached and may die.

During early 1983, when surface water temperatures around the Galápagos exceeded 30 °C, many coral colonies suffered bleaching; and by the end of the year, 95% of them had been destroyed. The fate of the stressed corals was greatly affected by the activities of other species. Bleaching corals were quickly colonized by filamentous algae which spread over corals and rocks, down to a depth of 20 m (Figure 5.31(b)). Sea-urchins fed on these algae, at the same time eroding the coral. As the ENSO event waned, new coral larvae began to settle and become established, but for a colony to survive, the recruitment of new coral organisms had to keep pace with their erosion by foraging sea-urchins, and by other coral grazers such as parrot fish (Figure 5.32), surgeon fish, and damsel fish. The activities of these grazers meant that Galápagos corals showed very little recovery until 1989; some reefs never recovered (Figure 5.33).

In general, the fish which were adversely affected during 1982–83 were endemic species and those species which originated in the waters off Peru and Chile. Some, such as the toothed parrot fish which normally hides and forages amongst the beds of brown macrophytic algae, disappeared

Figure 5.32 Parrot fish grazing on coral.

Figure 5.33 All that remains of Onslow Reef, near Floreana (taken in 1985).

along with its habitat. The rusty damsel fish, the only cool-water species of damsel fish, was devastated by lack of food and by fungal skin infections.

On the other hand, fish originating from the western Pacific or the Panamic Province increased in numbers and range: these included Moorish idols, filefish, and some species of parrot fish. Four species of fish appeared which had never before been seen in the archipelago. One extremely unusual visitor was the yellow-bellied sea-snake which is dispersed by surface currents and is thought to have been carried to the Galápagos in warm, low-salinity water from the region of Panama.

The changes in species distribution brought about by this ENSO event caused great problems for the archipelago's artisanal fishery. This is a seasonal fishery, active from October to April, based on demersal groupers, particularly 'bacalao', 'camotillo' and 'vieja'. Normally, brine-soaked fish flesh is dried on the hot coastal rocks during a fishing trip and then stored in the hold until the vessel returns to port. The product is mostly sent to Ecuador where it is eaten as a traditional dish during Easter Holy Week. In December 1982, despite high fishing effort, catches of the most important commercial species began to decline. It is thought that the fish migrated into deeper, cooler water, from where they returned when the ENSO event ended. In the high humidity and persistent rains, the fish which *were* caught proved difficult to dry sufficiently to store in the vessels' holds: the product was poor and profits slumped.

The shifts in the distribution of nektonic species around the islands had a drastic effect upon the Galápagos fur seal and the Galápagos sea-lion, which depend on cold-water species of fish and squid. These prey species were substantially reduced and pushed deeper by the influx of warm water and the resulting depression of the thermocline.

According to Figure 5.27(b), how deep would a seal or sea-lion have had to dive during November 1982 to encounter water cooler than 20 °C, the normal temperature of the sea-surface in March?

Figure 5.34 Young sea-lion suffering from pox.

They would have had to dive to a depth of about 130 m. This would be no problem for sea-lions, which can dive to a depth of about 200 m. However, fur seals can only dive to a depth of 20 m, and under normal conditions they feed at night on schools of fish and squid which have migrated up to the surface; so, for them, food was largely out of reach.

Despite the sea-lions' diving ability, colonies on the Galápagos appeared almost deserted in comparison with former years. They were stressed partly by shortage of food but mainly by the high sea and air temperatures, and became particularly susceptible to disease: it seems that many sea-lions died from sea-lion pox (Figure 5.34). Most of the young born in 1982 died, effectively abandoned by the adults who had to remain at sea much longer than usual, searching for food; the 1983 year class was also much smaller in number than usual.

Fur seals fared even worse during the 1982 reproductive season, their population declining by some 20%. As with the sea-lions, the mothers remained at sea for extended periods, and could not produce enough milk to keep their pups alive. Many of the 1–2-year-olds also starved to death, while the poor physical condition of the older animals made them more susceptible to disease and predation.

The effects on seabirds and marine iguanas

The Galápagos archipelago is a major breeding site for seabirds in the eastern equatorial Pacific—19 species of seabird breed there and in normal years their combined numbers amount to over one million individuals.

From what you have read in Section 5.3, which seabirds are likely to suffer most during an ENSO event?

Those which normally feed offshore in the cold upwelling regions on the western side of the archipelago. The Galápagos penguin and the flightless cormorant are particularly vulnerable because, when the upwelling fails, they cannot fly long distances in search of more productive waters.

Normally, Galápagos penguins breed throughout the year, but they are believed to suspend breeding when the sea-surface temperature goes above about 24 °C. On this basis, they would not have bred between October 1982 and July 1983, and indeed no juvenile penguins could be seen in August–September 1982–83. It seems that during 1982–83, the population fell from between 6 000 and 15 000 to about 400: a population decrease of at least 90%.

Flightless cormorants breed opportunistically throughout the year, and showed a high level of breeding activity during 1982–83. Although they managed to raise chicks for a month or so, none survived to young adulthood and during the ENSO event the population fell by about half.

One of the birds studied intensively during 1982–83 was the waved albatross (Figure 5.35(a)). This breeds almost exclusively on Española in the south-east of the archipelago, and has a fixed breeding season from April to mid-June. The albatrosses suffered from food shortage in 1982,

(a)

(b)

Figure 5.35 (a) Waved albatrosses; no pairs bred successfully in 1983. (b) Flightless cormorant nest made of stems of the salt bush, rather than the more usual *Sargassum* weed which had all died; later, iguana skeletons were also used.

and 60% of the eggs were deserted; the birds returned late in 1983, and 20% of them—the experienced breeders—did not return at all. Those that did return found their nesting areas covered by a thick cloak of vegetation; far fewer adults laid eggs than in 1982 and the eggs were smaller. The rain submerged nests, and adult birds were seen swimming around on pools frantically trying to retrieve their eggs from the bottom. Furthermore, the shortage of food meant that male birds stayed away at sea for much longer than usual; when they returned to relieve their partner from incubating the eggs, it was already too late. No chicks of the waved albatross hatched in 1983.

A few species of seabirds were successful at producing young during 1983; these were some of the pelagic feeders and scavenging foragers such as the lava or dusky gull, which had ample supplies of abandoned eggs, dead chicks and offal.

When the ENSO event ended in September 1983, thousands of birds began returning to their colonies; populations were depleted, particularly of juveniles, but the sizes of the returning populations suggested that many birds had found alternative feeding grounds away from the Galápagos. Birds which breed annually waited until the start of the breeding season, but the others began courting and bred within a remarkably short time. The opportunistic breeders—flightless cormorants, blue-footed boobies and brown pelicans—were the first to recover. Galápagos penguins have low fecundity which makes them very vulnerable to environmental changes, and so were very slow to recover;

Figure 5.36 One of the many marine iguanas which perished during the El Niño event.

nearly ten years later, populations were still at about 10–15% of their former numbers.

Like penguins and flightless cormorants, the marine iguanas live in fixed colonies and have narrowly defined feeding areas; they also suffered great losses during the 1982–83 ENSO event (Figure 5.36). Under normal conditions, as the tide recedes, iguanas move out onto exposed rocks to feed on leafy red algae. The low nutrient supply, high temperatures and lower than usual salinities during 1982–83 meant that most of these algae died and were replaced by a variety of new species. The main invading species, particularly in the intertidal and splash zones, was a filamentous brown alga which the iguanas found hard to digest.

The high sea-levels and heavy swell made intertidal feeding difficult, particularly for the juveniles. The young born in 1982 were the most affected, suffering 90% mortality; the few 1983 hatchlings mostly survived, probably because they lived off their yolk-sac reserves. Adult iguanas can dive to feed, and so in regions where subtidal food was available, more of them survived.

When dead iguanas were examined, it was found that their stomachs were either completely empty, or contained the largely undigested remains of the brown alga on which the animals had continued to feed (even though it tended to make them vomit), until eventually their guts became blocked with its undigested fibres. When they died, they were between 50 and 60% of their normal body weight. Stomach contents showed that the iguanas had also fed on carrion (e.g. dead crabs, sea-lions and other iguanas), as well as on iguana skin (often taken from a neighbour's flanks); they also consumed earth and stones.

The deaths of adults and juveniles, combined with the small number of young born in 1983, reduced marine iguana colonies to 30–50% of their former size. Iguanas have a very low reproductive rate, and it has been calculated that the population as a whole may not recover to its pre-1982 size for at least 100 years. On islands where there are introduced predators such as cats, rats and pigs, the rate of recovery could be even slower—and local populations could even become extinct.

ENSO events and evolution in the Galápagos

There are two distinct ways in which ENSO events may bring new species to the Galápagos. The first is simply that the strong, sustained eastward flows from the Indo-Pacific region may bring to the islands adults and dispersal stages of marine species not yet established there, which in time could evolve into new species, through isolation and genetic 'drift'.

It has also been speculated that by pushing certain species to the brink of extinction, extreme events like the ENSO of 1982–83 may set the stage for the evolution of new species. The idea is that after such an event, only small scattered outposts of a species might remain; being isolated one from another, these could evolve in different ways to produce distinct new species. As we noted in Section 5.3, each island tends to have its own set of species which seem to have evolved through several cycles of introduction and isolation.

5.5 GALÁPAGOS CORALS: A CLUE TO PAST CLIMATE

The large size of some of the massive coral heads killed by the 1982–83 ENSO event is a strong indication that there has not been an El Niño event of similar intensity over the past hundred years or so. Corals can provide quite detailed information about climatic fluctuations over the course of several centuries, in a variety of ways. The volcanic nature of the Galápagos Islands has provided access to coral 'climate recorders' in a most dramatic manner.

Early in 1954, part of the western coast of Isabela was suddenly uplifted as a result of inflation of the western flank of the Alcedo volcano by injection of magma. This caused the shoreline of Urvina Bay to the west (Figure 5.1(a)) to move 1.2 km seaward, lifting several square kilometres of the shallow bed of the bay high and dry (Figure 5.37). That this uplift occurred almost instantaneously could be seen from the remains of organisms stranded on the upraised rock. One of the first visitors to the site wrote that he found 'in every niche and crack and cave in the rock . . . the skeleton of some sea animal—a crab, a starfish, a sea-urchin, a fish. In the depression below the rocks, the dried remains of lobsters, sea turtles, and marine worms lay mummified in the sun'. There have been other uplifts around the Galápagos but they have generally been smaller and more localized, and/or have occurred along more steeply sloping coastlines.

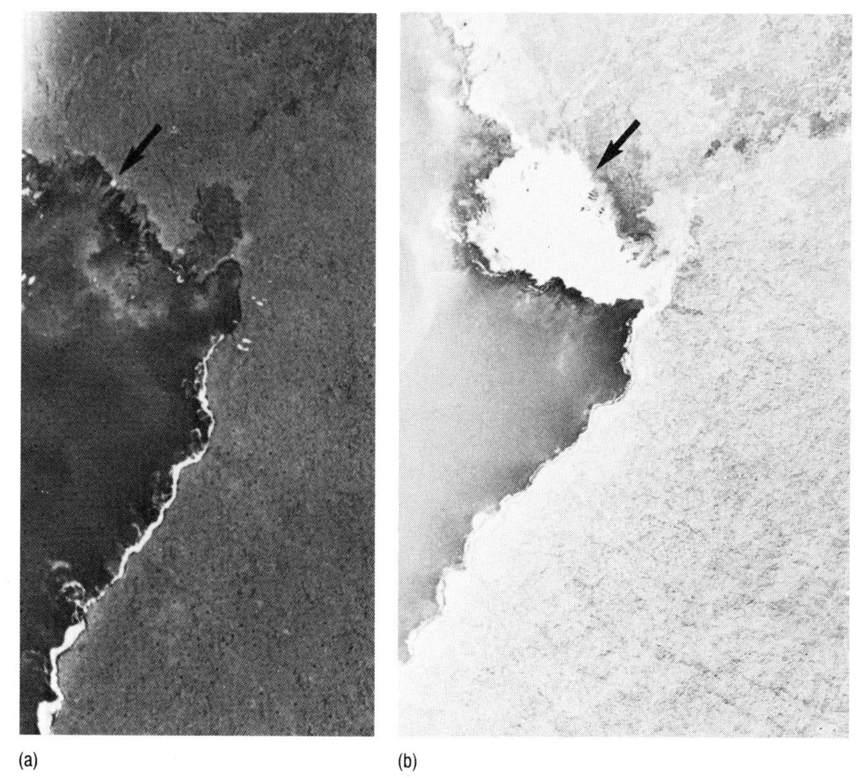

Figure 5.37 (a) Urvina Bay in 1946, eight years before the uplift. The arrow points to a white beach of carbonate sand, on either side of which there were mangroves.

(b) The same area in 1968; the beach, again shown by the arrow, is now more than a kilometre from the sea.

The Urvina Bay uplift is unusual because it elevated a shallow bay, so exposing an extensive area. On the uplifted shelf were eight species of coral, of which the star coral *Pavona clavus* and two head corals made

Figure 5.38 Uplifted star coral at Urvina Bay.

up monospecific reefs. One of these reefs, a giant colony of star coral, was 371 years old (Figure 5.38).

Given the seasonal changes in the oceanic environment around the Galápagos, can you suggest how this age might have been estimated?

Through seasonal growth bands, analogous to tree rings. The coral grows faster during the warm, sunny season of the year than during the cool, overcast season. As a result, new coral skeleton is laid down in alternating more dense and less dense layers. If vertical slices of coral are laid on film and illuminated with X-rays, the growth bands are recorded on the film. Figure 5.39 shows part of a section of the giant star coral colony photographed in this way. Knowing exactly when the coral died, and counting each pair of light and dark bands as one year, it was possible to estimate the year when the coral first became established—1583.

Growth bands in the lower sections of the coral are widely spaced and regular, indicating that for about the first 150 years of its existence the coral was flourishing. Then, in the mid-18th century, conditions for the coral seem to have deteriorated; the bands are closer together, and in some places there are hiatuses in growth indicated by stars on Figure 5.39. One of these occurred in 1747 (indicated by a red star), where two dark bands come together. This indicates a period when the coral was severely stressed—indeed, parts of the colony died at this time. It is thought that this disruption of coral growth was caused by a major ENSO event.

After about 1755, the coral again grew healthily, producing broad regular bands. Healthy growth continued until about 1890, when the coral became fragmented and from this time until the coral was killed by the uplift in 1954 the bands are indistinct and close together. This could indicate *either* an increase in ENSO events, with their influxes of warm water, *or*—bearing in mind the oceanographic setting of the Galápagos—water generally too cold for optimum coral growth.

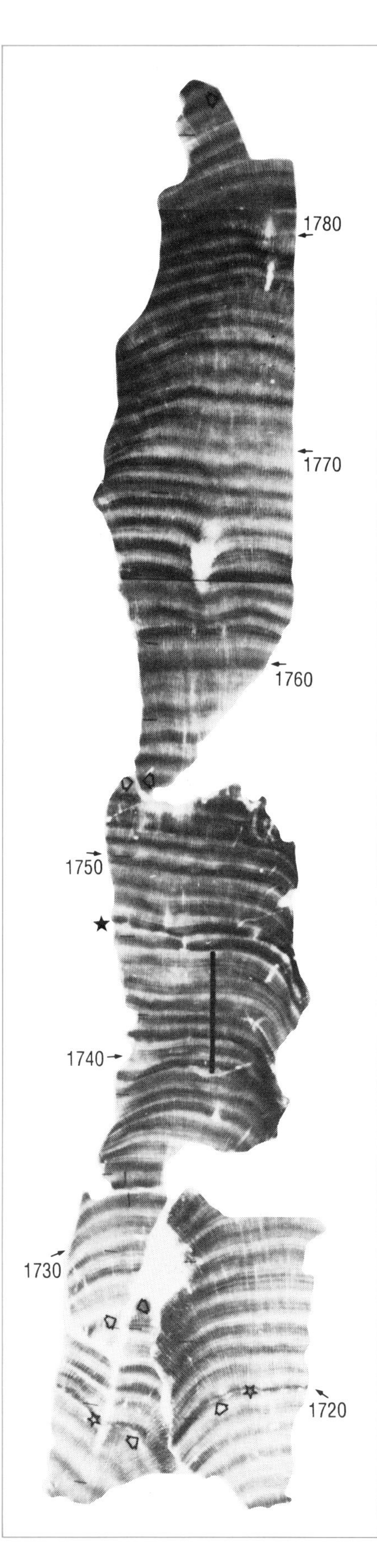

Figure 5.39 Part of a collage of X-ray prints of 5 mm-thick slabs cut from a 5.5 m-long section of an uplifted star coral *Pavona clavus* colony from Urvina Bay (pre-1715 and post-1785 portions are not shown). Each light and dark band corresponds to one year, and the average growth rate of the coral was about 10 mm per year. The red star indicates a possible ENSO event in 1747. Arrows indicate individual years. The vertical black bar indicates a transect selected for more-detailed study. Ignore the open arrows.

How might we be able to find out more about past water temperatures in Urvina Bay by analysing the carbonate of the coral skeleton?

Through the ratio in the skeleton of the stable isotopes of oxygen: ^{18}O and ^{16}O. When the ratio of ^{18}O to ^{16}O (expressed as $\delta^{18}O$) is relatively high, the coral was growing in relatively cool water, and *vice versa*. Figure 5.40 shows the variation in the oxygen isotope ratio of the coral skeleton over the life of the coral.

QUESTION 5.7 (a) From Figure 5.40, what can you say about the causes of disrupted coral growth in (i) the mid-1700s and (ii) the 1900s?

(b) Is there a possible alternative cause for the disruptions?

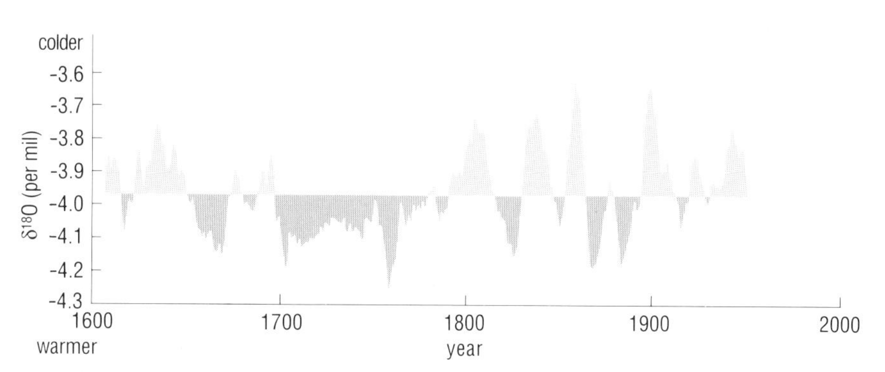

Figure 5.40 Variation with time/height of $\delta^{18}O$ in the Urvina Bay coral, with dates estimated by counting growth bands, but with data for groups of seven years averaged to show longer-scale trends more clearly.

It is interesting to see from Figure 5.40 that warm conditions prevailed in the Galápagos region for most of the 18th century, at a time when much of western Europe was in the grip of the 'Little Ice Age', during which major rivers such as the Thames regularly froze over completely in the winter. Changes in global climate do not necessarily affect all parts of the world in the same way.

5.6 EXPLORATION, EXPLOITATION AND CONSERVATION: CHANGING ATTITUDES

It is not known who originally discovered the Galápagos, but the first human visitors must have been carried there by the same strong westward currents which brought the islands their flora and fauna. No humans stayed to found a permanent population, but flints and other artefacts discovered around the islands suggest that in prehistoric times they may have been used as fishing outposts.

The Spaniards who invaded Peru in the 16th century heard stories of Iupac Yupanqui, a great Inca monarch who, with many men and a fleet of balsa rafts, had set out to explore the sea to the west, and reported having found two islands, one of which they named Island of Fire. Whether these islands were two of the Galápagos group is open to speculation, but given the prevailing winds and currents it is certainly possible.

The Spanish found the islands for themselves in 1535, by accident. A ship sailing from Panama to Peru became becalmed off the coast of South America; it was then swept westwards by a strong offshore current and, after a week of drifting, came to a group of barren volcanic islands. One of those who scrambled ashore after much-needed freshwater was Bishop Fray Tomás de Berlanga, who later sent a report of the circumstances of the discovery to Charles V, the Holy Roman Emperor. In his report, he described the strange volcanic topography, the arid coastal soil (they found little freshwater) and the wildlife—the great numbers of seals and iguanas, and, 'so large that each could carry a man on top', the giant tortoises or '*galopegos*'. The identification of the islands with the tortoises persisted, and on a Flemish map of 1570 they are shown under the name 'Insulae de los Galopegos'. The islands also acquired another name, Las Encantadas—the Enchanted, or Bewitched, Islands; this name may have arisen amongst mariners because the local winds and currents made the task of sailing around the islands very difficult. It is also said that as early navigational techniques were too imprecise to establish latitude and, particularly, longitude accurately, the position of the islands seemed to keep changing. Even today they can be elusive, with a tendency to disappear suddenly behind a veil of mist.

5.6.1 BUCCANEERS, WHALERS AND SEALERS

The Spanish made no attempt to settle on the Galápagos, and the islands remained uninhabited until the late 17th century. Ironically, it was Spain's enemy, England, which now made use of the islands—as a base for buccaneers. These 'pirates with patronage' sailed out to loot coastal towns and raid the Spanish ships which at that time were transporting large amounts of treasure along the South American coast between Panama and Callao. The Galápagos provided an ideal bolt-hole: they had safe and secluded anchorages, an abundance of fresh food (largely in the form of the easily caught tortoises), and several reasonably reliable sources of freshwater.

The buccaneering period of the Galápagos' history lasted until the early 18th century. Amongst those buccaneers who wrote about their experiences were William Dampier and Ambrose Cowley, whose chart of the islands is reproduced as Figure 5.41. The island names used on this chart of 1684 honour the Stuart Kings and various high officials, not a few of whom had turned a blind eye to the buccaneers' activities. Subsequently, the islands acquired other names, some of which are given in the caption to the chart.

The next major role for the Galápagos was as a base for the Pacific whaling fleets. When, in 1788, the British ship *Emilia* rounded Cape Horn in search of sperm whales, she began a century of commercial sea-faring which was to be on a scale never seen before. British whalers were soon joined in the Pacific by vessels from New England and France

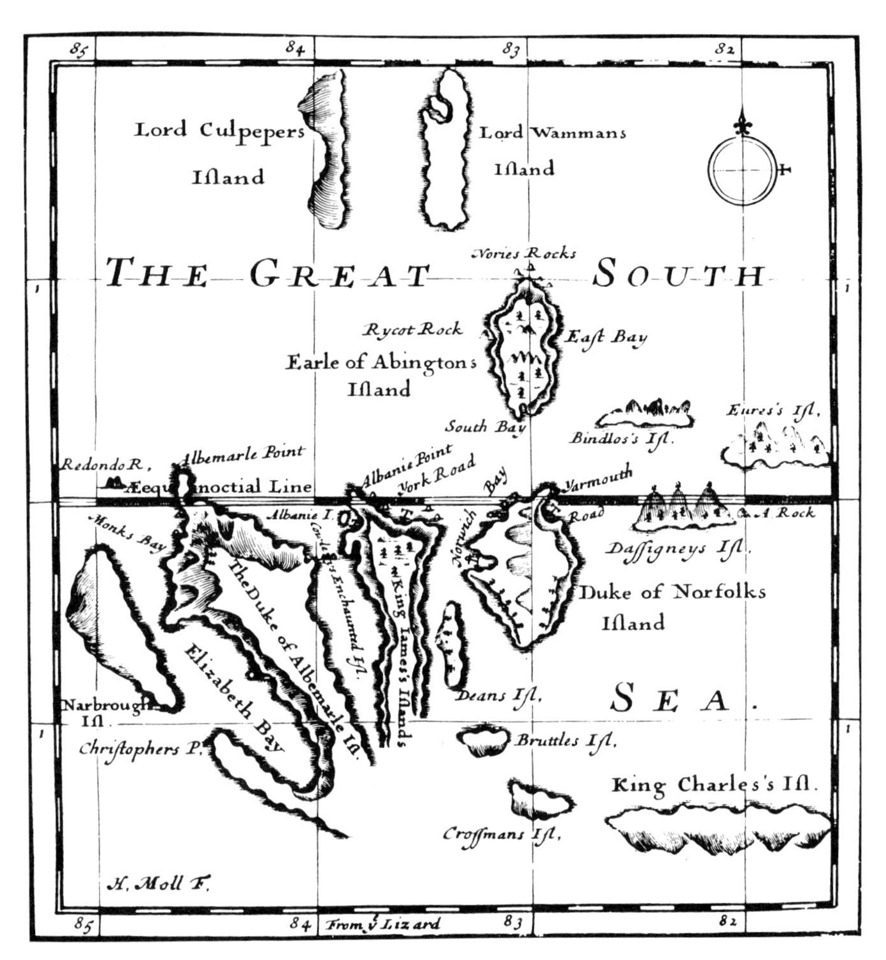

Figure 5.41 Ambrose Cowley's chart of the Galápagos Islands from his *Collection of Original Voyages*, 1699. Comparison with a modern chart such as Figure 5.1(a) shows discrepancies of scale (particularly for some smaller islands) and King James Island (San Salvador) and Eure's Island (probably Genovesa) have been displaced to the south. The anchorage on the western side of the Duke of Norfolk's Island (Santa Cruz) corresponds to Bahia de Conway. Names given here correspond to Ecuadorian equivalents (in italics) as follows (subsequent better-known names in parentheses): Narborough Island, *Fernandina*; Duke of Albermarle Island, *Isabela*; King James's Island, *San Salvador* (Santiago); Dean's Island (Duncan), *Pinzón*; Duke of Norfolk's Island (Indefatigable), *Santa Cruz*; Bruttles Island, *Tortuga*; Crossman's Island, *Crossman* (Los Hermanos); King Charles's Island, *Santa Maria* (Floreana); Dassigney's Island (Chatham), *San Cristóbal*; Eure's Island (Tower), *Genovesa*; Bindlos's Island, *Marchena*; Earl of Abington's Island, *Pinta*; Lord Wamman's (Wenman's) Island, *Wolf*; Lord Culpeper's Island, *Darwin*.

and, in smaller numbers, from Holland, Spain and Germany. They sailed up the coast of South America to the rich whaling grounds in the eastern Pacific and then westward along the Equator with the Trade Winds and prevailing currents to the whaling grounds around the Gilbert and Ellice Islands. From this 'Line fishery', they headed off to the other whaling grounds in the eastern Pacific.

The Galápagos were seen by whalers as the 'gateway to the Pacific' (Figure 5.42, overleaf). The first whaler to visit the Galápagos was the *Rattler*, which under Captain James Colnett surveyed the Pacific for the whaling industry. The *Rattler* returned with the first reliable information on the movement of sperm whales throughout the eastern Pacific. Between the 1790s and the early 1900s, whalers congregated about the Galápagos in huge numbers; not only were there good whaling grounds nearby but—as the buccaneers had discovered—the islands offered safe harbours, materials needed for refitting, freshwater and food. The tortoises in particular were a great attraction as a source of easily caught fresh meat for long periods at sea. As Admiral Porter wrote in 1813, after loading 14 tons of the animals from Santiago [San Salvador] onto his vessel the *Tartar*: 'They require no provisions or water for a year . . . they have been piled away among the casks in the hold of a ship, where they have been kept for eighteen months'. How many tortoises were taken away in the holds of whaling vessels is impossible to say—at least 100 000 and possibly several times that. First one race of tortoises and then another was brought to the edge of extinction.

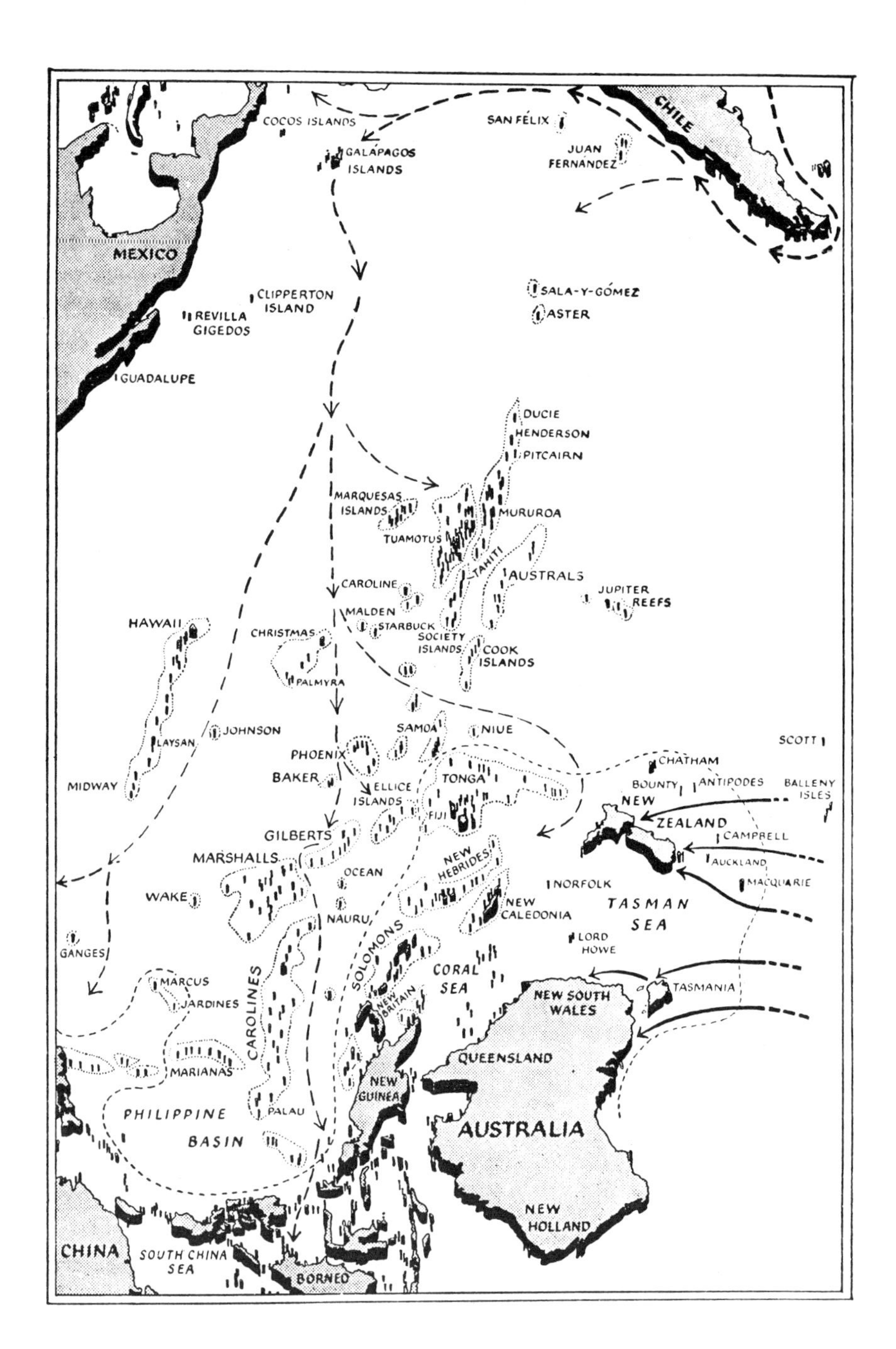

Figure 5.42 A whaler's view of the Pacific, showing how the Galápagos served as a convenient 'jumping-off point' for Pacific whalers. The dashed arrows indicate the advance of the whalers; and the thicker curved arrows the seasonal migration of the southern right whale.

The number of whales taken around the Galápagos is perhaps even harder to estimate. As an example, the British vessel *William*, the second whaling ship to enter Galápagos waters, took 42 sperm whales in only 18 days in 1797; shortly afterwards, the *Cyres* obtained a full load of sperm oil after only a year-and-a-half of sailing around the islands. Given that, at its peak, the American whaling fleet alone had 70 vessels visiting the Pacific (Figure 5.43), the total number of whales taken in more than a century of whaling must be enormous.

From the beginning of whaling around the Galápagos, sea-lions and, especially, fur seals were taken in great numbers. A log of the *Rattler* reads: 'We saw but few seals on the beach, either of the hairy [sea-lion] or furry [fur seal] species ... but a few hundred of them might at any

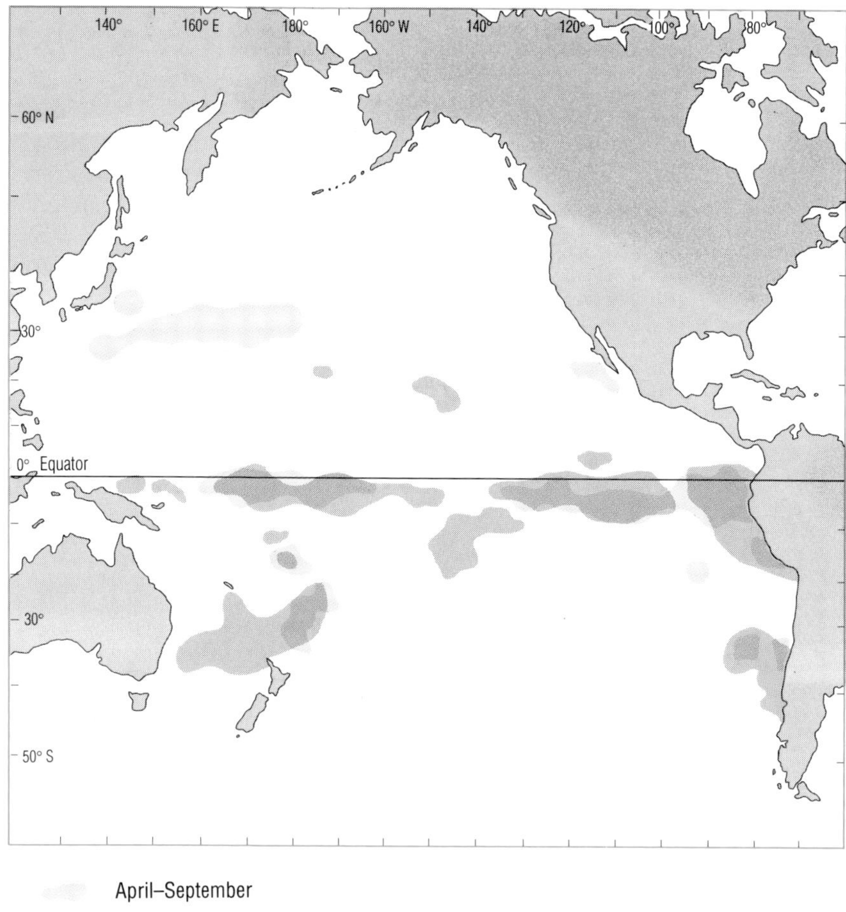

Figure 5.43 Areas in the Pacific where sperm whales were taken by whalers from Nantucket, New England, between 1729 and 1919. Red areas show regions where kills were made from October to March, blue areas show regions where kills were made from April to September. The 'Line fishery' in the South Equatorial Current (on the equinoctial line, *cf.* Figure 5.41) is one of the regions where kills were made in both seasons.

time be collected without difficulty, and form no inconsiderable addition to the profits of a voyage'. Sealers frequented the Galápagos throughout the 18th century although never in such great numbers as the whaling vessels. Initially, the numbers of seals and sea-lions taken from the sea or killed around the rocky shores was enormous: in a single voyage in 1816, 8 000 fur seals and 2 000 sea-lions were taken; in 1823, 5 000 fur seals were killed in two months. Over the course of the century, kills became smaller and smaller and by the 1890s Galápagos fur seals were thought to have been exterminated. In 1891, the *Albatross* found that 'a scattered remnant of a herd still frequented the more inaccessible rocks of the archipelago' but subsequent sealing voyages 'resulted in the killing of all seals that could be found'.

In 1905–06, an expedition from the California Academy of Sciences found only one fur seal during a comprehensive survey of the archipelago. Although fur seal numbers had risen to between 20 000 and 40 000 by the 1980s, the population still bears the marks of the whaling and sealing days and is made up of scattered small clusters. Sea-lions, whose pelts were of less value commercially, are widespread throughout the archipelago.

The Spanish never attempted to colonize the Galápagos, which were annexed by the newly independent Republic of Ecuador in 1832. Within a year, the first settlers—including many political and other prisoners who had exchanged a sentence of death for one of deportation—arrived on Floreana. They began to clear the natural vegetation and introduce animal husbandry, and for a few years the community prospered. However, Floreana is small and subject to drought, and morale and living standards declined. The Government began to use the island as a penal colony, and there followed years of strife and lawlessness, abandonment and recolonization. The history of the settlement of the other three islands to become inhabited—Isabela, San Cristóbal and Santa Cruz—is equally dark. As settlers returned to the mainland, or died, the domestic plants and animals they left behind ran wild and multiplied, becoming a serious threat to the natural environment. Meanwhile, as the Ecuadorian authorities continued to be inadequately represented on the islands, they could do little to prevent foreign vessels coming to plunder the stocks of tortoises, seals and whales.

5.6.2 THE GALÁPAGOS MARINE RESOURCES RESERVE

On May 13, 1986, the Government of Ecuador published a Decree which began:

> 'It is the duty of the National Government to protect those natural areas which stand out because of their ecological, scientific, educational, economic, and political value, in order to conserve their resources and furnish today's generations with an improved and sustained benefit from their use, while at the same time, maintaining their potential to meet the needs and aspirations of future generations.'

There followed articles concerning the extent, the management and the funding of the Galápagos Marine Resources Reserve.

That this Reserve has come into being around islands once regarded as fit only for a penal colony is the result of a remarkable cooperation between Science on the one hand and Government on the other.

The chain of events which led to the establishment of the Reserve can be said to have begun when Captain Fitzroy of the *Beagle* asked Charles Darwin to join him on an investigative voyage to the Pacific. Darwin's vivid and enthusiastic descriptions of the wildlife in his *Journal* as the naturalist of the expedition, and the development of his ideas about evolution, stimulated by what he saw in the islands, kept the Galápagos in the minds of at least some scientists.

Darwin and his companions visited the islands in 1835 (three years after Floreana was first colonized) and they were amazed by birds and animals with no fear of humans, and impressed by ecosystems not yet altered by human interference.

For nearly a century after the visit of the *Beagle*, life for the people on the islands remained hard, the poor conditions often aggravated by tyranny and bloodshed. Meanwhile, the natural environment continued to deteriorate. During the 1930s, small groups of scientists in the United States and Britain, as well as a band of Ecuadorian activists, campaigned both for the islands to be protected and for the establishment of a Galápagos research station. Although they succeeded in that the Ecuadorian Government declared large parts of the archipelago to be nature reserves, it was a hollow victory because no agency was put in place to implement protective regulations.

In 1959, great steps were made towards the eventual establishment of a national nature reserve with an international centre for scientific research. The Government of Ecuador declared all unsettled areas of the archipelago (nearly 97% of the total land area) to be a National Park. With the encouragement and backing of a number of international bodies, a group of able and committed scientists and conservationists came together to set up the Charles Darwin Foundation for the Galápagos Islands. The Charles Darwin Research Station was officially inaugurated in 1964 and an agreement was signed between the Government and the Foundation empowering the latter to maintain and manage the Research Station, and formally recognizing its role as advisor to the Government on all matters affecting science and conservation in the archipelago. This alliance between international Science and Government has continued, and eventually led to the setting up of the Marine Resources Reserve.

Despite the public's fascination with the land creatures of the Galápagos, and the obvious dependence of some of the more famous of them on the sea for food, almost nothing was known about the islands' marine life until the early 1970s. The first systematic study of subtidal plants and animals was published in 1974 by Gerard Wellington, an American marine biologist. Wellington noted in particular the large number of marine species endemic to the Galápagos (*cf.* Table 5.1), and stressed that the archipelago—the islands and the surrounding seas—should be seen as an integral ecosystem. In the same year that Wellington's report was published, the National Park put forward a 'Master Plan' proposing an extension of the Park boundaries one kilometre seaward of all uninhabited islands, some of which involved defining zones where human activities would be restricted to a greater or lesser extent.

Recognizing that many islanders felt their livelihoods would be threatened by restriction of activities around the coast, the President of Ecuador declared in 1982 that the Galápagos would not simply be a conservation area but would also be a Marine Resources Reserve, protecting the resources of the sea and sea-bed for the exclusive use of the local inhabitants. The idea of a Marine Resources Reserve persisted and the first article of the Government's 1986 Decree reads as follows:

> 'That the water column, the seabed, and the marine subsoil of the sea located within the interior of the Galápagos Archipelago—which is understood to be the area within the baselines used to measure the territorial sea of the Galápagos Archipelago, according to the Supreme Decree No. 959-A of June 28th 1971 as proclaimed in Official Register No. 265 of July 13th 1971—are declared to be a resources reserve, along with a band of 15 nautical miles surrounding the said baselines, and fall under the exclusive domain of the State.'

Article 2 lists those ministers and council representatives who would make up the Inter-institutional Commission charged with 'the management and vigilance of the marine reserve', and states that

'The Commission may request the assistance and collaboration of the Charles Darwin Research Station and national and international organizations as deemed necessary.'

The total area of the Reserve is about 70 000 km^2: about 50 000 km^2 of internal waters within the archipelagic baselines (solid red line in Figure 5.44) and 20 000 km^2 in the 15-(nautical)-mile buffer zone. The length of protected coastline is greater than the total coastline of continental Ecuador.

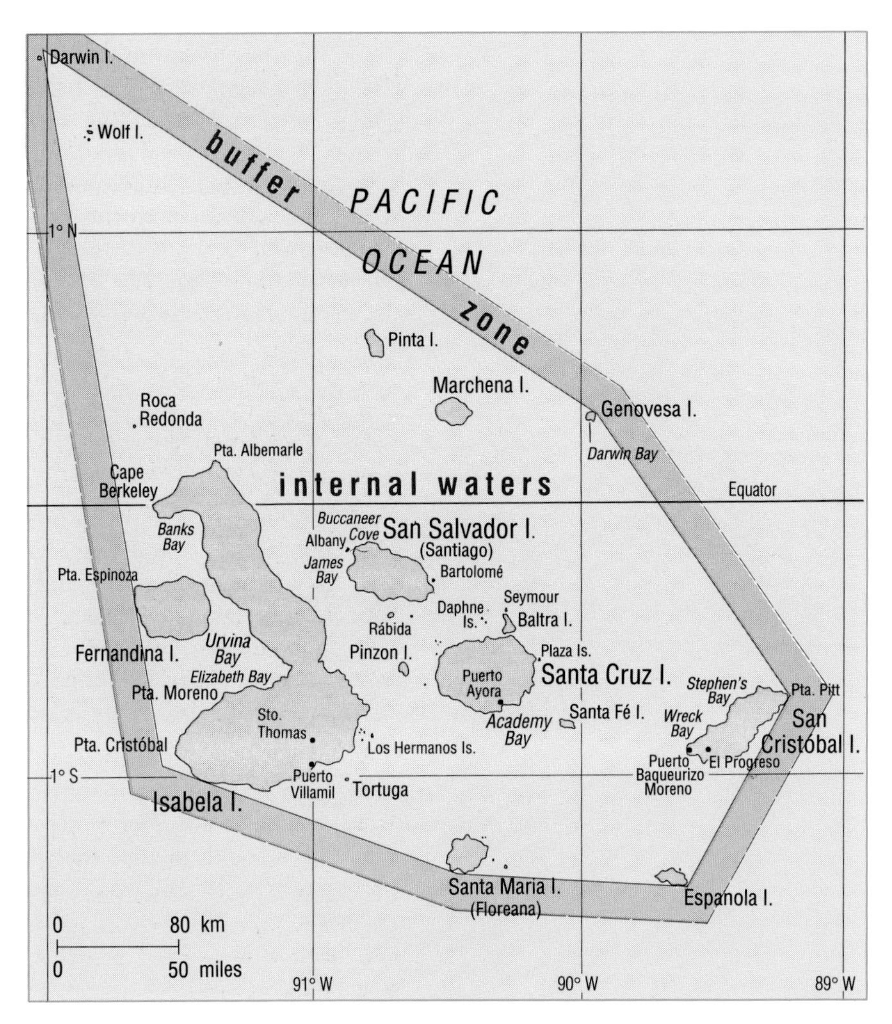

Figure 5.44 The extent of the Galápagos Marine Reserve. The inner straight red lines are the archipelagic baselines, and the 'buffer zone' between these and the outermost boundary is 15 nautical miles wide.

The strong links with international scientific institutions have continued. For example, in 1983 the Woods Hole Oceanographic Institution Marine Policy Center was asked by the Government of Ecuador to provide a report and recommendations concerning management of coastal and marine resources, and later provided advice on the management of the Reserve.

High on the management agenda is the issue of tourism. From at least the 1960s it was realized that tourism could play a large role in the development of the Galápagos National Park: if conservation was supported, there was a strong chance that foreign tourists would come to see what was being conserved. Foreign exchange could be Ecuador's reward for what has been described as a 'peculiar, generous and unprecedented deal with international science'.

The growth in tourism has been enormous, and between 20 000 and 30 000 visitors come to the islands each year. To minimize the need for hotels or tourist facilities, visitors generally sleep on board the boats by which they tour the islands. As yet, the environmental impact of so many visitors seems to be small, but new problems are arising. The tourist industry attracts new settlers to the islands and the permanent population is now about 10 000, and growing. One serious problem is that, both by accident and design, settlers continue to introduce new species of plants and animals which threaten the naturally occurring species. Also, there is potential for conflict between the desire of the Galápagos population for an improved standard of living and the requirements of the National Park/Reserve, which of necessity limit opportunities for economic development.

One region of particular concern is the fishing industry. In 1989, all non-traditional methods of fishing, including the use of monofilament nets and spear-fishing, were banned from the Reserve, as was the killing of sharks for their fins. These measures have been reasonably successful, although enforcement remains a problem. It is not clear how the exclusion of commercial fisheries from the Reserve should be interpreted in the light of Ecuadorian Fisheries Law, nor how it will affect the (declining) national tuna fishery. Local artisanal fisheries, on the other hand, are regarded as an integral aspect of the Reserve, and until at least the late 1980s their activities were not thought likely to threaten any marine species. However, as fishing gear improves and vessels become modernized, the situation could change. Would relatively small-scale but no longer strictly 'traditional' fishing operations be stopped?

Regulations may also be needed to stop the ceiling on tourist numbers being pushed even higher. More and more tourists are taking part in offshore leisure activities such as diving and windsurfing, and these also need to be controlled if the environment is to be protected.

5.6.3 WIDER PERSPECTIVES

The establishment of the Galápagos Marine Resources Reserve raises some interesting issues so far as international law is concerned. For instance, we have seen that the inner boundary of the Reserve (Figure 5.44) corresponds to the archipelagic baselines (although being Ecuadorian territory, the Galápagos do not count as an 'archipelagic state' under UNCLOS).

What, then, is the implication of the 15-nautical-mile width of the 'buffer zone', given the terms of the Law of the Sea Convention?

As the internationally recognized territorial sea limit is twelve nautical miles, the implication is that Ecuadorian sovereignty is being extended seawards by three nautical miles. However, Ecuador and other South American states have claimed 200-mile territorial seas, which under the terms of the 1982 Convention (which Ecuador has not signed) could be seen to correspond, in effect, to exclusive economic zones (EEZs)/continental shelves; from this point of view, the international legal status of the buffer zone is rather confusing. To add to the confusion the Ecuadorian Constitution has always allowed (in theory, at least) for legislation to be applied somewhat differently on the Galápagos.

As far as EEZs are concerned, the 1982 Convention requires states to take measures necessary to protect and preserve rare or fragile ecosystems, as well as the habitat of depleted, threatened, or rare species, or other forms of marine life. (Ecuador introduced such a measure in 1990 when it declared the area within 200 nautical miles of the Galápagos to bc a whale sanctuary.) The Convention also states that if there are 'recognised technical reasons' why certain areas of their EEZs need special protection, they may take appropriate special measures. This provision is generally taken to refer to threats from pollution by vessels. It has been argued that the foraging ranges of certain protected Galápagos-based seabirds require an extended area to be protected, and that this constitutes a 'recognised technical reason'. Whether or not this interpretation is upheld, it does seem that the requirements of the 1982 Convention for states to conserve marine species, combined with the right of coastal states to conserve and manage the natural resources in their EEZs, together provide a reasonable basis for the 'buffer zone', under international law.

Finally, remember that the Galápagos Reserve is a *Resources* Reserve, and encompasses the sea-bed and subsoil.

Given the tectonic setting of the archipelago (Figure 5.3), what sea-bed resources might fall within the Reserve, and in what sense could they need management or protection?

Mineral deposits precipitated from hydrothermal solutions some time in the past are the most likely sea-bed resources. They would need management in the sense that exploitation of such resources would involve considerable disturbance of marine ecosystems.

Various areas of the Marine Reserve have been identified as in need of special protection, and suitable to be made into a Marine Park. These include nursery grounds for commercial species, areas providing habitats for certain species, areas for tourist education and areas of special interest to science.

5.7 SUMMARY OF CHAPTER 5

1 The Galápagos are geologically young volcanic islands close to the Equator, just south of the east–west Galápagos spreading axis and about 1 000 km east of the East Pacific Rise. They lie at the apex of a 'v'-shaped hot-spot trace formed by the Carnegie and Cocos Ridges. Detailed mapping of the sea-bed along the spreading axis has provided new information about processes at overlapping spreading centres and about the mechanism of rift propagation at ridge offsets.

2 The Galápagos typically experience annual alternations of a cool dry season with strong South-East Trade Winds, and a warm wet season with tropical storms and weaker South-East Trades. The seasons are controlled by the position of the ITCZ, and by that of the Equatorial Front (locally known as the Galápagos Front), which separates warmer, less saline water to the north from cooler, more saline water to the south: both lie further north in the Southern Hemisphere winter than in that Hemisphere's summer. The Equatorial Front is also the northern

boundary of the westward-flowing South Equatorial Current, fed mainly by the Peru Current. North of the Front, the eastward-flowing (North) Equatorial Counter-Current brings warm water from the western Pacific—average sea-surface temperatures in the western half of the equatorial Pacific are typically a degree or two higher than those in the eastern half. For part of the year, the Panama Current brings warm low-salinity water southwards and westwards towards the Galápagos. In the eastern Pacific, wind-induced upwelling of cool nutrient-enriched sub-surface water is most intense during the southern winter, and around the Galápagos there is also intermittent localized upwelling where the eastward-flowing Equatorial Undercurrent is deflected upwards by the steep topography on the western slopes of the islands.

3 Most of the plant and animal species that have colonized the Galápagos originated in South and Central America, carried to the islands by prevailing winds and currents; only a small proportion originated in the western Pacific. In terms of sea-surface temperatures, Galápagos waters in the north and north-east—nearest to the Equatorial Front—are most 'tropical'; they are cooler in the south and south-east, furthest from the Front; seasonably variable in the central zone; and coolest in the west where upwelling is most frequent and strongest. Coral reef communities are more abundant on east-facing coasts; and macrophytic algal (seaweed) communities are more abundant on west-facing coasts where most upwelling occurs. The resulting high productivity supports a great diversity of invertebrates, fishes, birds and mammals, as well as the renowned Galápagos turtles and iguanas. The Equatorial Divergence, just south of the Equatorial Front, is another highly productive region near the islands, and an important feeding ground for whales and for the wider-ranging fishes and birds. The relatively high diversity of Galápagos flora and fauna is attributable also to the variety of environments which the islands offer and to the way in which each island seems to have evolved its own set of species.

4 The onset of the warm, wet season at the turn of the year was formerly known locally as El Niño. Superimposed on the seasonal cycle is an irregular (2–10-year) cycle of more extreme warm and wet conditions, called El Niño–Southern Oscillation (ENSO) events, which generally last about a year. They occur when there is a reduction in the atmospheric pressure difference between the high pressure region over the South (East) Pacific and the low pressure region over Indonesia. The South-East Trade Winds slacken, and equatorial winds may even become westerly for a time. The normal west-to-east sea-surface slope of the equatorial Pacific relaxes and the 'pool' of warm water in the west flows eastwards, depressing the thermocline there, so that the already weakened upwelling now brings only nutrient-depleted water to the surface. Higher sea-surface temperatures lead to increased incidence of tropical storms and higher than average rainfall. ENSO events are commonly associated with climatic perturbations in other parts of the world.

5 The 1982–83 ENSO event was the longest and most intense on record; equatorial westerly winds were stronger and more prolonged than previously observed; sea-surface temperatures in the eastern tropical Pacific were higher, and the ITCZ was further south; the Galápagos experienced nine times their normal wet-season rainfall; the Galápagos Front all but disappeared and water warmer than 20 °C extended down to 125 m, compared with less than 50 m normally. Stratification of

surface waters by high temperature and high rainfall further reduced supplies of nutrients. Primary productivity slumped and the species composition of phytoplankton changed; populations of many animals were decimated, including several species of birds, sea-lions, seals and iguanas; corals were killed, probably by a combination of high temperature and reduced salinity; and storm waves combined with high sea-levels to cause havoc among intertidal and coastal populations.

6 Progress of the 1982–83 ENSO event could be traced by changes of mean sea-level across the tropical Pacific. Mean sea-level was more than 30 cm above normal in the east, during the peak of the event. The transfer of warm water from west to east across the Pacific under the influence of the (temporary) westerly winds occurred as a series of eastward-propagating Kelvin waves—a succession of 'bulges' of warm water travelling east along the Equator at about $3\,\mathrm{m\,s^{-1}}$. Equatorial Counter-Currents also increased in strength, aiding the advection of warm water eastwards; but the Equatorial Undercurrent virtually disappeared, as a consequence of relaxation of the west-to-east slope of the sea-surface.

7 Old coral reefs can be dated by counting annual growth bands, and the thickness and spacing of the bands provides information about changing environmental conditions. This information can be supplemented by analysing the oxygen isotopes in the calcium carbonate of which the reefs are made—higher values of δ^{18}O indicate lower temperatures. Results suggest, for example, that warm-water conditions prevailed around the Galápagos for most of the 18th century, but were more variable thereafter.

8 The Galápagos were probably known to the Incas, but were not discovered by Europeans (Spaniards) until the 16th century. Initially used as a base by buccaneers, from the mid-18th century until the early 20th the islands were visited annually by whalers and fur-sealers, who brought populations of some animals to the verge of extinction. Ecuador claimed the islands in 1832 and used them sporadically as a penal settlement; animals and plants introduced by the settlers ran wild and adversely affected indigenous communities. International efforts to introduce conservation measures began in the 1930s and took a major step forward with establishment of the Charles Darwin Research Station in 1964. In 1986, the Government of Ecuador decreed that the islands should become a Marine Resources Reserve, and there is now good international cooperation over policies to protect the unique Galápagos ecosystems. Problems remain, however, for a balance must be struck between the needs of conservation on the one hand, and those of fisheries, tourism and the local population on the other.

Now try the following questions to consolidate your understanding of this Chapter.

QUESTION 5.8 Examine Figure 5.6(b). How does the oceanic crust on either side of the 'pseudofaults' differ in terms of age? What techniques would you therefore expect researchers to use in trying to locate the pseudofaults?

QUESTION 5.9 Examine Figure 5.45, which shows measurements of temperature and salinity at the surface and 60 m depth in Academy Bay during the 1982–83 ENSO event (Figure 5.29).

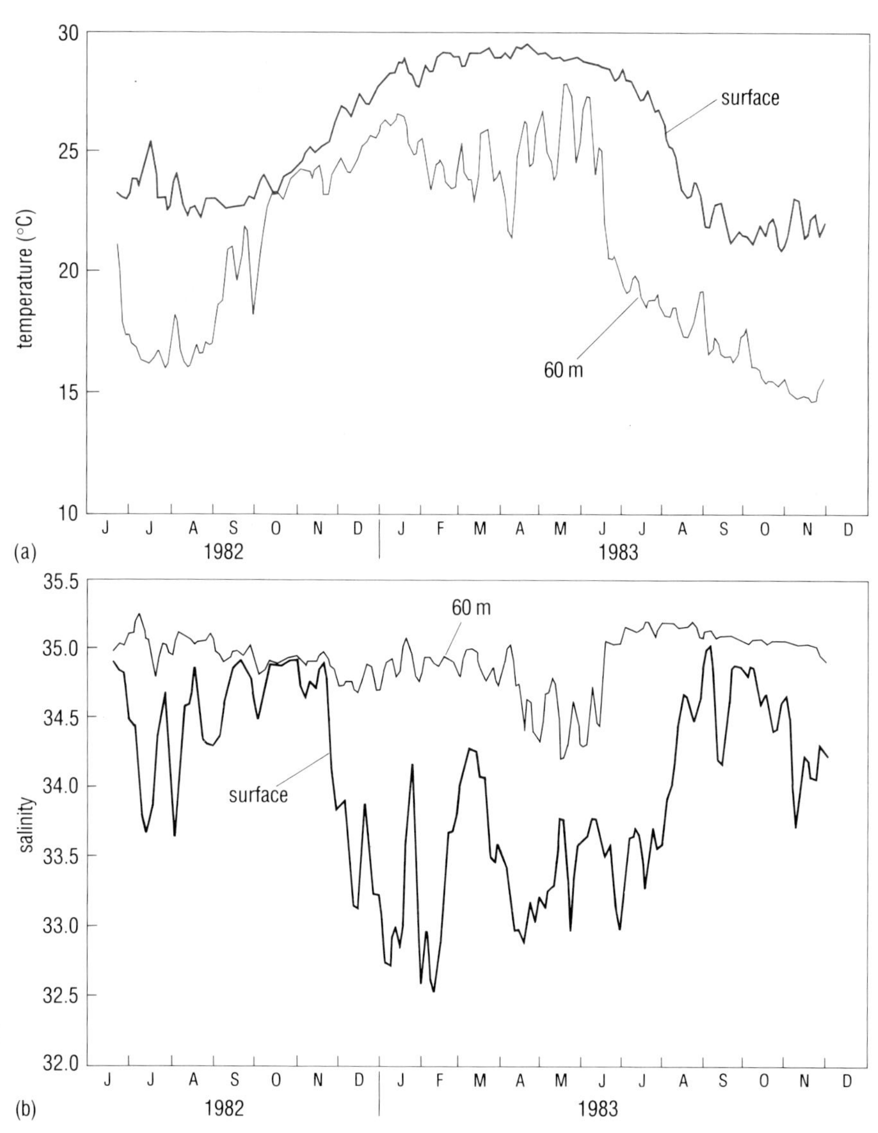

Figure 5.45 The variation of (a) temperature and (b) salinity at the surface and at 60 m depth in Academy Bay, during the 1982–83 ENSO event. For use with Question 5.9.

(a) On the temperature record, do (i) the 'arrival' of the ENSO and (ii) the return to normal conditions, show up first at the surface or at depth? Is this what you would expect?

(b) How would you account for the salinity changes over the same period?

(c) By comparing the curves for the surface with those for a depth of 60 m, when would you say that the waters in Academy Bay were (i) most homogeneous (well-mixed) and (ii) most stratified?

QUESTION 5.10 Examine Figure 5.46 (overleaf). For each of the variables shown, explain briefly why it provides an index for ENSO events.

QUESTION 5.11 In Figure 5.3, the red ring shows the position of active hydrothermal vents on the Galápagos spreading axis. Hydrothermal vents are relatively short-lived environments for marine organisms, lasting only a few thousand years at most. In what sense do processes influencing colonization of a new hydrothermal vent resemble those affecting colonization of an oceanic island group such as the Galápagos?

Figure 5.46 For use with Question 5.10. Time series of selected variables whose interannual variability is related to the ENSO phenomenon. Vertical lines indicate ENSO events. Letters represent:

A Sea-surface temperatures in the eastern equatorial Pacific (South American coast to dateline, 180°), average for April to March.

B Rainfall index for central equatorial Pacific island stations, average for April to March.

C Atmospheric pressure at sea-level, Darwin, Australia, average for April to March.

D Atmospheric pressure at sea-level, Tahiti, average for April to March.

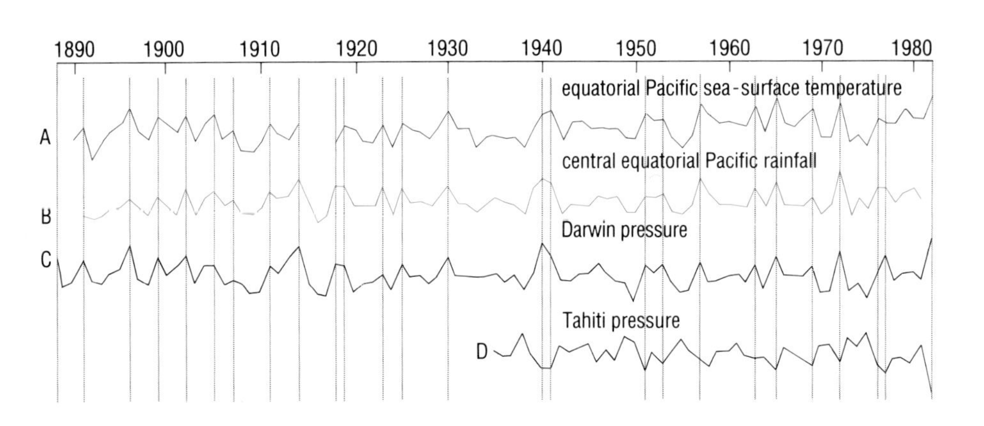

APPENDIX 1 TABLES I–III

Note: the names and status of several countries have changed since Tables I–III were compiled.

Table I The width of the territorial sea and the extent of fishing limits claimed by various countries (1972).

Country	Territorial sea (miles)	Fishing limit (miles)
Albania	12	12
Algeria	12	12
Argentina	200	200
Australia	3	12
Bahrain	3	—
Barbados	3	—
Belgium	3	12
Brazil	200	12
Bulgaria	12	12
Burma	12	12
Cambodia	12	12
Cameroon	18	18
Canada	12	12
Ceylon	12	6
Chile	200	200
China	12	12
Colombia	12	12
Congo (Brazzaville)	12	3–15
Costa Rica	3	—
Cuba	3	3
Cyprus	12	12
Dahomey	12	12
Denmark	3	12
Dominican Republic	6	12
Ecuador	200	200
Egypt	12	12
El Salvador	200	200
Ethiopia	12	12
Finland	4	4
France	3	12
Gabon	25	25
Gambia	50	18
Germany, East	3	3
Germany, Federal Republic of	3	12
Ghana	12	12
Greece	6	6
Guatemala	12	12
Guinea	130	12
Guinea, Equatorial	12	—
Guyana	3	3
Haiti	6	6
Honduras	12	12
Iceland	4	12
India	12	12
Indonesia	12	12
Iran	12	12
Iraq	12	12
Ireland	3	12
Israel	6	6
Italy	6	12
Ivory Coast	6	12
Jamaica	12	12
Japan	3	3
Jordan	3	3
Kenya	12	12
Korea, North	12	—
Korea, South	20–200	20–200
Kuwait	12	12
Lebanon	—	6
Liberia	12	12
Libya	12	12
Madagascar	12	12
Malaysia	12	12
Maldive Islands	3–55	100–150
Malta	6	12
Mauritania	12	12
Mauritius	12	—
Mexico	12	12
Monaco	3	12
Morocco	12	12
Netherlands	3	12
New Zealand	3	12
Nicaragua	3	200
Nigeria	30	30
Norway	4	12
Oman	3	—
Pakistan	12	12
Panama	200	200
Peru	200	200
Poland	3	12
Portugal	—	12
Qatar	3	—
Romania	12	12
Saudi Arabia	12	12
Senegal	12	18
Sierra Leone	200	12
Singapore	3	3
Somali Republic	12	12
South Africa	6	12
Spain	6	12
Sudan	12	12
Sweden	4	12
Syria	12	12
Taiwan	3	3
Tanzania	12	12
Thailand	12	12
Togo	12	12
Trinidad and Tobago	12	12
Tunisia	6	12
Turkey	6	12
Ukrainian SSR	12	12
Union of Arab Emirates:		
Abu Dhabi	3	—
Ajman	3	—
Dubai	3	—
Fujairah	3	—
Ras al Khaimah	3	—
Sharjah	12	—
Umm al Qaiwain	3	—
USSR	12	12
United Kingdom	3	12
Overseas areas	3	3
USA	3	12
Uruguay	200	12
Venezuela	12	12
Vietnam, North	12	20 (km)
Vietnam, South	3	20 (km)
Yemen Arab Republic	12	12
Yemen, People's Democratic Republic of	12	12
Yugoslavia	10	10
Zaire	3	3

Table II The ratio of shoreline length to land area for 141 countries. The units are reciprocal miles (miles^{-1}).

Country	Ratio	Country	Ratio	Country	Ratio
Monaco	5.000 00	Honduras	0.008 64	Pakistan	0.001 41
Malta	0.409 83	Sierra Leone	0.007 90	Libya	0.001 33
Barbados	0.331 32	Sweden	0.007 82	Romania	0.001 23
Bahrain	0.294 37	Liberia	0.006 74	Togo	0.001 20
Trinidad and Tobago	0.128 28	Thailand	0.006 54	Nigeria	0.001 16
Singapore	0.125 00	France	0.006 53	Ethiopia	0.001 15
Mauritius	0.110 26	Somalia	0.006 48	Syria	0.001 14
Cyprus	0.081 18	Turkey	0.006 37	Brazil	0.001 12
Jamaica	0.066 16	Mexico	0.006 36	Kenya	0.001 09
Philippines	0.060 40	North Vietnam	0.006 23	Cameroon	0.001 01
Haiti	0.054 50	South Yemen	0.005 88	China	0.000 94
Denmark	0.041 25	Finland	0.005 64	Mauritania	0.000 90
Cuba	0.039 50	Bangladesh	0.005 62	Algeria	0.000 64
Japan	0.033 00	Morocco	0.005 19	Congo	0.000 63
Taiwan	0.033 84	Australia	0.005 08	Sudan	0.000 40
Panama	0.033 51	Burma	0.004 69	Jordan	0.000 39
Greece	0.032 29	East Germany	0.004 57	Iraq	0.000 05
UK	0.029 61	Yugoslavia	0.004 31	Zaire	0.000 02
Iceland	0.027 13	Guatemala	0.004 23	Afghanistan	0.0
New Zealand	0.026 70	Uruguay	0.004 22	Andorra	0.0
Lebanon	0.026 58	Ecuador	0.004 18	Austria	0.0
Indonesia	0.026 26	Gabon	0.003 86	Bhutan	0.0
Sri Lanka	0.025 65	UAR	0.003 38	Bolivia	0.0
Ireland	0.024 43	Yemen	0.003 24	Botswana	0.0
Qatar	0.024 00	West Germany	0.003 20	Burundi	0.0
Costa Rica	0.022 69	Senegal	0.003 17	CAR	0.0
Italy	0.021 07	USA	0.003 16	Chad	0.0
Portugal	0.020 89	Bulgaria	0.003 12	Czechoslovakia	0.0
El Salvador	0.019 85	Ghana	0.003 09	Hungary	0.0
South Korea	0.018 72	South Africa	0.003 09	Laos	0.0
Kuwait	0.018 54	Venezuela	0.003 06	Lesotho	0.0
Dominican Republic	0.017 27	Cambodia	0.003 00	Liechtenstein	0.0
Equatorial Guinea	0.016 98	Belgium	0.002 88	Luxembourg	0.0
Israel	0.015 46	Canada	0.002 88	Malawi	0.0
Malaysia	0.014 42	Guyana	0.002 79	Mali	0.0
Albania	0.014 09	USSR	0.002 68	Mongolia	0.0
Netherlands	0.014 00	Peru	0.002 53	Nepal	0.0
Norway	0.013 19	Colombia	0.002 32	Niger	0.0
South Vietnam	0.012 88	India	0.002 24	Paraguay	0.0
North Korea	0.012 41	Ivory Coast	0.002 20	Rhodesia	0.0
Oman	0.012 25	Guinea	0.002 00	Rwanda	0.0
Spain	0.010 45	Poland	0.001 99	San Marino	0.0
Chile	0.009 86	Argentina	0.001 97	Swaziland	0.0
Madagascar	0.009 50	Tanzania	0.001 84	Uganda	0.0
Nicaragua	0.008 86	Saudi Arabia	0.001 58	Upper Volta	0.0
Tunisia	0.008 75	Iran	0.001 55	Vatican City	0.0
Gambia	0.008 71	Dahomey	0.001 49	Zambia	0.0

Table III Area of the legal continental shelf (to the 200 m isobath) for various countries, according to the Convention on the Continental Shelf (1958). Units are square nautical miles.

Country	Area	Country	Area	Country	Area
Canada	846 500	Cambodia	16 200	Congo	2 600
Indonesia	809 600	Bangladesh	16 000	Cyprus	1 900
Australia	661 600	Honduras	15 600	Gambia	1 700
USA	545 400	South Yemen	15 000	Albania	1 600
USSR	364 300	Tunisia	14 800	Bahrain	1 500
Argentina	232 200	Turkey	14 700	Israel	1 300
China	230 000	Guyana	14 600	Lebanon	1 300
Brazil	224 100	Ethiopia	13 900	Western Samoa	1 200
UK	143 500	Ecuador	13 700	Syria	1 100
Japan	140 100	Nigeria	13 500	Belgium	800
India	131 800	Gabon	13 400	Fiji	600
Mexico	128 900	North Korea	13 200	Dahomey	500
Malaysia	108 900	Mauritania	12 900	Togo	300
South Vietnam	95 600	Tanzania	12 000	Zaire	300
Thailand	75 100	West Germany	11 900	Iraq	200
South Korea	71 300	Jamaica	11 700	Jordan	200
New Zealand	70 800	Portugal	11 400	Barbados	100
Burma	66 900	Guinea	11 200	Nauru	100
Madagascar	52 600	UAR	10 900	Singapore	100
Philippines	52 000	Yugoslavia	10 700	Monaco	95
Spain	49 700	Senegal	9 200	Afghanistan	0
Sweden	45 200	Trinidad and Tobago	8 500	Andorra	0
France	43 005	Poland	8 300	Austria	0
Italy	42 000	Chile	8 000	Bhutan	0
South Africa	41 800	Sri Lanka	7 800	Bolivia	0
Iceland	39 000	Qatar	7 700	Botswana	0
Ireland	36 700	Sierra Leone	7 700	Burundi	0
Iran	31 200	Greece	7 200	CAR	0
Norway	30 000	Yemen	7 200	Chad	0
Finland	28 600	Romania	7 100	Czechoslovakia	0
Mauritius	26 700	Sudan	6 500	Hungary	0
Venezuela	25 700	Ghana	6 100	Laos	0
Netherlands	24 700	Liberia	5 700	Lesotho	0
Libya	24 400	Dominican Republic	5 300	Liechtenstein	0
Peru	24 100	El Salvador	5 200	Luxembourg	0
Taiwan	23 500	Costa Rica	4 600	Malawi	0
Cuba	23 300	Kenya	4 200	Mali	0
North Vietnam	22 200	Tonga	4 200	Mongolia	0
Saudi Arabia	22 100	Algeria	4 000	Nepal	0
Nicaragua	21 200	Malta	3 800	Niger	0
Denmark	20 000	Bulgaria	3 600	Paraguay	0
Colombia	19 800	Equatorial Guinea	3 600	Rhodesia	0
Morocco	18 100	Guatemala	3 600	Rwanda	0
Oman	17 800	Kuwait	3 500	San Marino	0
Somalia	17 700	Cameroon	3 100	Swaziland	0
UAE	17 300	Haiti	3 100	Switzerland	0
Pakistan	17 000	Ivory Coast	3 000	Uganda	0
Panama	16 700	Maldives	3 000	Upper Volta	0
Uruguay	16 500	East Germany	2 800	Vatican City	0
				Zambia	0

APPENDIX 2 SUMMARY OF THE UN CONVENTION ON THE LAW OF THE SEA

Note: You are not expected to remember all the details of this summary.

The Convention as a whole is in 17 parts, which are summarized below. It will come into force only when ratified by 60 States, and even then will only be binding on those States which have ratified or acceded to it.

1 Use of terms The international sea-bed area, activities in the area, pollution of the marine environment, and dumping, are here given specific definitions for the purposes of the Convention.

2 Territorial sea and contiguous zone Every coastal State has sovereignty over its territorial sea, up to 12-miles wide as from straight baselines running along the coast. Foreign vessels, including merchant ships and warships, are permitted 'innocent passage' through these waters, defined as navigation that does not prejudice the coastal State's peace, good order or security. In a further 12-mile area, called the contiguous zone, the coastal State can exercise the control needed to prevent infringement of its customs, and of its fiscal, immigration and sanitary laws, and can punish violators.

3 Straits used for international navigation Vessels and aircraft of all nations may exercise 'transit passage' through these straits. This is defined as freedom of navigation and overflight solely for the purpose of continuous and expeditious transit, without threat to the States on either side, which are entitled to regulate navigation and other aspects of passage.

4 Archipelagic States States made up of a group or groups of closely related islands and interconnecting waters have sovereignty over a sea area enclosed by straight lines drawn between the outermost points of the islands, while the ships of all other States enjoy the right of passage through sea lanes designated by the archipelagic State.

5 Exclusive economic zone Coastal States have sovereign rights up to 200 miles from shore with respect to resources and certain economic activities, and also have specified types of jurisdiction over scientific research and environmental preservation, while all other States have freedom of navigation and overflight, as well as freedom to lay submarine cables and pipelines. States are to co-operate for the conservation of highly migratory species including marine mammals. Land-locked States and 'States with special geographical characteristics' have the right to participate in exploiting part of the zone's fisheries when the coastal State cannot harvest them all itself. Delimitation of economic zones between States will be 'effected by agreement on the basis of international law . . . in order to achieve an equitable solution'.

6 Continental shelf Coastal States have sovereign rights over this national area of the sea-bed for the purpose of exploring and exploiting it, without affecting the legal status of the water or air space above. The shelf extends at least to 200 miles from shore, and out to 350 miles or even beyond, depending on the nature and configuration of the continental margin. Coastal States will be required to pay to the International Sea-Bed Authority (see item 11) part of the revenue they derive from exploiting oil and other resources from any part of their shelf beyond 200 miles. Delimitation of overlapping shelves is to be on the same basis as for the exclusive economic zone. A Commission on the Limits of the Continental Shelf will make recommendations to States on the shelf's outer boundaries.

7 High seas All States enjoy the traditional freedoms of navigation, overflight, scientific research and fishing in this area of the oceans beyond national jurisdiction. They will be obliged to adopt, or co-operate with other States in adopting, measures to manage and conserve living resources.

8 Regime of islands The territorial sea, contiguous zone, exclusive economic zone and continental shelf of islands are to be determined in the same way as for other land territory, but uninhabitable rocks will have no economic zone or continental shelf.

9 Enclosed or semi-enclosed seas States bordering seas such as the Caribbean and the Mediterranean should co-operate on management of living resources and on environmental and research policies and activities.

10 Right of access of land-locked States to and from the sea, and freedom of transit The people and goods of States with no sea coast must be allowed to move through a neighbouring coastal State to reach and return from the sea, under mutually agreed terms.

11 International Sea-Bed Area All activities on the sea-bed area beyond national jurisdiction will be controlled by an International Sea-Bed Authority, following the principle that the area and its resources are the common heritage of mankind. A parallel system will be established under which exploration and exploitation of deep-sea minerals will be carried out by the Authority as well as by States and private and public corporations and consortia under contract with the Authority. The Authority will establish rules, regulations and procedures for sea-bed mining, in accordance with basic conditions for prospecting, exploration and exploitation set out in the Convention. The Authority will have an Enterprise for Mining, as well as an Assembly, a Council and a Secretariat. A Sea-Bed Disputes Chamber of the International Tribunal will settle disputes and issue advisory opinions on request.

12 Protection and preservation of the marine environment States are obligated to use 'the best practical means at their disposal' to prevent and control marine pollution from any source. They are to co-operate globally and regionally, notify one another of imminent or actual damage, and develop contingency plans against pollution. Technical assistance, monitoring and environmental assessment will be promoted. International rules and national legislation will be devised to prevent, reduce and control pollution of the marine environment from land-based sources and activities on the oceans and sea-bed, including dumping. Enforcement will be the responsibility of coastal States, port States, and flag States, depending on the nature, source and location of the offence. Safeguards can be invoked against inappropriate enforcement actions. Ice-covered areas may be protected by special rules against pollution from vessels. States will be liable for damage caused by violation of their international obligations to combat marine pollution. Warships will have sovereign immunity from environmental regulations imposed by other States, but the States operating them must ensure that they act in a way that is consistent with the Convention as far as practicable. Obligations under other Conventions on the protection and preservation of the marine environment will not be prejudiced by the new Convention.

13 Marine scientific research All States have the right to conduct ocean research for exclusively peaceful purposes. International co-operation in this area is to be promoted through such means as the publication and dissemination of information and knowledge. All such research in the exclusive economic zone and on the continental shelf is subject to the consent of the coastal State, but consent must be given when the research is for peaceful purposes and fulfills other criteria laid down in the Convention. Scientific research installations or equipment in the marine environment must not interfere with shipping routes and must bear identification markings and warning signals. States and international organizations are to be held responsible for damage caused by their own research activities or for action they take against the research conducted by others when such action contravenes the Convention. In the event of certain disputes the researching State may require the coastal State to submit to international conciliation on the grounds that it is not acting in a manner compatible with the Convention.

14 Development and transfer of marine technology States are bound to promote marine technology 'on fair and reasonable terms and conditions', with due regard for all legitimate interests, including the rights and duties of holders, suppliers and recipients of technology. International co-operation will be promoted through the establishment of guidelines, criteria and standards for technology transfer, co-ordination of international programmes, and co-operation with international organizations. The establishment and strengthening of national and regional marine scientific and technological centres are to be promoted.

15 Settlement of disputes States are obligated to settle by peaceful means any disputes over the interpretation or application of the Convention. When they cannot reach agreement on a bilateral basis, they will have to submit most types of disputes to a compulsory procedure entailing a decision binding on all sides. They will have four options: the International Tribunal for the Law of the Sea, to be established under the Convention; the existing International Court of Justice; arbitration; and special arbitration procedures for particular categories of disputes. Certain types of disputes will have to be submitted to conciliation, a procedure whose outcome is not binding on the parties. States will have the option of declining to accept compulsory settlement procedures for certain types of disputes on especially sensitive matters such as boundaries and military activities.

16 General provisions States undertake to discharge their obligations under the Convention in good faith and without abusing their rights, and to refrain from threatening or using force contrary to international law. They are not bound to disclose information contrary to their essential security interests. Coastal States have jurisdiction over archaeological objects and objects of historical origin found at sea up to the outer edge of their contiguous zone (24 miles from shore).

17 Final provisions The non-sea-bed provisions of the Convention can be amended by two-thirds of the States which are party to the Convention, but the amendments will apply only to those States which ratify or accede to them. The sea-bed provisions can be amended with the approval of the Assembly and the Council of the Authority, but only if they do not prejudice the system of exploration and exploitation pending a review Conference to be held 15 years after commercial exploitation begins. Sixty States must adhere before the Convention enters into force. Intergovernmental organizations to which States have transferred legal competence over matters governed by the Convention can sign and accede to it under specified conditions.

APPENDIX 3 THE COD WAR IS NOT ONLY ABOUT FISH

By David Spanier, Diplomatic Correspondent of *The Times*
(from *The Times* of 23 December 1975).

The cod war, which looks like going on well into 1976, is not altogether about fish. It is also about national psychology, and that is why it is so difficult to settle.

Anyone who has been to Iceland is aware of the peculiar strength of the Icelanders. The will to survive, in their remote, freezing and, at this time of the year, mostly dark, outcrop of rock is altogether marvellous. But as Newton's third law of motion instructs us, to every action there is an equal and opposite reaction.

The result of battling to survive in this isolated, volcanic and inhospitable environment has given the Icelanders a stern, aggressive, inward-looking patriotism which is rarely touched by consideration of life outside. (It is characteristic that the Icelanders spend lovingly and lavishly in decorating their own homes, but leave the roadway outside unmade and unpaved.)

The struggle to make a living depends, of course, on cod and the struggle against Britain—the present dispute is the third round—awakens the fiercest strains of nationalist sentiment. 'People say we are selfish,' Mr Einar Agustsson, the Foreign Minister, remarked the other day, 'and I admit it, of course we are selfish'. Yet Mr Agustsson, for all his outspokenness, ranks as a 'dove' in Icelandic politics.

He is one of the Icelandic leaders who would like to make a settlement with Britain, on something approaching the sort of catch which the British industry could live with. Yet at the very moment he was discussing figures with the British, albeit figures which seemed far too low to our side, other members of his own party were going on television and denouncing him as a traitor.

That illustrates the depth of Icelandic feeling about cod, and there is no doubt that the majority of the people support the hard line. Why should the British come in and take their precious cod? When the British carry on fishing, and send in Royal Navy frigates to protect their trawlers, is it any wonder that Icelandic blood rises to fever pitch, in resisting the invader?

Icelandic politics is a complicated balance of forces in which the desire of each group to hold its place with the electorate is the paramount motive of policy. Well, we know something about that in Britain, too.

But in the closed circuit of Icelandic politics, with its concentration on fishing, policy inevitably comes out harder and harder. Rival Icelandic politicians—this is the point—see no advantage in telling their supporters that there is a world outside, and that Iceland has some responsibilities in it. The public, on the contrary, has its patriotism—hostility continually stoked up.

It is useless for Britain to protest that Iceland's attitude in the present cod war is ungenerous, unneighbourly and above all unnecessary. Feelings run too deep. The fact that British and Icelandic marine biologists more or less agree on the case for conservation, and that a way of implementing some protective measures in which the British catch would be significantly reduced could be agreed in a morning, is all beside the point.

The question now is whether the external environment may itself have some influence in changing Icelandic attitudes. In the last two rounds of the cod war, the two countries were at loggerheads, then the dispute was raised in NATO and the United Nations, and finally they found a motive for coming together and settling. The same pattern is being repeated this time, though at a faster pace, but with two important differences.

The first is that with the Law of the Sea Conference rapidly moving towards general approval of a 200-mile limit, there is no motive for the Icelanders to modify their stand; and in the new situation, Britain might look rather ridiculous in maintaining naval support.

The second difference is that Britain is now a member of the European Community. So far, Iceland has been denied the tariff benefits she is entitled to in exporting her fish to Community countries, which bothers her a good deal. Moreover, her recent agreements with West Germany and Belgium will come up for review this spring, if there is no settlement with Britain.

There is a possibility, therefore, if the dispute has to drag on through the winter, that Iceland will come to realize that she is standing to lose more than she is to gain by refusing the British a fair, though comparatively small, share of the market. When that realization comes, the saga could be settled before you can say fish fingers.

* * *

Elizabeth Young comments: In the event, the writing on the wall in 1975 addressed not so much the Icelanders as the British Government and people, who had failed to inform themselves about what was happening and had indeed become foolishly over-excited. There were four lessons there for the British to learn, the first being about fish stocks. As early as 1972 a working party chaired by Dr Garrod of the Ministry of Agriculture, Fisheries and Food (MAFF) Laboratories at Lowestoft had produced a report on the state of cod fisheries in the North Atlantic. This report had concluded that in many areas stocks of cod had already been seriously depleted by overfishing and that, although the stock in the Iceland area was not yet overfished, it was nevertheless at immediate risk. The working party recommended that the cod-fishing effort in the whole of the North Atlantic should be reduced by half immediately to allow the fishery to recover, with luck within some five years. The longer the delay in putting this reduction into effect, the report said, the slower and the less certain would be the recovery of cod stocks.

The report did not exclude the Iceland area from its recommendations because—and this was the second lesson to be learnt—the development of large, fast freezer-trawlers in the late 1960s had made it possible for a fishing fleet, foiled in one area by the absence of fish, to move swiftly to another: North Atlantic cod, in the days of modern high-powered fishing fleets, had become a single 'target'. The Icelanders, having already fished out their local herring, were determined to save their cod from a similar fate—hence the attitude that Mr Spanier, no doubt echoing Whitehall, described as 'ungenerous' and 'unneighbourly'.

For reasons which no doubt seemed good at the time, the Garrod Report was kept 'Confidential', and it seems likely that its unequivocal recommendations never travelled very far within the British Government machine. It spelt out bad news for the policy of the MAFF, whose job was the health and well-being of the hard-pressed British fishing industry—a job which the MAFF fostered mainly by way of subsidies of various kinds. The last thing the Ministry wanted was to see British fishermen excluded from traditional fishing grounds. Whether the Foreign and Commonwealth Office (FCO) saw the report itself, or only a report of the report from the MAFF, is not known. But just as the MAFF could deplore the practical effects of the report's recommendations, so the FCO could deplore the report's cavalier attitude to international law; British and all other fishermen had a complete right in international law to fish in Icelandic waters, and if necessary the Royal Navy would defend that right. Ministers were advised to this effect and journalists were briefed. National feeling was whipped up, the implications for NATO of Britain sending warships against an unarmed ally were ignored, and anyone pointing out that we too needed to protect *our* fish had their views dismissed as demented or disloyal or both.

The first two lessons have now been learnt. The third—that suppressing the unwelcome results of research does not make the facts uncovered by that research go away—has also been learned; but the temptation to suppress the unwelcome is perennial. The fourth lesson—that the Icelandic civil-controlled coastguard was a more suitable arm of government to use in the policing role than the Royal Navy's warships—still remains unlearned in Whitehall.

SUGGESTED FURTHER READING

BADGER, G. (1989) *Explorers of the Pacific*, Kangaroo Press. A vivid historical account which emphasizes the importance of Polynesian voyages and navigation as a prelude to the saga of European exploration, with its hardships, privations and triumphs.

COMMITTEE ON EXISTING AND POTENTIAL USE OF THE SEAFLOOR (1990) *Our Seabed Frontier*, National Research Council, USA. An assessment of sea-floor properties and processes as they relate to utilization of the United States' EEZ sea-bed: mineral and energy resources, waste disposal, pipelines, cables, etc.

CRIBB, J. (1986) *Subtidal Galápagos*, Camden House Publishing, Canada. A record of impressions of a diving expedition around the Galápagos. Its main feature is the stunning underwater photographs, but the text also deals in a simple way with the oceanographic reasons for the striking differences in biota between one island and another.

ENGELHARDT, F.R (ed.) (1985) *Petroleum Effects in the Arctic Environment*, Elsevier. A collection of in-depth articles covering the Arctic marine ecosystem, the fate of spilled oil and its effects on different groups of organisms.

GLASSNER, M.I. (1990) *Neptune's Domain: A Political Geography of the Sea*, Unwin Hyman. A short but wide-ranging book, written in an accessible style, dealing with the Law of the Sea and its application to certain areas (e.g. the eastern Pacific, Rockall, and the Arctic) and specific problems (e.g. pollution).

MITCHELL, A. (1989) *A Fragile Paradise*, Collins. A highly readable account of how animals and humans have colonized the volcanic and coral islands of the Pacific, and the effects of human activity on the fragile ecosystems there.

PRESCOTT, J.R.V. (1975) *The Political Geography of the Oceans*, David & Charles. A readable historical account of the evolution of maritime law.

REY, L. (1982) *The Arctic Ocean: The Hydrographic Environment and the Fate of Pollutants*, Macmillan. A collection of articles on topics ranging from the discovery of the Arctic basin to the physical oceanography of Arctic waters and their marine ecosystems. The articles are variable in style: some are difficult to digest, but others are very readable.

'The Galápagos Marine Resources Reserve', *Oceanus*, **30,** No. 2, Summer 1987. This special Issue contains articles on the setting up of the Galápagos Reserve, its marine flora and fauna, the effects of the 1982–83 El Niño, and the history of the islands.

ANSWERS AND COMMENTS TO QUESTIONS

CHAPTER 1

Question 1.1 In both cases, the flux is important. (a) If huge tonnages of dredged sediment are dumped offshore daily, they may affect coastal sediment transport regimes; they may overwhelm bottom communities at the dumping site; finer-grained components will settle slowly, reducing the transmission of light through the water, and so cutting down primary production by photosynthesis. (b) Huge tonnages of organic matter will require correspondingly large amounts of oxygen in order to decay, which may lead to serious local depletion of oxygen. Nitrogen and phosphorus in the wastes may provide additional nutrients and lead to increased productivity, thus increasing the amount of organic matter and accelerating the oxygen depletion.

In both cases, the effects can be mitigated by dumping at sites where fast currents can transport and disperse the materials at rates comparable with input fluxes. The selection of such sites requires good oceanographic intelligence.

(It is of course incorrect to assume either that harbour sediments are totally inert or that domestic/agricultural wastes are wholly organic. There will be contaminants, such as heavy metals (e.g. lead, zinc, copper) pesticides, germicides, and so on.)

Question 1.2 Not more than a few per cent of the world's annual industrial production can be lost to the environment, for it is wasteful and expensive to lose manufactured metal. Even if as much as 10% is lost (i.e. 9×10^2 tonnes), that can still be only about one-quarter of the riverborne flux, and about 1% of the atmospheric flux. It follows:
(a) that the main source of oceanic mercury is not industrial, and
(b) that the most likely pathway for mercury into the oceans is atmospheric fall-out, which is consistent with Table 1.1.

Question 1.3 (a) Quantitatively, the most important product obtained from seawater is freshwater. This is especially important in arid regions where the rainfall is low.

(b) Magnesium metal. About half of the world's total production of this comes from the sea.

Question 1.4 The answer is a matter of simple division:

$$\frac{500 \times 10^3\,\text{kW}}{50\,\text{kW m}^{-1} \times 0.1} = 100 \times 10^3\,\text{m}$$

$$= 100\,\text{km}$$

Maintenance of such a length of wave converters would be difficult and costly. There would be considerable environmental impact and potentially significant interference with shipping. Trebling the conversion efficiency (which is not likely) would reduce the total length to about 30 km, which is still very long. None of this, however, precludes the utilization of wave energy on a smaller local scale.

Question 1.5 The remaining (approximately) 25% of the global fish catch comes mainly from areas of upwelling (Figure 1.4) in tropical and sub-tropical latitudes, along eastern boundaries of the Atlantic and Pacific Oceans. Water wells up from deeper levels to replace surface waters moved away from the land in surface currents driven by the prevailing (Trade) winds. The upwelling water is rich in nutrients derived from the decomposition of organic matter sinking from the surface; so upwelling regions are characterized by high biological production.

Question 1.6 (a) In the strictest sense, CO_2 is not a pollutant gas. It is essential for photosynthesis and the production of plant tissue and carbon is a ubiquitous constituent of all life on the planet. However, its capacity to adsorb infra-red radiation means that increased concentrations of atmospheric CO_2 lead to global warming. As it is emitted in huge quantities by the combustion of fossil fuels, which form the basis of our energy-hungry civilization, it is difficult to reduce CO_2 emissions substantially.

(b) Only about half of the CO_2 liberated by human activities stays in the atmosphere. Most of the rest probably enters the oceans, and a major focus of oceanographic research is to find out about the manner and rate of its sequestration in the marine environment; how much is physically dissolved in the cold waters of polar latitudes; and how much is biologically removed by organisms in surface waters. This is the function of international research projects such as the Joint Global Ocean Flux Study (JGOFS).

Question 1.7 (a) High concentrations of copper along the south Wales coast are due to the industrial complexes of Cardiff (and Avonmouth) and Swansea. The only other heavy industrial conurbation is round Liverpool. High copper values off west Wales and along the coast of south-west England therefore have other causes. These two areas are regions where copper was mined (along with other metals) until the end of the 19th century. It is not possible to know whether the copper comes from soil tips, mine waters, or simply groundwater that has dissolved metal from the mineralized rocks themselves.

(b) These are some of the questions to which answers might be sought. What is the tolerance of local marine organisms to copper? These are winter and spring data: do copper concentrations change in the drier summer months? What is the local residence time of copper? In what form is the copper dissolved in seawater, and is this form absorbed by marine plants and animals? No doubt you can think of other questions.

Question 1.8 (a) Sands and gravels, which potentially amount to nearly 7×10^{11} tonnes. They will be found mostly on continental shelves in higher northern latitudes. Most of them were deposited by large rivers transporting glacial debris from melting glaciers onto coastal plains that subsequently became submerged by post-glacial sea-level rise.

(b) Calcium carbonate (shell) deposits, which potentially amount to some 9×10^{10} tonnes. Here, too, the deposits would be mostly in shallow (continental shelf) waters, where corals and shellfish grow, but this time in lower latitudes, where the warm water temperatures encourage precipitation of $CaCO_3$.

Question 1.9 Although marine resources *are* tangible, not all are fixed, especially biological resources. Moreover, ownership of a resource is of no value unless exploitation is possible—and exploitation is an activity, whether it be drilling for oil or gas, fishing, or harnessing wave power.

CHAPTER 2

Question 2.1 (a) The passage lies within Irish territorial waters because Tory Island itself has a territorial sea three miles wide. In fact, the outer limit of Irish territorial waters must extend to seven miles (4 + 3) from the mainland, opposite Tory Island

(b) Irish territorial waters would extend fourteen miles (4 + 10) from the mainland opposite Tory Island, because of the island's own territorial sea of ten miles width.

(c) No, because the island lies within the system of straight baselines. (This is a small price to pay for the huge areas of internal water enclosed within the baselines on Figure 2.5. What is more, Ecuador actually claims a 200-mile wide territorial sea.)

Question 2.2 The claims could have been made as a means of preventing stronger maritime nations, notably the USA, USSR and Japan, from exploiting the offshore fishery resources of western South America. Fish are a major source of revenue for poor Latin American countries, which are bordered by rich upwelling zones (Figure 1.4). These upwelling zones are associated with the Peru Current, the width of which was estimated to be 200 miles.

Question 2.3 (a) Yes, Mandalya Gulf is a legal bay according to the criteria laid down in the Convention on the Territorial Sea. It is certainly a 'well-marked indentation whose penetration is in such proportion to the width of its mouth as to contain land-locked waters and constitute more than a mere curvature of the coast'. Furthermore, its area is larger than a semicircle whose diameter is the line A–B (see Figures 2.3, 2.6 and related text).

(b) No, this bay is *not* a legal bay. A semicircle of diameter CD would be *bigger* than the area of the bay.

(c) According to the scale bar, the distance between A and B is less than 24 miles long, and therefore the line joining them would be the closing line for the Mandalya Gulf. On the other hand, even if the bay with headlands C and D *were* a legal bay, its closing line would *not* be the line joining C and D, because these points are more than 24 miles apart. Instead, the baseline would be a line 24 miles long drawn so as to enclose the maximum possible area of water.

(d) There are two main points. First, the closing line of a bay separates *internal waters* (over which the coastal state has full sovereignty and in which there is *no* right of innocent passage) from territorial waters in which there *is* right of innocent passage. Secondly, the closing line forms the *baseline* from which maritime zones—territorial sea, exclusive economic zone, etc.—are measured, in the vicinity of the bay. However, given the geography of the region, there is not even room for a full-width territorial sea.

Question 2.4 The Dardanelles must be a strategically important international strait, even though it does not strictly speaking join two areas of the high seas, but rather two areas of territorial sea, in the Aegean and Black Seas. It is the only maritime link between the Black Sea and the Mediterranean (and hence the rest of the world's oceans). It obviously lies wholly within Turkish internal waters, but Turkey presumably does not treat it as such.

Question 2.5 (a) (i) Territorial sea—it lies within three miles of St. Kilda, which, as an offshore island, has its own territorial waters.

(ii) Internal waters—it lies inside the straight baselines from which the territorial sea is measured.

(iii) Contiguous zone—it lies in the middle of straits about 11 miles wide, and therefore outside the (three-mile) territorial sea measured from either shore but inside their overlapping (nine-mile) contiguous zones.

Jurisdiction over the territorial sea is essentially complete, apart from the right of innocent passage (i). Indeed, apart from this provision, a state's sovereignty over its territory extends to its territorial waters.

Internal waters are sovereign territory (ii).

In its contiguous zone, a coastal state may only 'prevent and punish infringement of its customs, fiscal, immigration or sanitary regulations committed within its territory or territorial sea' (iii).

Important: This example is a fictitious one; in fact, the UK has not declared a contiguous zone and its territorial sea has been extended from three to twelve miles wide.

(b) Because it is a 'well-marked indentation whose penetration is in such proportion to the width of its mouth as to contain land-locked waters and constitute more than a mere curvature of the coast' whose 'area is as large as, or larger than, that of the semicircle whose diameter is a line drawn across the mouth of that indentation'. Its seaward limit is defined by 'a straight baseline of 24 miles . . . drawn within the bay in such a manner as to enclose the maximum area of water that is possible with a line of that length'.

CHAPTER 3

Question 3.1 (a) The outer edge of the continental margin is indeed at the slope base, coinciding with the line which defines the landward limit of the boundary zone in Figure 3.2(a). However, a state can claim 350 miles of shelf, or even more, if it can establish that it has a naturally prolonging shelf extending to that distance.

(b) The Convention on the Continental Shelf set the limit at 200 metres depth, or beyond that 'where the depth of the superadjacent waters admits of exploitation'; so the depth limit was somewhat open-ended (Section 2.2.2). Two states with similarly shaped coastlines could thus have very differing widths of shelf per unit length of coastline, if one had a shelf that sloped gently for a long distance, and the other had a narrow shelf that sloped steeply for a short distance. The 200-mile minimum limit proposed in UNCLOS gives states with similarly shaped coastlines an equivalent area of shelf per unit length of that coastline. Even where coastlines have very dissimilar shapes, the new proposals are still more equitable in these terms than under the Convention on the Continental Shelf.

(c) The EEZ is 200 miles wide, while the legal continental shelf can extend out to 350 miles or more. There is some overlap: both stipulate national jurisdiction over resources of the sea-bed and sub-soil. However, the articles and clauses relating to the EEZ limit jurisdiction over the waters above to 200 miles from shore; while those relating to the legal continental shelf can give jurisdiction over sea-bed and sub-soil out to the edge of the geological continental margin—but *not* over the waters above (*cf.* exchange of letters at the end of Chapter 2).

Question 3.2 The Japanese must be assuming that if they preserve North and East Dew Rocks they will also secure the right to a 200-mile wide exclusive economic zone/continental shelf around the rocks, if and when the UN Convention on the Law of the Sea comes into force. Otherwise, the rocks would be eroded to become low-tide elevations which would not even have a territorial sea

Question 3.3 The only maritime zones to which the United Kingdom could lay claim would be a twelve-mile territorial sea plus a twelve-mile contiguous zone around the rock (see items 2 and 8 of Appendix 2). In the absence of UK ratification of UNCLOS (Table 3.1), these zones could still be claimed under the provisions of the Convention on the Territorial Sea (1958)—see Section 2.2.1.

Question 3.4 (a) Apart from the obvious point that 'miles' means 'nautical miles', it is the term 'shore' that is the problem. The distance is measured from the baseline that is used to determine the territorial sea boundary; this is twelve miles out, and the EEZ boundary is a further 188 miles beyond that. Bearing in mind the provisos of the answer to Question 3.1(b), the same applies to the second case: '. . . per unit length of baseline. . . .' would be more correct, because shorelines/coastlines have a habit of being somewhat convoluted.

(b) Off western South America, the width of the continental shelf (distance to the base of the continental slope) is nowhere more than about 100 miles and is generally much less. Thus, even with the possibility of claiming '100 miles beyond the 2 500 m isobath', the legal shelf width would in most cases not even reach 200 miles, let alone exceed it.

Off Newfoundland, the geological continental shelf extends in places as much as about 500 miles offshore, and nowhere does it extend for less than about 200 miles. By judicious use of the clause relating to the 2 500 m isobath, Canada's legal continental shelf in this region can be as wide as 350 miles or more.

Question 3.5 Table III of Appendix 1 shows the 'ranking' according to the 1958 Convention. For some coastal states with wide geological shelves, there would not be much change under UNCLOS (e.g UK, USA, at least on the eastern coasts). For others, the change would be significant (e.g. Spain and Portugal, with only a narrow geological shelf off the Iberian peninsula) and they would probably go further up the Table. So would many island states, which typically have narrow geological shelves.

Question 3.6 (a) The 200-mile line defines the limit of the EEZ, so both states have the right to explore and exploit the deposits lying within that line.

(b) If the Canada/USA boundary is projected south-westwards, it cuts across this southern segment of the Juan de Fuca Ridge. Both the USA and Canada *could* claim their respective portions of this segment to be part of their legal continental shelves (i) under the Convention on the Continental Shelf, because although the normal depth limit of jurisdiction is 200 metres, this would be subsumed within the 'limit of exploitability' clause (Section 2.2.2): 'if they can mine it, they can have it'. It is doubtful whether they could also justifiably claim it (ii) under the terms of UNCLOS. The continental shelf off the western USA is narrow, and it is difficult to see that the Juan de Fuca Ridge could be regarded as a prolongation of the continental shelf, however close to the coast it is. Also, the sea-bed between the 200-mile limit and the rise crest not only lies below the 2 500 m isobath, it is well beyond that isobath running along the coast. (Beyond the 200-mile limit there would be a requirement to pay part of the revenue to the International Sea-Bed Authority (Appendix 2, item 6)).

(c) As outlined in Section 2.2.2, with this criterion there was a risk that the 'shelf region could expand, as new technology gave access to ever greater depths, until it covered the entire ocean floor. . . .' Hence, the need for a distance limit under UNCLOS.

(d) The short answer is no, because the USA did not sign and has not yet ratified or acceded to the Convention. Nor can it readily appeal to customary law in this case, because there is as yet no precedent for deep-sea mining—a few pilot operations have been done for nodule mining, but that is all. However, there is nothing to prevent the USA from enacting its own legislation. Indeed, for your information:

Under the Deep Seabed Hard Minerals Resources Act of 1980, the National Oceanic and Atmospheric Administration (NOAA) is authorized to license or permit private firms to explore or exploit areas seaward of the continental shelf on an interim basis, pending entry into force of the Law of the Sea Treaty (UNCLOS) or alternative reciprocating states arrangement. UNCLOS has not come into force, and anyway the USA is not a signatory; but who could stop them?

Question 3.7 According to the criteria set out in Section 3.1.2, spider crabs, clams and razor shells would qualify as exploitable; lobsters, swimming crabs and scallops would not, because at the harvestable stage they are not 'in constant physical contact with the sea-bed'. Accordingly, the coastal state does not have the exclusive right to exploit these stages.

CHAPTER 4

Question 4.1 (a) The islands certainly form an archipelago or closely related group of islands; furthermore, some of the more southerly ones have their southern shores within the Canadian territorial sea (even when this was three miles wide; it is now twelve miles). As will be discussed later, the year-round ice cover 'fusing together' many of the islands makes them even more 'closely related' than they would otherwise have been.

Set against this, the Parry Channel seems somewhat to disrupt the coherence of the 'island group'; and the most north-easterly islands appear geographically more related to Greenland. Add to this the difficulties of administering such inhospitable areas and perhaps the best we can say is that, using contiguity, Canada has only a moderately good case to claim the islands as sovereign territory.

(b) Territorial sovereignty over sea areas can only be generated through contiguity to adjacent land territories, and then only out to a width of twelve miles (although under the 1982 Convention some degree of control may be exercized out to the edge of the exclusive economic zone and/or continental shelf, i.e. out to a distance of 200 miles or more). If the case for Canadian sovereignty over the islands is only moderately strong, that for sovereignty over the intervening waters must, of necessity, be weaker still. Sovereignty over waters all the way to the North Pole is out of the question for Canada (or any other state).

(c) The arguments in (a) and, especially, (b) indicate that contiguity *cannot be used as a justification for the application of the sector theory in the Arctic.*

Question 4.2 Clearly, the two situations are very different: in the Arctic, the region concerned is largely ocean; while in the Antarctic, a continent and its adjacent waters are being divided up. Furthermore, in the Arctic, the sector lines can be drawn using territorial boundaries of the surrounding land (all poleward of 60° of latitude). In the Antarctic, it is not so easy. If the polar circle were taken as the latitude from which territories were to be 'extended', there would be no sectors; if 50° S were used, only Argentina and Chile could claim; for 40° S, New Zealand and Australia could be added; if the Equator were used, there would be a multitude of claimants.

Question 4.3 (a) Hudson Bay. Other bays claimed by Canada as historic bays include the Bay of Fundy, the Gulf of St. Lawrence, and a number of bays off the coasts of Newfoundland, Nova Scotia, Prince Edward Island, New Brunswick and Quebec. Sir Richard Cartwright described Hudson Bay as *mare clausum* and 'as belonging to' Canada. Application of the concept of *mare clausum* (or *res nullius*) has generally resulted in claims to *territorial* waters rather than internal waters (see Section 2.1) and, in fact, the notion of internal waters had not been clearly formulated in 1907.

(b) The important (and only) difference is that foreign vessels have the *right* of innocent passage through territorial waters. No such right exists with regard to internal waters; the coastal state may *permit* innocent passage, as a privilege. The significance of this, as far as Arctic waters are concerned, will be explored further in Section 4.4.

Question 4.4 The introduction of straight baselines (Section 2.2.1), particularly for coasts which are deeply indented and/or have lots of islands along them. The straight-baseline system would effectively have legitimized the tendency of such a coastal state to regard the waters within its inlets, and between its islands, as internal waters.

Question 4.5 (a) In general, apart from the north-eastern coast of Baffin Island, the characteristics of the Canadian Arctic Archipelago do not closely resemble those of the Norwegian coast. However, the geographical relationship of the islands to Canada is such that one might justifiably feel that 'what really constitutes the *Canadian* coastline is the outer line of the *Archipelago*'.

(b) (i) Figure 4.15 indicates that the baselines chosen follow the general direction of the coast closely, so criterion 1 is fulfilled. Furthermore, the complex pattern of channels weaving between the islands, and the area of water forming bays and inlets, also demonstrates a close linkage between land and water. From a purely geographical point of view, the most important exceptions to this are probably the western entrance to the Parry Channel and Amundsen Gulf. Criterion 2 is, by and large, fulfilled. (ii) Yes, the baselines do reflect the spirit of the law: for example, none of them is exceptionally long, or chosen so as to enclose as large an area as possible.

Question 4.6 The flat topography of the continental shelves must have been imposed at the time of the last glaciation, approximately 25 000 years ago. During this period, the sea-level was about 200 m lower than at present, and the sediments would have been exposed to the action of waves, surface currents and, during the coldest periods, erosion by ice.

Question 4.7 Figure 4.18 shows that the two main features of the surface circulation of the Arctic Ocean are the Transpolar Current and the Beaufort Gyre. Hence:

(a) Pack-ice is constantly being carried in the Transpolar Current from the Chukchi Sea (or even the Bering Sea) across the ocean towards the Fram Strait, and then southwards along the eastern coast of Greenland.

(b) By contrast, ice circulating in the Beaufort Gyre tends to be trapped in the Arctic for many years, during which successive ridging and rafting of ice-floes can build up great thicknesses of multi-year ice, only some of which melts each summer. (There is no theoretical limit to the time a floe can survive in the Arctic, although successive episodes of surface melt in summer and bottom freezing in winter mean that after 20 years none of the original ice is left. One particular ice-island is known to have lasted 34 years before escaping from the Arctic in the East Greenland Current.)

Question 4.8 Continental ice-sheets form from precipitation, and the lead in them comes from *atmospheric* pollution, notably that from vehicle exhausts. The lead concentration in Arctic sea-ice has also increased to some extent because each winter it is augmented by a covering of snow which becomes incorporated into the multi-year ice. (Lead also reaches Arctic waters by rivers, but there are a number of pollutants which enter them entirely via the atmosphere. These include chlorofluorocarbons (CFCs) and polychlorinated biphenyls (PCBs).)

Question 4.9 (a) Having failed to complete her voyage via Route 2, the *Manhattan* was forced to use the other deep-water alternative, Route 1, through the Prince of Wales Strait (see Figure 4.34(b)). The Prince of Wales Strait is very narrow: its minimum width is 6.5 nautical miles and, as shown by Figure 4.37(a), in the vicinity of the (Canadian) Princess Royal Islands the width of the seaway is further reduced to 5.7 miles on one side of the islands and 4.65 miles on the other. Both of these widths are less than $3 + 3 = 6$ miles so even with the three-mile territorial sea which Canada then had, the Strait would be completely closed off to international navigation, having no strip of high seas (or exclusive economic zone) down the middle.

(b) By increasing the width of the territorial sea from three to twelve miles (now, in any case, the generally accepted width under the 1982 Law of the Sea Convention), Canada was effectively shutting off the North-West Passage at a number of other places, including the Barrow Strait between Bathurst Island and Cornwallis Island where there are a number of smaller islands (Figure 4.37(b)).

Question 4.10 Via the Fram Strait, which, as comparison of Figures 4.17 and 4.18 shows, is the only deep-water passage out of the Arctic Ocean. The main warm current affecting the region is the Norwegian Current, the downstream continuation of the Gulf Stream.

Question 4.11 (a) The United States was allocated a sector because of its Alaskan territory.

(b) Denmark (in respect of Greenland).

Question 4.12 Coastal states are entitled to their territorial seas (and contiguous zones) purely on the basis of contiguity to their land territories. Under the 1982 Law of the Sea Convention, the same is true of the continental shelf.

Question 4.13 The International Court awarded the islands to Britain, because they belonged to it on the basis of an ancient title. As far as contiguity was concerned, the islands were not only closer to Jersey than to the French coast, but they were part of the 'natural unity' of the British Channel Islands.

Question 4.14 (a) The Aleutian arc. This generates an enormous area of (200-mile plus) continental shelf (and 200-mile) exclusive economic zone. It may not surprise you to learn that, prior to a USA–USSR agreement in 1990, the Soviet Union was unhappy about how the Bering Sea was divided between itself and the United States, according to the criteria used to define exclusive economic zones. United States rights over the oil-bearing basins in the southern and central Bering Sea (see Figure 4.17) were particularly contentious.

(b) (i) The oilfields shown extend into the United States (twelve-mile) territorial sea. (ii) and (iii) Most of the fields shown fall within the United States EEZ, although it is hard to tell that from Figure 4.17, and they therefore also fall within the continental shelf, according to the geometrical (as well as the geological) criterion. (iv) None of the fields extends into the high seas.

(c) As the oilfields shown are within the United States EEZ, and none is on the US continental shelf *outside* the 200-mile limit, the United States will not be asked to pay royalties to the International Sea-Bed Authority (see Section 3.1.2).

Question 4.15 (a) Heat. Warm air or water from industrial complexes, or extra heat generated simply through buildings and installations absorbing solar radiation, could (and does) lead to melting of snow, ice and permafrost, at times of year when it normally would be stable.

(b) Oil is dark in colour (i.e. its albedo is low) and a spill can cover a large area of ice or snow. The capacity of the surface to absorb solar radiation could therefore increase significantly, and in spring or summer this could encourage increased melting of snow and ice. The fear is that because of the feedback loop between absorption of solar radiation and ice cover, regions where the ice-cap is now stable would start to melt, and this could bring about significant changes in the Arctic environment.

Question 4.16 (a) By absorbing and scattering both incoming sunlight, and that re-radiated and reflected by the snow and ice, Arctic haze could significantly alter the Arctic radiation budget. At present it is not known how the net radiative heat flux is being affected.

(b) With the melting of the sea-ice cover, there will be much more evaporation and the moisture content of the atmosphere will rise, causing increased precipitation. Land ice is formed from precipitation.

CHAPTER 5

Question 5.1 The reason is that as well as the north–south divergence of the Cocos and Nazca Plates caused by spreading at the Galápagos spreading axis, both these plates are moving eastwards relative to the 'hot spot', as a consequence of spreading at the East Pacific Rise.

Question 5.2 The period of gentle local winds and calm weather (1) is typical Doldrums weather and corresponds to the summer period of the year when the ITCZ has moved south and the South-East Trades are weak (December–March). The period of strong local winds and rough seas (2) corresponds to the winter period when the Trade Winds are strong (June–September).

The weather described in (1) is typical of that resulting from vigorous upward convection over a warm sea (*cf.* Figure 5.8(b)); by contrast, the weather in (2) is what might be expected when a cold sea-surface (*cf.* Figure 5.8(a)) cools the overlying air.

Question 5.3 (a) Perhaps the most striking aspect of the distributions shown in Figure 5.12 is that whereas the coast of Fernandina and the western side of Isabela are characterized by macrophytic algae, the eastern side of Isabela and the coasts of several other islands are characterized by coral reefs. Comparison with Figure 5.11 shows that the coasts subjected to upwelling—i.e. those supplied with cold, upwelling nutrient-rich water—are fringed with macrophytic algae rather than coral reefs. This is to be expected, given that to thrive algae need high concentrations of dissolved nutrients, while coral reefs can flourish in barren tropical waters, because they are ecosystems which efficiently recycle nutrients. They do, however, need surface water temperatures to be warm.

(b) Reef-building corals have a symbiotic relationship with zooxanthellae, algae which live sheltered by the coral skeleton and whose metabolic activities aid the coral to deposit calcium carbonate. Zooxanthellae need fairly high light levels to photosynthesize, which is why reef-building corals only flourish within about 70 m of the surface.

Question 5.4 (a) No; for example, there were no ENSO events in either 1968–69 or 1979–80.

(b) The answer is different for the two cases shown. Before the 1976 ENSO event, easterly winds over the western equatorial Pacific *were* unusually strong, and there was a dramatic drop in average easterly wind speed between mid-1975 and mid-1976. In the case of the 1982–83 event, there had been strong easterly winds during the first half of 1981 but by the beginning of 1982 they had weakened considerably.
ENSO events occur when warm water that has been piled up in the western Pacific surges eastwards. As more water is piled up in the west when the easterly winds are strong (*cf.* Figure 5.19), we might expect a bigger ENSO event—i.e. a bigger surge of warm water—to occur after a period of strong easterly winds.

(c) On the basis of the answer to (b) we would not have expected the 1982–83 ENSO event to be particularly severe. However, the extremely sharp drop in easterly wind strength (which in mid-1982 was a reversal to westerly winds, as we shall see later) could indicate that the event would have been very severe—as was indeed the case.

Question 5.5 Considering all the information in Figures 5.24 and 5.25, it must have been clear that an ENSO event was underway by October–November. Temperatures had been higher than normal for some months (Figure 5.24), and rainfall had been more frequent (Figure 5.25 (b)), but the actual *amount* of rainfall began to increase significantly in October–November; normally, the wet season would not become established until the Southern Hemisphere spring.

Question 5.6 The low-salinity surface water would be less dense than more saline water beneath (which is probably also slightly cooler). The water column would be stable and the steep halocline would also be a steep density gradient (**pycnocline**), acting as an effective barrier to vertical mixing. The nutrients used up by phytoplankton which on death would have sunk into deeper water would not easily have been replenished from below.

Question 5.7 (a) (i) Bearing in mind that lower $\delta^{18}O$ values mean warmer water, it is possible that the period of poor coral growth in the mid-1700s was related to too-warm water around the Galápagos, either because of frequent ENSO events (not necessarily as severe as the one in 1747) or because water temperatures in the equatorial Pacific were generally higher then. (ii) The reverse seems to have been true during much of the 20th century, with water temperatures generally cooler than required for optimum coral growth.

(b) An alternative explanation is, of course, volcanic or tectonic activity, which could have caused at least some of the disruption in growth, particularly from the late 19th century onwards. This is not likely to have caused the hiatus in 1747 (Figure 5.39), because that lasted only about a year, and the effects of geological disturbance would have lasted longer.

Question 5.8 The crust on one side was formed at the propagating spreading axis and is younger than that on the other side, which was formed earlier at the failing spreading axis. The junction can be identified by its bathymetric expression as a zone of linear disturbance which would show up on side-scan sonar and narrow-beam echo-sounding maps (*cf.* Figure 5.6(b)). Magnetic surveys might also help.

Question 5.9 (a) The temperature (i) began to rise between one and two months earlier at depth than at the surface (the beginning of September 1982, rather than mid-October). It also (ii) began to fall one to two months earlier at depth (early June as opposed to mid-July). In other words, the 'signal' of the ENSO was 'felt' first at depth and only later at the surface, and was stronger and more sharply defined at depth. This is consistent with it being propagated as a Kelvin wave in the thermocline.

(b) Variable but marked decreases in salinity affected mainly surface waters, and occurred during the period of high rainfall indicated by Figure 5.25; so the decrease can be attributed mainly to rain. At 60 m depth, salinity mostly stayed between 34.7 and 35, except for the small dip between April and June 1983. This could be related to the influx of a low-salinity water mass from the north, but there is no conclusive evidence that this was the case.

(c) At times when the upper water column was (i) well-mixed, the temperature and salinity values would have been similar at the surface and at a depth of 60 m. Both the temperature and salinity records show the greatest correspondence between surface and 60 m values in the period October–November 1982. After that the curves diverge, indicating (ii) increasing stratification. The large differences of both temperature and salinity between the surface and 60 m depth in June–July 1983 indicate that stratification was greatest at that time.

Question 5.10 A Average eastern equatorial Pacific sea-surface temperatures would be higher during ENSO events, because the area of warm water normally in the west spreads all the way along the Equator then.

B Accompanying the warm surface water is a region of vigorous atmospheric convection, so that the area of heavy rainfall, usually over Indonesia, moves eastwards. Also, the zone of high precipitation associated with the ITCZ moves south, and the net result is greatly increased precipitation in the central and eastern Pacific.

C and D The difference between these two is the Southern Oscillation Index, which is low during ENSO events (Figure 5.21).

Question 5.11 A new hydrothermal vent can only be colonized through currents carrying organisms—probably as eggs or larvae—from another, perhaps distant, hydrothermal vent. So colonization is strongly influenced by current patterns; also, the long stretches of cold, deep water between vents could act as a 'barrier' preventing warm-water vent species travelling long distances between vents (*cf.* the East Pacific Barrier), allowing vent communities to develop with different species composition in different parts of the ocean.

ACKNOWLEDGEMENTS

The Course Team wishes to thank the following: Martin Angel, the external assessor; Quentin Huggett for providing manganese nodules from which Figure 1.14(a) was derived; Professor Pat Birnie for valuable comment and advice on Chapters 3 and 4; Peter Wadhams for advice and comment on Chapter 4; Mike Davey, Günther Reck and members of the Charles Darwin Foundation for the Galápagos Islands for advice and comment on Chapter 5, Gary Robinson for his generosity in supplying photographs for Chapter 5; and Mary Llewellyn for comments on the whole Volume. The structure and content of the series as a whole owes much to our experience of producing and presenting the first Open University course in Oceanography (S334), from 1978 to 1987. We are grateful to the Course Team who prepared and maintained that Course, to the tutors and students who provided valuable feedback and advice, and to Unesco for supporting its use overseas.

Grateful acknowledgement is also made to the following for material used in this Volume.

Figure 1.1 John Ross; *Figure 1.2* Post Office; *Figure 1.3(a),(b)* Norman Etherson, Loch Etive Farmed Shellfish; *Figure 1.3(c)* J. B. Wright; *Figure 1.5* G. E. R. Deacon (1962) *Oceans*, Hamlyn; *Figure 1.7(c)* Department of Transport; *Figure 1.11* P. H. Zaalberg (1970) in *Trans. Inst. Min. Metall.*, Sect. A, **79**; *Figure 1.12* J. Ben Carsey (1950) in *Bulletin of the American Association of Petroleum Geologists*, **34**, AAPG; *Figure 1.13* NERC; *Figures 1.15, 1.19(a)* B. J. Skinner and K. K. Turekian (1973) *Man and the Ocean*, Prentice-Hall; *Figure 1.16* John Edmond/Woods Hole Oceanographic Institution; *Figure 1.17* E. T. Degens and D. A. Ross (eds) (1969) *Hot Brines and Heavy Metal Deposits in the Red Sea*, Springer Verlag; *Figure 1.19(b)* French Embassy, London; *Figure 1.20* Department of Energy; *Figure 1.21(a)* NERC; *Figure 1.21(b)* M. Kenward (1976) in *New Scientist*, **70**, May; *Figure 1.22* US Department of Energy; *Figure 1.23* S. Rusby (1975) in *New Scientist*, **68**, November; *Figure 1.25* M. I. Abdullah *et al.* (1972) in *Nature*, **235**, January; *Table 1.3* J. M. Broadus (1987) in *Science*, **235**, AAAS; *Figure 2.9* L. F. Laporte (1975) *Encounter with the Earth*, Canfield Press; *Shark cartoon* Nick Hobart, Ontario, Canada; *Figure 3.1* D. A. Ross and Therese A. Landry/Woods Hole Oceanographic Institution; *Figure 3.4* The Admiralty; *Rockall collage* Lt. Cdr. Desmond Scott/Trustees of the Imperial War Museum/Crown copyright; *Figures 4.7, 4.13, 4.15, 4.34(b), 4.37* D. Pharand (1988) *Canada's Arctic Waters in International Law*, Cambridge University Press; *Figures 4.9 and 4.32* Dome Petroleum Ltd., Esso Resources Canada Ltd. and Gulf Resources Inc.; *Figure 4.11* National Archives of Canada; *Figures 4.16 and 4.41* (1979) *Polar Regions Atlas*, C.I.A.; *Figures 4.19, 4.26, 4.31(b)* Peter Wadhams; *Figure 4.21* British Petroleum; *Figures 4.22 and 4.23* Gulf Canada Resources Ltd.; *Figures 4.24(a), 4.30, 5.1(b)* NASA; *Figure 4.24(b)* Tromsø Satellite Station; *Figures 4.27–4.29* F. R. Engelhardt (ed.) (1985) *Petroleum Effects in the Arctic Environment*, Elsevier; *Figure 4.31(a)* Bernie McConnell; *Figure 4.34* Department of Energy, Mines & Resources, Canada; *Figure 4.35* National Maritime Museum; *Figure 4.36* Dan Guravich; *Figure 4.38* *Toronto Star*;

Table 4.1 C. Lamson and D. Vanderzwaag (eds) (1988) *Transit Management in the Northwest Passage*, Cambridge University Press and Petro-Canada Explorations; *Figure 5.6(b)* R. N. Hey and M. C. Kleinrock *et al.*/Woods Hole Oceanographic Institution; *Figure 5.9* S. P. Hayes *et al.* (1986) in *Progress in Oceanography*, **17**, Pergamon; *Figure 5.11* G. T. Houvenaghel; *Figure 5.12* P. W. Glynn and G. M. Wellington (1983) *Corals and Coral Reefs of the Galápagos Islands*, University of California Press; *Figure 5.13* Zoological Society of London; *Figures 5.14–5.16 and 5.31–5.36* Gary Robinson; *Figure 5.20* M. A. Cane (1983) in *Science*, **222**, AAAS; *Figure 5.21* A. E. Gill and E. M. Rasmusson (1983) in *Nature*, **306**, Macmillan; *Figures 5.22 and 5.26* K. Wyrtki in G. Robinson and E. M. del Pinto (1985) *El Niño in the Galapagos Islands*, Charles Darwin Foundation; *Figures 5.24 and 5.25* M. Robalino in Robinson and del Pinto *op. cit.*; *Figure 5.27* S. P. Hayes in Robinson and del Pinto *op. cit.*; *Figures 5.28, 5.29, 5.45* J. Kogelschatz in Robinson and del Pinto *op. cit.*; *Figure 5.30* G. Feldman *et al.* (1984) in *Science*, **226**, AAAS; *Figures 5.37, 5.38, 5.39* Rob Dunbar; *Figure 5.40* R. B. Dunbar *et al.* in J. L. Betancourt (ed.) (1990) *Proc. 7th Annual Pacific Climate (PACLIM) Workshop*, California Dept of Water Resources; *Figure 5.42* I. T. Sanderson (1956) *Follow the Whale*, Cassell, by permission of John Farquharson Ltd.; *Figure 5.46* E. M. Rasmusson and J. M. Wallace (1983) in *Science*, **222**, AAAS; *Table 5.1* G. M. Wellington in R. Perry (ed.) (1984) *Key Environments: Galapagos*, Pergamon; *Appendix 1 (Tables I and III)* J. K. Gamble (1974) *Global Marine Attributes*, Ballinger Publishing; *Appendix 3* D. Spanier (1975) in *The Times*, 23 December, Times Newspapers Ltd.; *letters* Greek Embassy and Turkish Embassy.

INDEX